2025 41st Semiconductor Thermal Measurement, Modeling & Management Symposium (SEMI-THERM 2025)

San Jose, California, USA
10-14 March 2025

IEEE Catalog Number: CFP25SEM-POD
ISBN: 979-8-3315-9936-2

Copyright © 2025, SEMI-THERM Educational Foundation
All Rights Reserved

*** *This is a print representation of what appears in the IEEE Digital Library. Some format issues inherent in the e-media version may also appear in this print version.*

IEEE Catalog Number: CFP25SEM-POD
ISBN (Print-On-Demand): 979-8-3315-9936-2
ISBN (Online):   978-1-7355325-5-4
ISSN: 1065-2221

Additional Copies of This Publication Are Available From:

Curran Associates, Inc
57 Morehouse Lane
Red Hook, NY 12571 USA
Phone: (845) 758-0400
Fax: (845) 758-2633
E-mail: curran@proceedings.com
Web: www.proceedings.com

TABLE OF CONTENTS

SEMICONDUCTOR THERMAL MEASUREMENT, MODELING AND MANAGEMENT SYMPOSIUM

Welcome to SEMI-THERM 41

Committees

Short Courses

Keynote Speaker: Mark Shultz, PhD, IBM

Thermi Award: Rajiv Mongia, Senior Principal Engineer, Intel

Rosten Award: Szilárd Zsigmond Szőke, Henrik Sebők

Thermal Hall of Fame Award: Jim Wilson, Raytheon (retired)

Session 1 - AI Accelerators, Data Centers, HPC
Session Chair: Marcelo del Valle

Dielectric Two-Phase Cooling Solution for High Power Compute Systems * 1
Devin Pellicone, Advanced Cooling Technologies, Inc.

Integration of Two-Phase Flow Boiling in Future Power Packages 7
Ralph Schacht[1], B. Majed[1], T. Grün[2], D. May[2,3], M. Abo Ras[2], B. Wunderle[3], [1]Brandenburg University of Technology, [2]Berliner Nanotest and Design, [3]University of Technology Chemnitz

A TinyML Model Driven Approach To Dynamic Fan Power Control For Energy Efficiency In Edge Devices 13
Ashok Kumar Sankaran, Mukul Golash, Damaruganath Pinjala, Mohan K, Manikandan R, Cisco Systems

Session 2 - AI Accelerators, Data Centers, HPC:
Immersion Single-phase and Two-phase Liquid Cooling
Session Chair: David Saums

A Practical Metric for Cold Plate Thermal Performance in Two-Phase Direct-to-Chip Cooling * 19
Qingyang Wang, Serdar Ozguc, Akshith Narayanan, Jacob D. Moore, Richard W. Bonner III, Accelsius

Validation of 1D Thermo-Fluid Analysis Tool for Liquid Cooling Systems in Data Center Product: A Comparison Study with 3D Simulations and Experimental Data 24
Jiwon Yu, Bharath Ram Ravi, Stephen Keefe, Celestica

Design Optimization of Manifold Integrated Skived Cold Plates for Two-Phase Flow-Boiling 30
Serdar Ozguc, Qingyang Wang, Akshith Narayanan, Jacob Moore, Richard W. Bonner III, Accelsius

Superposition Method of Predicting Junction Temperature of Multi Chip Packages - Important Thermal Parameters 37
Kamrul Russel, Adeel Ahmad, Muhammad Ahmad, Otto Joe, Advanced Mico Devices Inc.

Experimental Study for Single-Phase High Heat Flux and High Coolant Temperature Spray Cooling * 42
Mengyuan Yuan, Nan Chen, Zhipeng Zhao, Zhibin Yan, Advance Liquid Cooling Technologies

Session 3 - Testing and Measurement Methods
Session Chair: Mohamad Abo Ras

Thermal Characterization Testing with a Three-Resistor Finite Difference Model 47
Adolfo Lozano III, Collins Aerospace

* -- Extended Abstract only

Calibration of the Performance Measurement for Thin Vapor Chambers by Photonics Technologies 54
Kuang-Yu Hsu, Yi-Jing Chu, Andhi Indira Kusuma, Krishn Patel, Rajveer G. V., Ming-Hsien Hsiao, T-Global Technology

Thermal Impedance Measurement of SiC MOSFETs Under Strong Negative Gate Bias * 59
Szilard Zsigmond Szoke, Mark Lazar, Henrik Sebok, Robert Bosch Kft.

Thin Film Thermal Conductivity Measurements * 63
V. Linseis, S. Feulner, Heng Wang, Linseis Inc.

Thermo-Optical Plane Source (TOPS) for Thermal Property Characterization and Package-Level Failure Analysis * 65
Jeffrey L. Braun, Bryan N. Baines, and John T. Gaskins, Patrick E. Hopkins, Laser Thermal

Session 4 - Liquid Metals
Session Chair: Navid Kazem; Co-Chair: Yasmin Khakpour

Heterogeneous Liquid Metal – Silver – Polymer Composites for Thermal Interface Materials * 67
Aastha Uppal, Wilson Kong, Ashish Rana, Jae Sang Lee, Matthew Green, Konrad Rykaczewski, Robert Y. Wang , Arizona State University

Design and Evaluation of Liquid Metal-based TIM for Laptop Application * 69
Hing Jii Mea, Benjamin F. Dorau, Navid Kazem, Arieca

Anisotropic And Heterogeneous Thermal Conductivity in Programmed Liquid Metal Composites Through Direct Ink Writing * 72
Ohnyoung Hur[1], Eric J. Markvicka[2], and Michael D. Barlett[1], [1]Virginia Tech, [2]Univ of Nebraska-Lincoln

Ultra-Low Resistance Bonded Flexible Metal (FlexiMetal) Thermal Interface Materials for Fast Transient Response In Mobile Computing Devices 74
Himanshu Pokharna, Deep Materials, Inc.

Acoustic Properties of Stretchable Liquid Metal-Elastomer Composites for Matching Layers in Wearable Ultrasonic Transducer Arrays * 79
Ethan Krings, Eric Markvicka, University of Nebraska Lincoln

Session 5 - Single-phase and Two-phase Liquid Cooling & Heat Sinks and Substrates
Session Chair: Sridhar Nagarajan, Coherent

A Multi-Method Approach for Optimizing and Accelerating Thermal Conductivity Characterization of Thermal Interface Materials * 81
Arya Hakimian, C-Therm Technologies Ltd.

Multi-Zone Thermoelectric On-Chip Cooling Under Different Heat Load Scenarios on a Thermal Test Vehicle * 83
Dan Ralf Wargulski, Torsten Nowak, Mohamad Abo Ras, Berliner Nanotest und Design GmbH

Impedance Matching and Flowrate Requirements for Heterogeneous Liquid Cooling Servers and Racks 85
Javier Avalos, Travis Gaskill, Umut Z. Uras, Nvidia

Development of Dielectric Oscillating Heat Pipes * 92
Corey Wilson, Patrick Margavio, ThermAvant Technologies

Unified Approach to Model Single-Phase Open-Loop Liquid Cooling 94
Albert Chan, Don Nguyen, Michael Brooks, Cisco Systems, Inc.

* -- Extended Abstract only

Session 6 - Liquid Metals
Session Chair: Navid Kazem, Arieca; Co-Chair: Yasmin Khakpour, RTX

Bi-Phasic Nanocomposites with Liquid Metal for Thermal Management and Soft Electronics * 100

Carmel Majidi, Carnegie Mellon University

Developments with Indium-Gallium Thermal Interface Materials for TIM1 Bare-die Server Processors * 102

Tim Jensen[1], Bob Jarrett[1], Ricky McDonough[1], Dave Saums[2], [1]Indium Corporation, [2]DS&A LLC

Evaluation of a Graphene-Enhanced Thermal Interface Material in Air and Immersion Cooling Systems 106

Henrik Barestrand[1], Markus Enmark[2], Jonas Gustafsson[1], Tina Stark[1], Hakan Fredriksson[1], Johan Liu[2], Jon Summers[1], [1]RISE ICE Data Center, [2]Chalmers University of Technology

Session 7 - Thermal Interface Materials
Session Chair: Jason Strader, DuPont

Die Warpage as a Critical Factor for Thermal Interface Material (TIM0) Development and Selection * 113

Dave Saums, DS&A LLC

Novel Metal Compressible Thermal Interface Materials (TIMs) for Non-Planar Surfaces * 118

Miloš Lazić, Bob Jarrett, Ricky McDonough, Indium Corporation

MXene-Liquid Metal Embedded Elastomer Aggregate Composites * 123

Mason Zadan[1], Yafeng Hu[1], Jeremiah Lipp[2], and Carmel Majidi[1], [1]Carnegie Mellon University, [2]Air Force Research Laboratory

Overcoming Thermal and Electromagnetic Challenges with a Single Dispensable Multifunctional Material for Consumer Electronics 127

Arun Raghupathy[1], Robin Huang[1], Judy Guu[1], Sathya Kaliyamoorthy[1], Jay Lee[1], RZ Chiang[1], Himanshu Pokharna[2], [1]Google, [2]Deep Materials Inc

Thermal Interface Materials for Power Electronics Application 132

Debabrata Pal, Mark Severson, Collins Aerospace

Session 8 - Emerging Technologies
Session Chair: Sai Kiran Hota; Co-Chair: Hamid Karani, RTX

Optimizing Structural Rigidity and Thermal Performance of Vapor Chambers via Pillar Distribution and Filling Ratio 140

Nawaf Khalid Y Rasheed[1], Dhruvil Prajapati[1], Chengyan Li[1], L. Winston Zhang[1,2] [1]University of Illinois Urbana-Champaign, [2]Novark Technologies, Inc.

Transforming Battery Thermal Management through the Design and Optimization of High-Heat Transfer Lithium-Ion Batteries 147

Alfred J. Piggott, Jeffrey S. Allen, Ahmad A. Pesaran, Applied Thermoelectric Solutions, LLC

Experimental Evaluation of Thin Oscillating Heat Pipes: A Comparison with Commercial Thermal Solutions * 151

Corey Wilson, ThermAvant Technologies

High Performance 3U-Form Factor Pulsating Heat Pipe Heat Spreader 153

Sai Kiran Hota, Kuan-Lin Lee, Ramy Abdelmaksoud, Srujan Rokkam, Advanced Cooling Technologies

---

* -- Extended Abstract only

Session 9 - Thermal Interface Materials
Session Chair: Robert Wang, ASU

Effect of Surface Roughness on the Performance of Thermal Interface Materials 162
Roshan Sameer Annam, Loren Russell, Cara Rossetti, Dylan Shah, Navid Kazem, Arieca

Vertically Aligned Flexible Insulation Thermal Conductive Sheet * 167
Tomofumi Watanabe, Fumihiro mukai, Hiroki Naito, Bando Chemical Industries

LIME: Liquid Metal Interconnections for Power Semiconductors * 169
Nicholas Baker, University of Alabama

Non-thermal Plasma Activation to Increase Surface Area of Carbon-Rich Materials * 172
Hector Gomez, Gerardo Diaz, University of California - Merced

---

SUBMIT A PAPER FOR SEMI-THERM 42!

As you further develop a technique or application, consider documenting it
for the thermal community. **SEMI-THERM 42** will begin accepting
abstracts during the summer (deadline is October 1, 2025).
We welcome your submissions! Visit us at **www.semi-therm.org**.

SEMI-THERM 42 is March 2026 – be there!

---

* -- Extended Abstract only

Forty First Annual

SEMICONDUCTOR THERMAL MEASUREMENT, MODELING AND MANAGEMENT SYMPOSIUM

PROCEEDINGS 2025

San Jose, CA   USA
March 10-13, 2025

Forty First Annual

SEMICONDUCTOR THERMAL MEASUREMENT, MODELING AND MANAGEMENT SYMPOSIUM

PROCEEDINGS 2025

San Jose, CA   USA
March 10-13, 2025

The 2025 Semiconductor Thermal Measurement, Modeling and Management (SEMI-THERM) Symposium is an annual international forum for the presentation of new developments in and applications relating to generation and removal of heat within semiconductor devices, and measurement of junction temperatures under various application and environmental conditions.

Attendance at the Symposium is in-person this year. The format of the symposium couples nine sessions of selected technical papers, keynote and luncheon talks, a panel discussion, vendor workshops, and a how-to session.

This year, the Symposium is preceded by five Short Courses: "**Efficient Thermal Simulations Using Compact Models**", "**Recent Use-Inspired Research Breakthroughs in Semiconductor Thermal Management Technologies and Metrologies**", "**Thermal Management and Thermomechanical Reliability of Automotive Power Electronics and Electric Motors**", "**Innovative Two-Phase Cooling Solutions: Shaping the Future of Data Center Thermal Management**", and "**Heat Pipes – Fundamentals and Applications**"

We trust you will take advantage of the rich array of information and experiences developed by this year's Steering and Program Committees, and consider submitting an abstract for next year's SEMI-THERM.

General Chair
Lieven Vervecken,
  Diabatix

Program Chair
Navid Kazem,
  Arieca

Vice Program Chair
Claire Wemp,
  DuPont

International Liaisons
John Parry,
  Siemens
Robin Bornoff,
  Siemens
Sobo Sun,
  Celsia Inc.
Winston Zhang,
  Novark

Symposium Marketing
Sarah da Silva Andrade,
  Diabatix

Symposium Management
Bonnie Crystall,
  C/S Communications, Inc

Proceedings
Paul Wesling,
Hewlett Packard (retired)

2025 PROCEEDINGS, Forty First Semiconductor
Thermal Measurement, Modeling and Management Symposium

PERMISSION TO REPRINT OR COPY

Copyright and Reprint Permission: Abstracting is permitted with credit to the source. Libraries are permitted to photocopy beyond the limit of U.S. copyright law for private use of patrons those articles in this volume that carry a code at the bottom of the first page, provided the per-copy fee indicated in the code is paid through Copyright Clearance Center, 222 Rosewood Drive, Danvers, MA 01923 USA. For other copying, reprint or republication permission, write to STEF Copyrights Manager, STEF, 3287 Kifer Road, Santa Clara CA 95051 USA. All rights reserved.

Copyright ©2025 by SEMI-THERM EDUCATIONAL FOUNDATION

**SEMI-THERM™, and the SEMI-THERM logo are registered trademarks of
SEMI-THERM Educational Foundation, Santa Clara, CA.**

PRINTED IN THE UNITED STATES OF AMERICA

Welcome Message

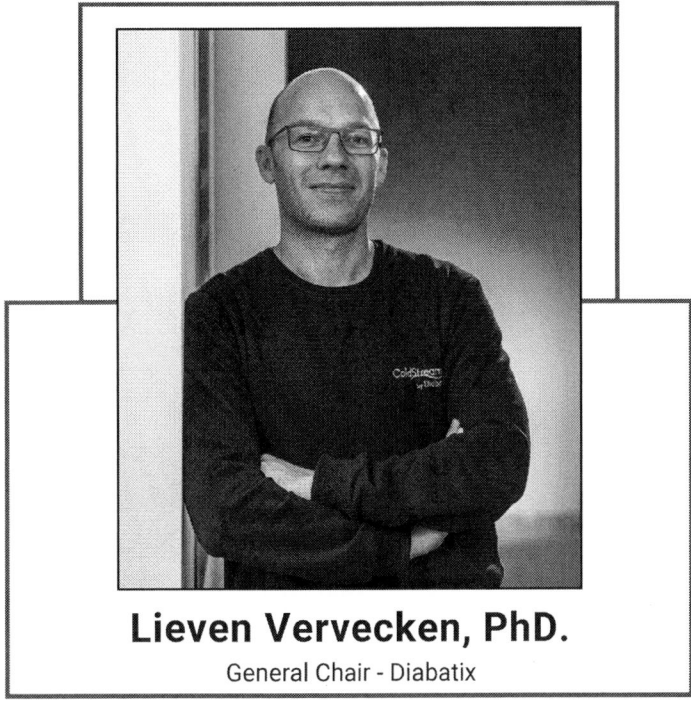

Lieven Vervecken, PhD.
General Chair - Diabatix

Dear Colleagues,

It is my great pleasure to welcome you to SEMI-THERM 41, our annual symposium dedicated to the thermal management and characterization of electronic components and systems. SEMI-THERM has long been a key platform for industry professionals and academia to exchange ideas, showcase innovations, and advance the field of thermal engineering. As this year's General Chair, I am honored to continue this tradition and bring together a diverse community of engineers, researchers, and thought leaders.

This year's program is again packed with exciting content, including an insightful keynote, technical paper presentations, an engaging panel discussion, vendor workshops, and hands-on short courses. Under the theme of "Thermal Innovations that Keep the World's Technology Cool," our sessions will explore cutting-edge topics, from the latest in the field of Thermal Interface Materials to emerging cooling technologies. In addition, we continue our tradition of recognizing excellence in the field with the Best Paper Award, the Thermi Award, the Harvey Rosten Award, and the Thermal Hall of Fame Award.

A special thanks goes out to our Program Chair, Dr. Navid Kazem (Arieca) and Vice Program Chair, Dr. Claire Wemp (DuPont), our Topic Champions, the dedicated reviewers, and all other members of the SEMI-THERM committees for their tireless work in organizing this year's outstanding symposium. I am also grateful to our sponsors and exhibitors for their invaluable support. Finally, I want to acknowledge all of you, our attendees, for playing a crucial role in making SEMI-THERM a vibrant community.

I hope you find this year's conference both insightful and inspiring. Whether you're here to present, learn, or network, may it be a time of valuable discussions, meaningful connections and exciting new collaborations.

Welcome to SEMI-THERM 41!

Dr. Lieven Vervecken,
General Chair, SEMI-THERM 41

WWW.SEMI-THERM.ORG

Personnel

Lieven Vervecken, PhD.

General Chair - Diabatix

lieven.vervecken@diabatix.com

Navid Kazem, PhD.

Program Chair - Arieca

navid@arieca.com

Claire Wemp, PhD.

Program Vice Chair - DuPont

claire.k.wemp@dupont.com

Personnel

Technical Committee

	Emails
Ross Wilcoxon, Collins - Chairman	Ross.Wilcoxon@collins.com
Marcelo del Valle, Infinera Corporation	mvalle@infinera.com
Pablo Hidalgo, AMD	pablo.hidalgoardana@amd.com
Dave Saums, DS&A	dsaums@dsa-thermal.com
Alex Ockfen, Meta	alex.ockfen@meta.com

STEF Board Members

	Emails
Dereje Agonafer, University of Texas, Arlington	agonafer@uta.edu
Alfonso Ortega, Villanova University	alfonso.ortega@villanova.edu
Jim Wilson, Raytheon (Ret.)	jswilsonrtx@gmail.com
Bernie Siegal, TEA	bsiegal@thermengr.net
George Meyer, Celsia	gmeyer@celsiainc.com
Dave Saums, DS&A	dsaums@dsa-thermal.com
Ross Wilcoxon, Collins	Wilcoxon@collins.com

Topic Champions

	Topics
Marcelo del Valle	AI Accelerators and Data Centers, HPC
Jason Strader	Thermal Interface Materials
Navid Kazem	Liquid Metals
Sridhar Nagarajan	Heat Sinks and Substrate
Dave Saums	Immersion Single-phase and Two-phase liquid cooling
Sridhar Nagarajan	Automotive Powertrain & ADAS
Azita Soleymani, Dave Saums	Thermal Management of Battery Energy Storage Systems
Claire Wemp	Testing and Measurement Methods
Azita Soleymani	CFD / Numerical Methods
Sai Kiran Hota	Emerging Technologies and Additive Manufacturing
Yasmin Khakpour, and Hamid Karani	Thermal Management of Electronics in in Aerospace

WWW.SEMI-THERM.ORG

Personnel

Symposium Management

Bonnie Crystall - Symposium Manager
Paul Wesling - IEEE Proceedings
Laura Dobbs - Exhibits, Marketing

Session Chairs	Sessions
Marcelo del Valle, Infinera	Session 1
Dave Saums, DS&A	Session 2
Mohamad Abo Ras, Nanotest	Session 3
Navid Kazem, Arieca	Session 4, 6
Sridhar Nagarajan, Coherent	Session 5
Jason Strader, DuPont	Session 7
Sai Kiran Hota, ACT	Session 8
Robert Wang, ASU	Session 9

Program Review Committee	Company
Azita Soleymani	Bamboo Charge
Claire Wemp	DuPont
Dave Saums	DS&A
David Nelson	Nelson Acoustics
Genevieve Martin	Signify
George Meyer	Celsia
Hamid Karani	RTX
Jaana Behm	Dell
James Petroski	Design by Analysis
Jason Strader	DuPont
Jim Wilson	Raytheon (ret)
Lieven Vervecken	Diabatix
Mohamad Abo Ras	Nanotest
Navid Kazem	Arieca
Pablo Hidalgo	AMD
Ross Wilcoxon	Collins
Sai Kiran Hota	ACT
Sridhar Nagarajan	Coherent
Valerie Eveloy	AE
Wendy Luiten	
Yasmin Khakpour	RTX
Prathiksha Ramprasad	Dhanpal Ampere Computing

Sponsors

Platinum

Diabatix is a pioneering Belgian software company at the forefront of generative design for thermal management. In an era of increasingly compact electronic devices, escalating computing power, urgent reduction of greenhouse gases, and the rapid adoption of renewable energy technologies, the demand for super-efficient cooling components has become paramount for streamlined product development. Diabatix rises to this challenge with its groundbreaking software solution, ColdStream.

Platinum

Celsia specializes in custom heat sink design and manufacturing using liquid two-phase devices: heat pipes and vapor chambers. Through its US headquarters and Taiwan design & production facility, the company's goal is to deliver fast, affordable, and reliable thermal solutions for the most demanding applications including high density electronics, performance CPU / GPU, amplifiers, HBLEDs, ASICS, and rugged systems. In recent years, Celsia has shipped over 2.5 million thermal assemblies to a global custom base in the telecommunications, computer, test equipment, defense, laser, and medical markets.

Platinum

SIEMENS

Siemens Digital Industries Software is driving transformation to enable a digital enterprise where engineering, manufacturing and electronics design meet tomorrow. The Xcelerator portfolio helps companies of all sizes create and leverage digital twins that provide organizations with new insights, opportunities and levels of automation to drive innovation. For more information on Siemens Digital Industries Software products and services, visit sw.siemens.com or follow us on LinkedIn, Twitter, Facebook and Instagram.

Sponsors

Gold

Arieca is an advanced materials start-up that is pushing the boundaries of materials functionalities in the most demanding applications. Our proprietary Liquid Metal Embedded Elastomer (LMEE) technologies allow for unprecedented performances in applications across semiconductor, automotive, and healthcare industries.

Silver

CoolIT systems™

CoolIT Systems specializes in scalable liquid cooling solutions for the world's most demanding computing environments. In the desktop enthusiast market, CoolIT provides unparalleled performance for a range of gaming systems. In the enterprise data center and high performance computing markets, CoolIT partners with global leaders in OEM server design to develop the most efficient and reliable liquid cooling solutions for their own leading-edge products. Through it's Direct Liquid Cooling technology, CoolIT enables dramatic increases in rack densities, component performance and power efficiencies. Together, CoolIT and its partners are leading the way for widespread adoption of high-performance computing.

Silver

Novark's team of highly skilled experts and nearly 1000 employees focus on the custom design, development, and manufacturing of Novark's three product families. Novark supports thermal solutions in a wide variety of markets, including PC, telecom, industrial power, servers, data centers, transportation, LED Lighting, and many more. Novark also supports scientific research at many universities, and frequently supplies materials and prototypes to researchers.

Media	Technical
electronics COOLING	

WWW.SEMI-THERM.ORG

Short Courses

Monday, March 10

7:00 AM - 12:00 PM • REGISTRATION OPEN • San Jose & Santa Clara

---

8:00 AM - 12:00 PM
Carmel

10:00 AM - 10:20 PM
Break

Efficient Thermal Simulations Using Compact Models
Presented by: Tamara Bechtold, PhD.
Jade University of Applied Sciences

Using industry-relevant examples, this course demonstrates the significant advantages of compact thermal models and explains the underlying theory in an understandable way. It introduces the state-of-the-art software tool "Model Reduction inside Ansys" and teaches the workflow through several hands-on exercises.

8:00 AM - 12:00 PM
San Jose & Santa Clara

10:00 AM - 10:20 PM
Break

Innovative Two-Phase Cooling Solutions: Shaping the Future of Data Center Thermal Management
Presented by: Olivier de Laet, Ben Sutton
Calyos SA

The course will start by providing an explanation of the physics behind heat pipes, and then using this foundational knowledge to explain micro-channel, loop and pulsating heat pipes. The lecturer will then present the trade-offs between the different heat pipe technologies including both thermal, mechanical, and environmental constraints.

1:00 PM - 5:00 PM
Carmel

3:00 PM - 3:20 PM
Break

Thermal Management and Thermomechanical Reliability of Automotive Power Electronics and Electric Motors
Presented by: Paul Paret, Gilbert Moreno, Bidzina Kekelia
National Renewable Energy Laboratory

In this course, we present our research conducted at the National Renewable Energy Laboratory (NREL) on automotive power electronics thermal management, thermomechanical reliability of high-temperature bonded interface materials, and electric motor thermal management.

WWW.SEMI-THERM.ORG

Short Courses
Monday, March 10

1:00 PM - 5:00 PM
San Jose & Santa Clara

3:00 PM - 3:20 PM
Break

Recent Use-Inspired Research Breakthroughs in Semiconductor Thermal Management Technologies and Metrologies
Presented by: Jason A. Wiebel, Amy M. Marconnet
Purdue University

Following a brief introduction and overview of research topics in the CTRC, this tutorial will review several use-inspired research topics explored in the CTRC where researchers have made breakthrough contributions in the development of technologies and metrologies for semiconductor thermal management applications.

1:00 PM - 3:00 PM
Monterey

Heat Pipes – Fundamentals and Applications
Presented by: William Anderson, Calin Tarau
Advanced Cooling Technologies, Inc.

The two-hour heat-pipe course will include:

- Heat Pipe Basics
- Heat Pipe Limits
- Heat Pipe Applications
- Different Types of Heat Pipes
- Heat Pipe Working Fluids and Compatibility

- Heat Pipe Wicks
- Heat Pipe Modeling
- Pulsating (Oscillating) Heat Pipes
- Conclusions
- References

WWW.SEMI-THERM.ORG

Keynote Speaker

Pumped Two-Phase Cooling for Sustainable High Performance Computing

Recent technology trends in computer systems include movement toward higher compute module power and higher rack power with 3D chip stacks out at the leading edge. These trends result in systems that cannot be cooled with conventional air cooling. Single phase water cooling has been the technology of choice in addressing these trends, but it has risks associated with potential leaks, biological contamination, and freeze damage. It also does not scale down to the very small channels required to cool 3D stacks of large high power chips. An alternative approach is to utilize two-phase flow boiling of a dielectric fluid, which has both benefits and challenges.

This talk will present challenges, approaches and results for two phase flow boiling in a near term innovative cold plate implementation as well as in a farther future 3D chip stack embedded cooling configuration. The results demonstrate capability to effectively cool conventional 2D/2.5D systems at a much lower, more sustainable cooling power cost relative to air cooling and the capability to build and effectively cool high power 3D chip stacks while mitigating the risks associated with single phase water.

Mark Shultz, PhD. IBM

Mark Schultz received his B.S. in Engineering from Harvey Mudd College and M.S and PhD in Electrical and Computer Engineering from Carnegie Mellon University. He is currently a Senior Research Scientist at IBM's TJ Watson Research Center. His research interests include data storage systems, computer system packaging, and computer system cooling, with his inventions having been used across a wide range of IBM products. He holds 110+ US patents in these and related fields and has authored 40+ published technical papers. He led the experimental portion of IBM's DARPA ICEcool program and now leads the cold plate portion of IBM's ARPA-E CoolerCHIPS program and the IBM Research portion of IBM's System Z cold plate R&D.

WWW.SEMI-THERM.ORG

2025 THERMI Award

Presented Wednesday, March 12

Rajiv Mongia

Senior Principal Engineer in the Assembly Test
Technology Development, Intel Corporation

In recognition of significant
contributions to the field of
Semiconductor Thermal Management

Rajiv Mongia is Senior Principal Engineer in the Assembly Test Technology Development (ATTD) group at Intel Corporation. He leads the Thermal Core Competency team focused on all thermal aspects of Intel Foundry packaging.

Prior to ATTD, Rajiv has had several positions at Intel including leading the thermal architecture team i Altera (FPGA's and other custom ASICs), leading Experience and Maker Outreach in the Modular Innovation Group; SW Product Management and User Experience for Intel RealSense products; Program Director for the Social Computing Intel Science and Technology Center and the Sustainable and Connected Cities Intel Collaborative Research Institute; and leading an international thermal technology development team focused on pathfinding to technology enabling in the PC Client Group. Prior to joining Intel, Rajiv was an engineering consultant where he led fire and explosion investigations and provided expert witness testimony. He was also the lead engineer of a start-up company developing gas turbines for distributed power generation.

Rajiv has been awarded 52 United States patents in the areas of combustion, thermals, mechanicals, platform engineering and human-computer interface design. He also received a number of awards including co-recipient of the 2015 Intel Achievement Award "For developing and bringing to market the first Intel RealSense products".

He received his undergraduate in Mechanical Engineering from Purdue University and his Masters and Ph.D. specializing in Combustion from the University of California, Berkeley.

WWW.SEMI-THERM.ORG

2024 Harvey Rosten Award

Presented Thursday, March 13

Title: Static and Dynamic Thermal Modelling of Si Photonic Thermo-Optic Phase Shifter - Company: Imec

The Award recognizes outstanding work in thermal analysis or modeling of electronic equipment and components, including validation experiments. Established by Harvey Rosten's family and friends, it honors his contributions to the field. The annual award includes a plaque and a $1,000 cash prize, encouraging innovation and excellence. A committee of experts selects the recipient based on criteria such as advancing thermal analysis, practical application to electronics design, insight into thermal behavior, innovation, and a pragmatic approach.

Kristof Croes received an MSc in physics and biostatistics. He obtained a PhD concerning the development of statistical techniques for planning reliability experiments. For seven years, he was product and application manager of the package level reliability products of the Singaporean based company Chiron holdings. Beginning 2007, he went back to research, where he is currently scientific director working on the reliability of advanced interconnects, packages and silicon photonics devices. Kristof was an (invited/tutorial) speaker at several leading-edge semi-conductor conferences [IRPS, IEDM, IITC, IPFA, ADMETA, ...]. He also (co)authored more than 100 articles in the field of reliability.

Minkyu Kim received his BS degree and PhD in electrical and electronics engineering from Yonsei University, Republic of Korea, in 2015 and 2021, respectively. His doctoral thesis concerned the monolithic silicon photonics transmitter design with a ring modulator and its temperature controller. Since 2021, he has been an R&D engineer at imec, where he is currently working on high-speed silicon photonics modulator design as well as I/O system design

WWW.SEMI-THERM.ORG

2024 Harvey Rosten Award

Presented Thursday, March 13

Title: Static and Dynamic Thermal Modelling of Si Photonic Thermo-Optic Phase Shifter - **Company: Imec**

Herman Oprins is a principal member of technical staff and R&D team leader at imec, where he is leading the thermal modeling and characterization team. He received his MSc degree and PhD in mechanical engineering from the KU Leuven, Belgium. In 2003, he joined imec, where he has been involved in the thermal experimental characterization, thermal modeling, and thermal management solutions ranging from the device level to the chip level to the system level.

Joris Van Campenhout is a fellow silicon photonics and director of the optical I/O R&D program at imec. Prior to joining imec in 2010, he was a post-doctoral researcher at IBM's TJ Watson Research Center (USA). He received his PhD in electrical engineering from Ghent University, Belgium, in 2007.articles in the field of reliability.

Ingrid De Wolf received her PhD in physics from KU Leuven, Belgium, in 1989. In the same year, she joined imec, Belgium, where she worked in the field of microelectronics reliability. From 1999 to 2014, she headed the reliability research group. She has (co)-authored 14 book chapters and more than 350 peer-reviewed journal articles. She is a fellow at imec, IEEE senior member, and a full professor in materials engineering of the KU Leuven.

WWW.SEMI-THERM.ORG

2024 Harvey Rosten Award
Presented Thursday, March 13

Title: Static and Dynamic Thermal Modelling of Si Photonic Thermo-Optic Phase Shifter - Company: Imec

Peter De Heyn is a principal member of technical staff working on the silicon photonics platform development as device responsible. He received his PhD in photonics engineering from Ghent University, Belgium, in 2014.

David Coenen received his MSc degrees in aerospace engineering and mechanical energy engineering from KU Leuven and Ghent University, in 2018 and 2020, respectively. He received his PhD in materials engineering from KU Leuven in 2023, focusing on thermal management of Si photonic optical transceivers. He is currently a researcher at imec, with research interests in Si photonics and advanced cooling solutions.

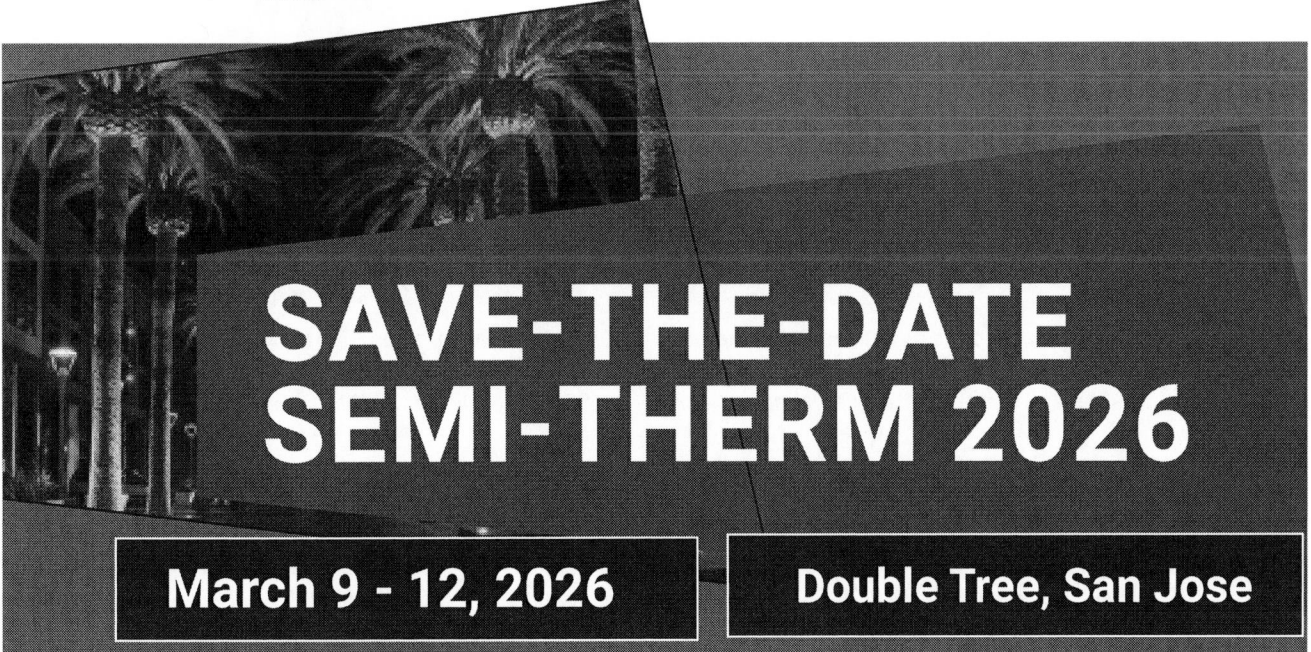

SAVE-THE-DATE
SEMI-THERM 2026

March 9 - 12, 2026   Double Tree, San Jose

WWW.SEMI-THERM.ORG

The SEMI-THERM Educational Foundation
2025 Thermal Hall of Fame
Presented Tuesday, March 11

Jim Wilson
Principal Engineering Fellow, ret., Raytheon

In recognition of significant contributions to the field of Electronics Thermal Management

Reflections on a Career in Electronics Cooling

Jim started a career in electronics cooling in the summer of 1985, a time when computer memory size limitations forced conscious decisions on how to simplify and construct a realistic simulation model, which nearly always required validation testing.

This presentation will cover some of the history of analog RF systems including Gallium Arsenide and Gallium Nitride high power amplifiers necessary for high power radars and jammers. Thermal issues associated with these devices helped mature device level measurement techniques and fund high heat flux cooling research. The talk will also include some examples of SEMI-THERM involvement over the years related to designing thermal architectures for RF systems.

Highlights from Jim's time as an editor for Electronics Cooling magazine will be covered including some career advice.

Biography

Jim Wilson retired as a Principal Engineering Fellow with Raytheon at the end of 2022 and currently consults part time. He has almost 40 years of experience in electronics thermal management designing cooling systems for a variety of defense and aerospace electronics.

Jim's work primarily focused on radar and RF related hardware including ground, airborne, and satellite applications. Jim developed packaging and thermal management capability for high heat flux Gallium Nitride semiconductors that enable high powered RF systems.

He has given invited presentations and short courses on defense electronics thermal management at ASME Interpack and IEEE Itherm symposia. Jim has been involved with SEMI-THERM since the mid-1990s and served on the general chair track for the SEMI-THERM conferences 16 through 18.

He has served as the Finance Chair for the past 20 years and helped transition the conference from IEEE sponsorship to being supported by the SEMI-THERM Educational Foundation. He was awarded the Thermi Award in 2011 for contributions to the field of electronics cooling. Jim was an editor for ElectronicsCooling magazine from 2005 to 2016. He is an ASME Fellow and has 15 patents related to thermal management. He received his undergraduate B.S. from Texas Tech University, M.S. from Stanford University and Ph.D. from Southern Methodist University, all in mechanical engineering.

WWW.SEMI-THERM.ORG

Exhibitors

All companies details: https://www.semi-therm.org/2025-exhibitors

ALPHA

Alpha Novatek
Booth #
400

Analysis Tech
Booth #
500

---

celsia

Celsia
Booth #
503

Element Six
Booth #
502

elementSIX
DE BEERS GROUP

INDIUM

Indium Corporation
Booth #
401

Laser Thermal
Booth #
302

SEKISUI

Sekisui Chemical Co. Ltd
Booth #
202

WWW.SEMI-THERM.ORG

Exhibitors

All companies details: https://www.semi-therm.org/2025-exhibitors

---

 **Sysmetric Thermal forcing Systems**

Sysmetric
Booth #
304

Thermal Engineering Associates
Booth #
501

T-Global Technology Co. Ltd
Booth #
103

Shin-Etsu MicroSi
Booth #
201

Ironwood ELECTRONICS

Ironwood Electronics
Booth #
205

Staubli Corporation
Booth #
200

Linseis Inc.
Booth #
308

Exhibitors

All companies details: https://www.semi-therm.org/2025-exhibitors

Nanotest

Booth #
508

LiSAT

Booth #
402

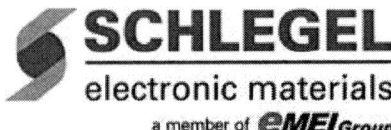

Schlegel Electronic Materials

Booth #
301

Parker Chomerics

Booth #
307

Shenzen FRD Science & Technology Co.

Booth #
209

CEJN

Booth #
505

Dongguan Sheen

Booth #
404

WWW.SEMI-THERM.ORG

Exhibitors

All companies details: https://www.semi-therm.org/2025-exhibitors

5N Plus
Booth #
102

Fujipoly America Corp
Booth #
300

Coherent
Booth #
506

Heatsync
Booth #
405

Honeywell
Booth #
303

m4 Engineering
Booth #
101

xMEMS Lab
Booth #
406

WWW.SEMI-THERM.ORG

Exhibitors

All companies details: https://www.semi-therm.org/2025-exhibitors

CPC
Booth #
206

Bando
Booth #
107

ebm-papst
Booth #
208

Readore Tech
Booth #
306

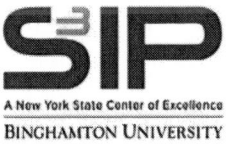

Arieca
Booth #
309

Binghamton University S3IP
Booth #
203

Deep Materials
Booth #
507

WWW.SEMI-THERM.ORG

Exhibitors

All companies details: https://www.semi-therm.org/2025-exhibitors

Malico Inc.
Booth #
509

Jones Tech
Booth #
407

SIEMENS

Siemens
Booth #
409

Cofan Thermal
Booth #
106

ThermAvant
Booth #
403

InfraTec
Booth #
408

INFRATEC.

WWW.SEMI-THERM.ORG

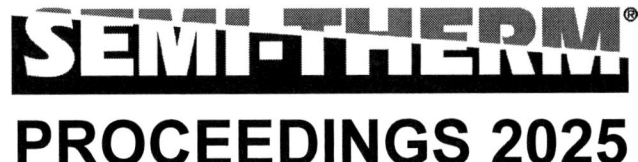

PROCEEDINGS 2025

The papers from **1988** through **2024** are available on the IEEE's **IEL/XPLORE** on-line system. Any researcher can use full-text search across the six million papers in the XPLORE database and access the abstracts of previous **SEMI-THERM** papers. Subscribers may download the PDFs of any **SEMI-THERM** papers. Non-subscribers may purchase single copies at a reasonable fee. To access this resource, please visit:

ieeexplore.ieee.org

DIELECTRIC TWO-PHASE COOLING SOLUTION FOR HIGH POWER COMPUTE SYSTEMS

Devin Pellicone
Advanced Cooling Technologies, Inc.
1046 New Holland Ave.
Lancaster PA, 17601
Devin.Pellicone@1-act.com
717-205-0643

SUMMARY
This paper will present the results of an internal development program where a pumped two-phase cooling system was demonstrated to cool a 200kW thermal load from numerous server blades and racks utilizing a centralized coolant distribution unit (CDU). The CDU was ultimately cooled using facility water and utilizes a dielectric two-phase cooling fluid for direct-to-chip cooling. Cold plate assemblies for each server blade and both in-rack and overhead fluid manifolding solutions have been demonstrated to reliably distribute fluid to each parallel server blade and rack.

1. INTRODUCTION
As the power density demand for high performance computing devices increases, dissipation of the waste heat generated during operation becomes critical. Existing air and liquid cooling solutions are approaching their usable limits and advanced cooling solutions are needed to continue to advance server and data center applications.

In a typical liquid loop, the fluid is pumped through a cold plate that is attached to the heat generating components. The heat from the components is conducted through the cold plate and absorbed by the working fluid as it flows through the plate. This heat is then absorbed by the sensible heat of the fluid, increasing its temperature in proportion to the amount of heat that was absorbed. The sensible heat of most cooling fluids is relatively low and therefore a large flow rate is required to maintain the components below the desired temperature. This requires significant pumping power and increases the cost and complexity of the cooling system.

Active two-phase cooling is a liquid cooling technology that utilizes the latent heat of vaporization of the working fluid to provide enhanced cooling capability. Similar to a traditional liquid loop, the working fluid is pumped through a cold plate to dissipate the heat from the device. However, in two-phase cooling the working fluid generates vapor along the length of the cold plate as the fluid boils, absorbing the waste heat by the latent heat of the fluid. The latent heat of a fluid is several orders of magnitude higher than the sensible heat, therefore, substantially less fluid flow is required as compared to a traditional liquid cooling system. The liquid/vapor mixture then travels on to a heat exchanger where the heat is dissipated and the vapor condensers. The liquid is then pumped back to the components to complete the loop, as illustrated in Figure 1.

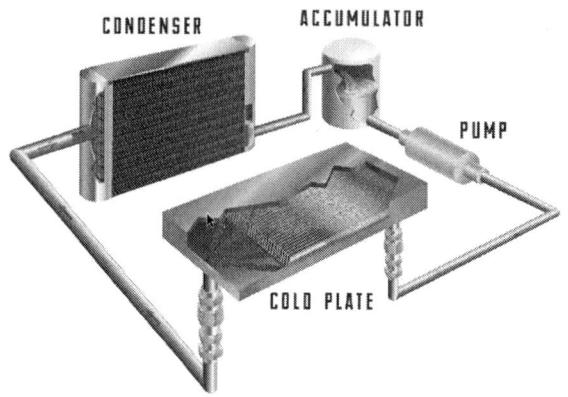

Figure 1. Basic 3D schematic of a pumped-two phase system and its components

2. EXPERIMENTAL/NUMERICAL METHODS

ACT utilized a custom developed pumped two-phase CDU for validating the performance of two-phase cooling in a high powered server application. The CDU hardware used for testing is shown in Figure 2 for reference.

Figure 2. Images of the CDU hardware used for testing

The CDU consists of a pump array, accumulator, liquid-to-liquid condenser, and all of the necessary controls and piping to reliably distribute the refrigerant to the adjacent server racks. A PnID for the CDU is shown in Figure 3, illustrating the numerous pressure, temperature, and flow sensors used to verify the performance of the system.

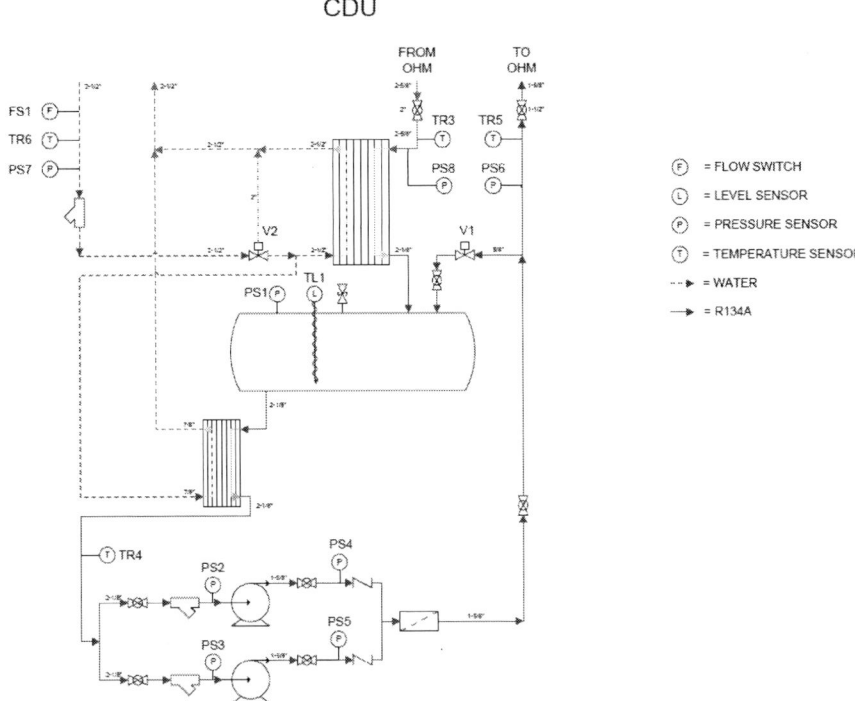

Figure 3. PnID for the CDU indicating the major components and instrumentation used for testing

Thermal test vehicles (TTVs) were used in place of real compute hardware to simulate the thermal load from each server blade, as shown in Figure 4. The TTVs were affixed with custom design two-phase cold plates and the server trays were instrumented with temperature and pressure sensors.

Figure 4. Detail of the thermal test vehicles and cold plates used to simulate the cooling on each server blade

An overhead manifold system was used to distribute the refrigerant to the server rack(s) and return it to the CDU for cooling, as shown in Figure 5. The test rack, shown in Figure 6 consisted of a total of forty (40) of these simulated server blades all connected to a manifold network in the back of the rack, illustrated in Figure 7, to distribute fluid to all of the parallel flow networks.

OVERHEAD MANIFOLD AND RACK MANIFOLDS

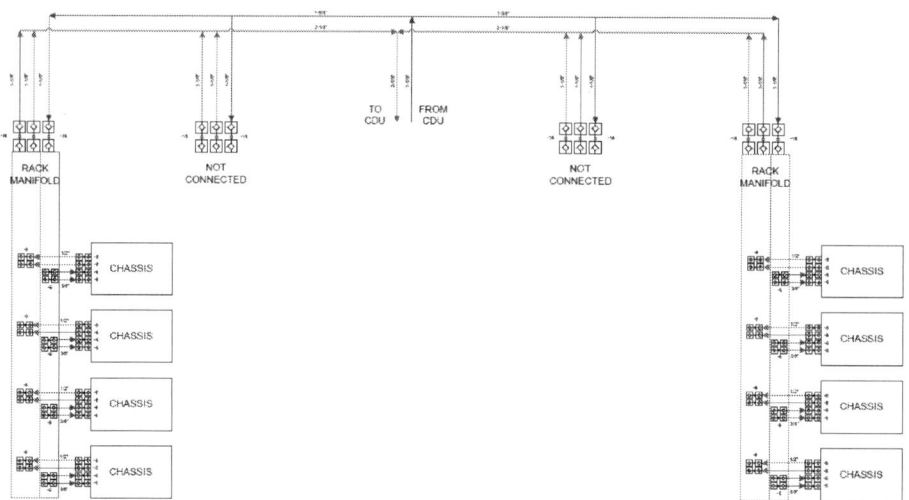

Figure 5. PnID of the overhead manifold network used to distribute flow to the servers

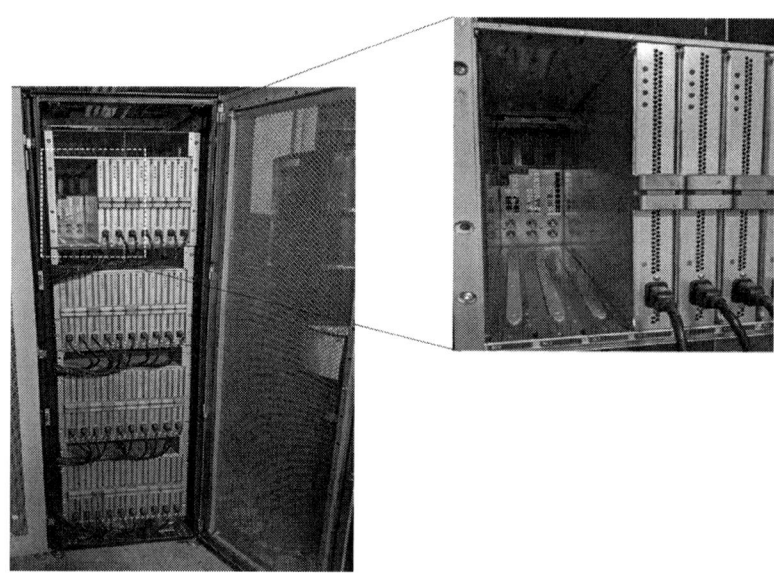

Figure 6. Image of the simulated server rack and blades with detail of the blind mate quick disconnects for each blade

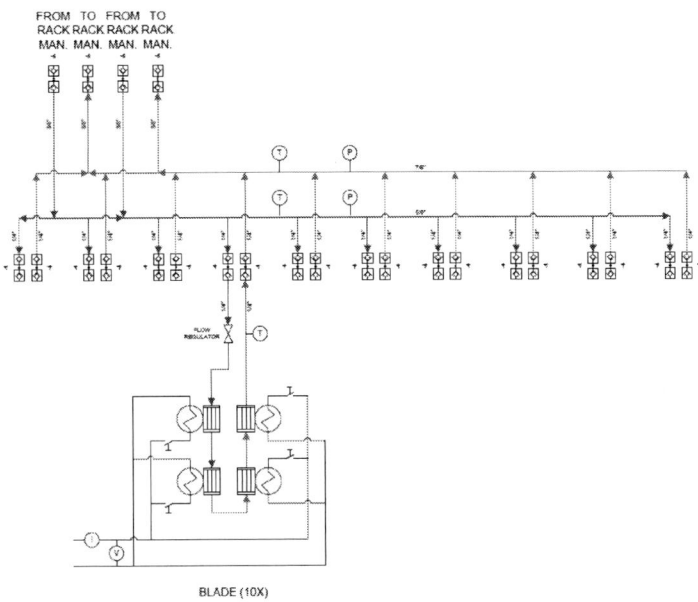

Figure 7. PnID of the in-rack manifold network used to distribute fluid to forty (40) individual server blades

3. RESULTS

Testing on the prototype CDU is still ongoing, but early results indicate the ability of the CDU to dissipate 200kW of thermal load from the server racks with a very low refrigerant-to-facility water temperature difference, as shown in Figure 8. It can be seen that even with 200kW of thermal load, the refrigerant can be condensed with as little as 10°C temperature difference to the facility water. This represents a very attractive thermal resistance for the CDU that will ultimately result in significantly reduced energy consumption from the facility chiller.

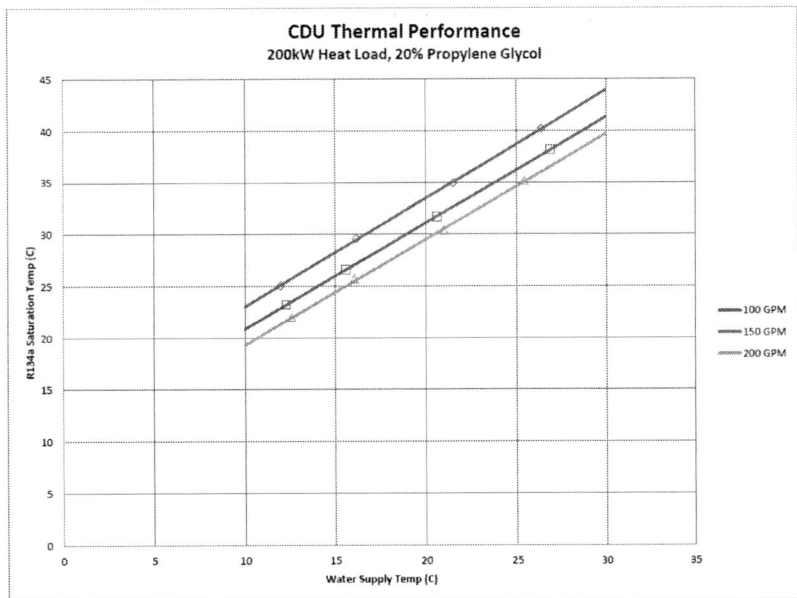

Figure 8. Preliminary CDU thermal performance indicating the ability to achieve 200kW of heat dissipation with low refrigerant-to-facility water delta T

Preliminary results for the individual server cold plate thermal performance are provided in Figure 9. It can be seen that a flux-based thermal resistance for the cold plate (fluid inlet temperature to cold plate base temperature) of less than 0.25°C-cm²/W has been achieved.

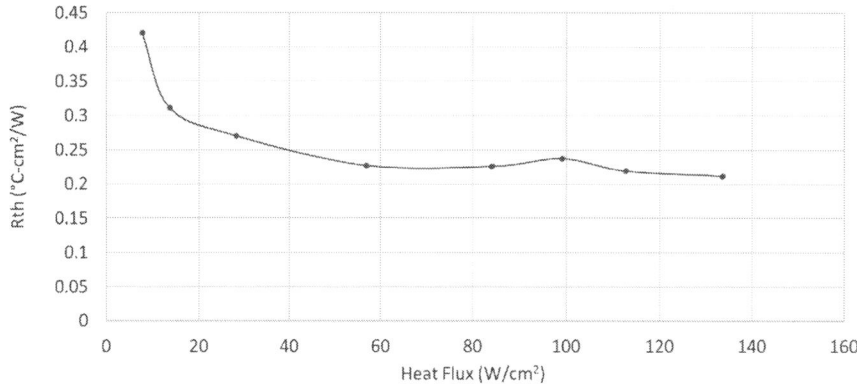

Figure 9. Preliminary cold plate data indicating the flux based thermal resistance of a single cold plate as a function of the thermal heat flux applied to each TTV

4. CONCLUSIONS

Initial test results obtained from a prototype two-phase CDU illustrate a very promising design for a novel cooling solution for high power server applications. The CDU has shown the ability to dissipate up to 200kW of thermal dissipation while maintaining a temperature differential of only 10°C above the supplied facility water. This technology would allow data center integrators to operate their facility water at significantly higher temperatures which would dramatically improve the efficiency of the system. Initial cold plate testing on simulated server nodes indicate that a low thermal resistance, fluid-to-case, is achievable which enables even higher coolant temperatures to be explored.

Integration of Two-Phase Flow Boiling in Future Power Packages

R. Schacht[1], B. Majed[1], T. Grün[2], D. May[2,3], M. Abo Ras[2], B. Wunderle[3]

[1]Brandenburg University of Technology Cottbus-Senftenberg, Germany; [2]Berliner Nanotest and Design GmbH, Germany, [3]University of Technology Chemnitz, Germany;
Email: ralph.schacht@b-tu.de

Abstract – **Flow boiling heat transfer is a topic that is of interest for the cooling of power electronics, high-performance computing in automotive applications, and other electronic packages with particularly high heat flux.**

Based on the investigations from [Schacht 2024], in which the effects on the high performance computing (HTC) with regard to the wetting behavior of de-ionised water (DI water) on silicon (SI) and gold (Au) surfaces during two-phase flow boiling were investigated, an initial design for the integration of two-phase cooling together with the power electronics in one housing (e.g. HPC or automotive) was introduced and the requirements for a controlled two-phase cooling system were discussed.

In this paper, the use of a Glycol-DI water mixture instead of DI water as a coolant is experimentally investigated with regard to heat transfer performance. In addition, the influence of the surface properties of matte and polished silicon on the heat transfer coefficient performance is investigated.

Glycol-DI-water mixture is used as a coolant in applications where ambient temperatures can be below 0°C (e.g. in the automotive industry) and which has no negative impact on global warming and is explosion-proof in useWith Glycol-DI-water mixture as coolant a maximum heat transfer coefficient HTC ~ 4.4 kW/(m²K) at a volume flow rate of $\dot{v}$ = 0.5 l/min (T_{in} = 60°C) with a maximum wall heat flux of $\dot{q}_{wall}$ ~ 38 W/cm² and a wall temperature of T_{wall} ~ 165°C could be achieved. An influence between matte and polished silicon surfaces on the heat transfer could not be determined.

Keywords— *Closed-loop two-phase flow boiling, future application-oriented flow boiling cooler concept, HPC, electronic power package.*

I. INTRODUCTION

The last years have again seen a further reduction of footprint in power electronics connected with an increase in power density, requiring more performant thermal management concepts. Whereas enforced (liquid) cooling has always been the last resort due to complexity, cost and reliability issues, sparking off progress in packaging technology and materials, power densities have reached levels where single-phase liquid cooling can just about guarantee the thermal budget available.

Single phase liquid cooling has for years been a reliable and available option for electronics cooling, mainly for inverter applications for electric vehicle (EV), making use of large pin-fin arrays for heat rejection. Still, with the advent of HPC for automotive and power losses of central processing units (CPU) reaching hundreds of watts and with the well-known Level 1 thermal interface between chip and lid as well as spreading limitations that have for decades been the bottleneck for effective and reliable thermal management of large CPU, there is a call for cooling concepts which have so far rather been confined to the space or other special applications (Figure 1a) [Wakil 2006]. So two-phase cooling concepts offer an interesting solution by enabling enhanced heat transfer by exploiting the latent heat of the working fluid. This cooling effect is larger by orders of magnitude, and it could therefore be exploited to cool effectively and locally where spreading is difficult and miniaturization is key.

This represents a potential solution for future highly integrated systems on an automotive platform, such as high-performance computing or power electronics integrated not only within electronic control units but also into structural components of electric vehicles—one of the key emerging trends. Integrated into the on-board liquid cooling system, this cooling approach would utilise flow boiling as a basis.

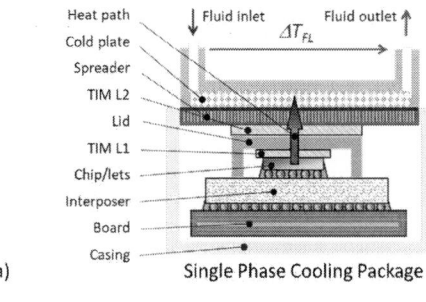

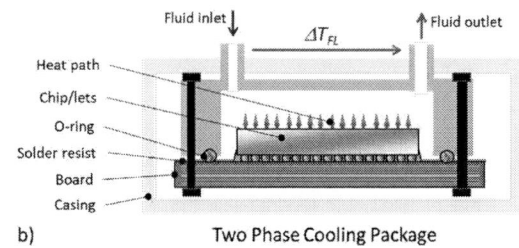

Fig 1: Schematic of (a) single-phase cooling packaging technology and (b) HPC application-oriented package concept with integrated two-phase cooling [Schacht 2024].

So, Figure 1b derives a possible future concept to integrate the two-phase flow boiling cooling together in one package with the power electronic. The advantages are (1) an integrability within casing which is general trend (2) no spreader is needed (3) the packaging technology becomes simpler (4) no TIM is needed (5) it has a smaller footprint (6) the weight can be lowered (7) the complexity of the cold plate can be lowered (8) no convective issues and (9) new design DOFs are possible and (10) concepts are based on the same closed fluid loop system [Schacht 2024].

In electric vehicles, Glycol-DI-water mixtures are used to cool batteries, power electronics and electric motors. The combination of DI water and Glycol combines the advantages of the high specific heat capacity of water with the antifreeze function and chemical stability of Glycol. [Schaeffer 2020]. The chemical stability and anti-corrosion properties of the Glycol-DI water mixture are essential to protect sensitive materials such as aluminium or copper in battery cooling plates and electric motors from damage [Kim J. 2019]. The electrical conductivity of the mixture is a critical factor in applications close to electrical components. Although the addition of Glycol reduces the thermal conductivity compared to pure DI water, it provides the necessary electrical insulation and increases operational safety [Lee 2017]. One of the main limitations of the Glycol-DI water mixture is the reduced thermal conductivity compared to pure DI water, which makes it necessary to adapt the flow rates and cooling systems [Zhao 2022]. Furthermore, Glycol residues can decompose after prolonged use, which reduces the anti-corrosion effect and requires regular maintenance [Bauer 2019].

The heat transfer properties of fluids during flow boiling depend significantly on the fluid composition and the surface properties of the substrate. Glycol-DI water mixtures and pure DI water exhibit different boiling behaviours, which are influenced by factors such as surface tension, thermophysical properties and wettability. Polished and matte silicon surfaces in particular exhibit different heat transfer and nucleation properties depending on the fluid.

Glycol-DI water mixtures reduce the surface tension of the fluid compared to pure DI water. This facilitates the formation of nucleation bubbles, which leads to an earlier start of flow boiling [Kandlikar 2002]. At the same time, the Glycol content increases the viscosity and reduces the thermal conductivity, which can limit heat transfer at high heat flux densities [Li 2007]. Pure DI water has a higher thermal conductivity and specific heat capacity, which makes it ideal for applications with high thermal requirements [You 1999].

Polished silicon surfaces have a low roughness, which limits the nucleation density. Pure DI water shows delayed boiling here due to the higher surface tension, which leads to lower bubble formation at moderate heat flux densities [Park 2011]. Glycol-DI water mixtures promote bubble formation due to the reduced surface tension, which improves heat transfer at low to moderate heat flux densities [Cornwell 1994]. At high heat flux densities, however, the lower thermal conductivity of the mixture can limit the heat transfer.

Matte surfaces with higher roughness promote nucleation bubble formation due to the presence of microcavities. Pure DI water benefits from this through improved heat transfer, as the surface tension is less critical and bubbles can detach more efficiently [Stephan 2003]. Glycol-DI water mixtures further enhance this effect, as the reduced surface tension and improved wetting enable a higher nucleation density. At high heat flux densities, however, the viscosity of the mixture can lead to increased pressure losses [Kreith 2001].

At low to moderate heat flux densities, Glycol-DI water mixtures are more efficient on both polished and matte silicon surfaces, as the reduced surface tension promotes bubble formation. At high heat flux densities, pure DI water performs better, especially on matte surfaces, as the higher thermal conductivity and lower viscosity of the fluid favour heat transfer [Bergles 1964].

The choice between Glycol-DI water mixtures and pure DI water depends on the specific operating conditions. Glycol-DI water mixtures are suitable for applications requiring improved bubble formation and lower freezing points, while DI water offers advantages for high heat flux densities and low flow resistance. The surface finish strongly influences both fluids, the polished surfaces favour a lower viscosity, while matte surfaces benefit from improved wetting properties.

However, using pure DI water or Glycol-DI water mixtures as coolant have also some disadvantages. The saturation temperature T_{sat} of pure water or Glycol-DI water mixtures are higher ($T_{sat,DI}$ =100°C, $T_{sat,mix}$ >105°C) compared to dielectric fluorinated coolants (R134a: T_{sat} = 26.3°C, HFE-7100: T_{sat} = 61°C [Hoang 2021]), which can lead to a higher wall temperature required for boiling. It has been noted that in all studies using pure DI water as a coolant, wall temperatures exceed 100°C, reaching up to 120°C. Elevated wall or chip temperatures can negatively impact the performance and lifetime of electronic devices [Hoang 2021]..

In order to be able to characterize such future power packages, the used closed loop cooler test stand is shortly introduced. A possible power packaging cooling solution with the thermal test vehicle used for an HPC application (Figure 1b) is then presented. Finally, the results are presented and discussed.

II. METHODOLOGY

A. Closed loop flow boiling cooler test stand

Figure 2 provides a schematic overview of the closed-loop flow boiling cooler test stand, designed for rapid experimental characterization and parameter variation. This setup enables the examination of various cooler designs, surface topographies, and surface chemistries to study nucleation behaviour and bubble formation, as well as the evaluation of other integrated power cooling concepts utilizing flow boiling.

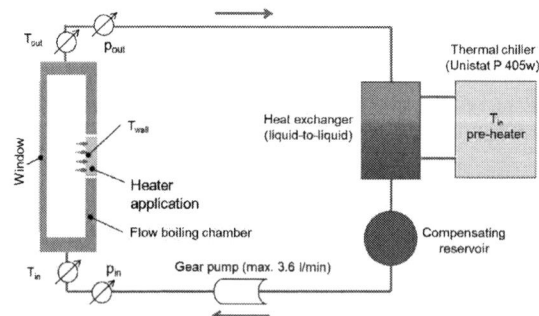

Fig. 2: Schematic of closed loop flow boiling cooler test stand.

More information is given in [Schacht 2024], as well as its proof of concept.

B. Flow boiling cooler concept

To emulate an HPC chip scenario a Thermal Test Vehicle (TTV) assembled on a printed circuit board (PCB) from *Berliner Nanotest and Design GmbH* is used. Figure 3a gives an overview of the assembly of the cooling concept design to integrate flow boiling cooling within the electronic component into one package. The TTV is cooled directly by the coolant without the need for a heat spreader. The PCB is protected against short circuits by the PCB solder resist. The cooler is mounted on top of the PCB, fixed by four screws and sealed by an O-ring.

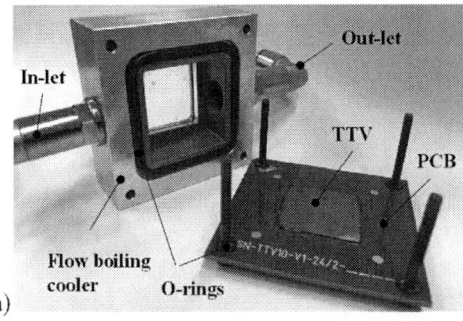

Fig. 3 a) Flow boiling cooler and TTV assembling concept and b) Flow boiling cooler package (with assembled TTV PCB) mounted on ETB.

The four O-rings at the screws ensure that the pressure is evenly distributed and that the PCB does not bend during fixing. A PMMA observation window is provided at the top, which is sealed with an O-ring and held in place by a fixing plate.

C. Thermal test vehicle (TTV)

The thermal test vehicle (TTV) is implemented as a monolithic die with a 10 × 10 matrix layout, covering an area of 24.9 × 24.9 mm². It is assembled using flip-chip technology with SAC305 solder bumps (300 µm diameter, 500 µm pitch) and underfill, mounted on a 60 × 60 mm² organic substrate

The TTV is available with a polished surface (Figure 4b), enabling a comparative analysis between polished (surface roughness ~7 µm) and matte (surface roughness ~20 µm) silicon surfaces using a laser texturing process (Figure 4c). This comparison facilitates the investigation of wetting behavior differences, as surface finish impacts fluid dynamics: polished surfaces reduce viscosity, whereas matte surfaces enhance wetting properties.

Structurally, the TTV comprises four independent quadrant power zones, each measuring 12.5 × 12.5 mm², along with six localized hotspots of 2.4 × 2.4 mm² (Figure 4a). The TTV can be powered up to 2000 W, with a resistance per zone of 7 ± 0.1 Ω at room temperature, a maximum voltage of 60 V, and a maximum current of 8.5 A per heater. The heater zones have a power density of 3.2 W/mm², while the hotspots reach 10 W/mm², with a maximum operating temperature of 125 °C.

Sixteen resistance temperature detectors (RTDs) are distributed across the TTV at central, edge, corner, and quadrant midpoints. These RTDs, connected using a Kelvin 4-wire configuration, exhibit an electrical resistance of 3.2 ± 0.1 kΩ at room temperature, a sensitivity of 10 Ω/K, and dimensions of 820 × 100 µm² [Sternberg 2023].

Fig. 4: TTV a) location of the four quadrant power zones (green, blue, yellow, grey), the six single cell hotspots (brown) and the sixteen temperature sensors (T1 – T16) b) TTV (polished silicion surface) assembled on organic substrate, in flip-chip technology with underfill and c) TTV with matte silicion surface.

The TTV and the cooler are mounted on an electronic test board (ETB) using four 26-pin connectors (Figure 3b). Independent power zones and hotspots are routed to designated power connectors on the ETB. The sixteen RTDs are multiplexed to two channels via an onboard multiplexer (MUX), with a constant current source providing the sense current for temperature measurements.

To manage the TTV's operation, a TTV Control Unit (TCU) is employed, comprising both hardware and software components. The hardware includes a 19" rack-mounted system with four programmable power supply units (PSUs) and a data acquisition (DAQ) system featuring eight analog channels. The TTV interfaces with the TCU via front-panel connectors (Figure 5).

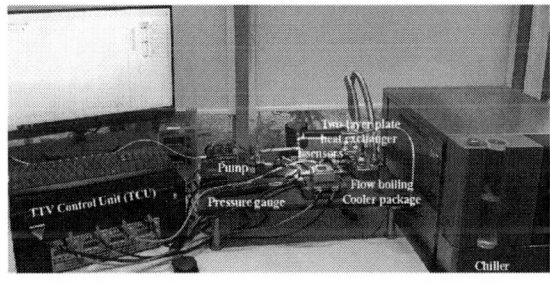

Fig. 5 Overview over test stand for the flow boiling investigation.

The TCU software handles power control, data acquisition from the RTDs, and the scheduling of power and temperature cycles. It enables real-time visualization and precise management of the TTV's thermal performance.

D. Data reduction

The wall heat flux $\dot{q}_{wall}$ [W/m²] defines Equation 1

$$\dot{q}_{wall} = \frac{P_D}{A} - \dot{q}_{loss} \left[\frac{W}{m^2}\right] \qquad (1)$$

where P_D [W] represents the generated the power dissipation by the TTV, and A [m²] the total chip surface area. The heat flux losses $\dot{q}_{loss}$ [W/m²], caused by conduction toward the PCB, are disregarded.

The mean fluid temperature is calculated using Equation 2

$$T_{fluid} = T_{in} + \frac{(T_{out} - T_{in})}{2} \ [K] \qquad (2)$$

where T_{out} [K] represents flow chamber outlet temperature and T_{in} [K] the flow chamber inlet temperature.

Equation 3 defines heat transfer coefficient HTC [W/(m²K)]

$$HTC = \frac{\dot{q}_{wall}}{(T_{wall} - T_{fluid})} \left[\frac{W}{m^2 \cdot K} \right] \qquad (3)$$

where T_{wall} [°C] represents the chip surface temperature.

The wall superheat ΔT_{sat} is calcuated by Equation 4

$$\Delta T_{sat} = T_{wall} - T_{sat} \ [K] \qquad (4)$$

where T_{sat} [K] represents fluid saturation temperature.

E. Operating conditions

The investigations were carried out at volume flow rates in the range of $\dot{v}$ = 0.3 – 1.0 l/min and inlet temperature of T_{in} = 60 °C, using 50% / 50% Glycol-DI water mixture (T_{sat} = 108 °C) [prokuehl 2003] and pure DI water (T_{sat} = 100 °C) as the coolant liquid.

The accuracy of the TTV power dissipation is specified as ± 1 W. Type K thermocouples with an accuracy of ± 1.5 °C are used to measure the inlet and outlet temperatures. The RTDs used to measure the TTV wall temperature have an accuracy of ± 1 °C.

This results in an uncertainty for the heat transfer coefficient of ΔHTC $\approx$ ± 10 W/(m²K) @ $P_{D,TTV}$ = 1 kW.

III. RESULTS AND DISCUSSION

This chapter presents and discusses the experimental results for the use of Glycol-DI water mixture instead of pure DI water as a coolant in terms of heat transfer performance using the industrial two-phase flow boiling cooler concept for an HPC package.

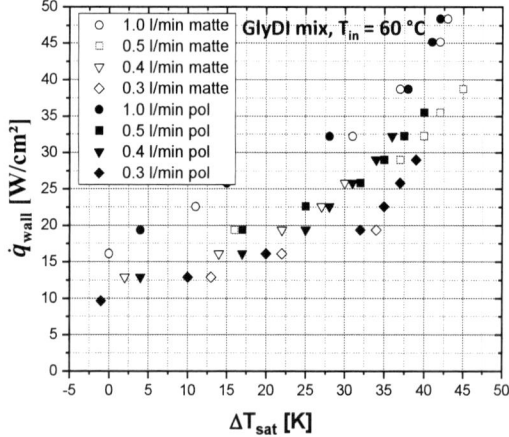

Fig. 6 Influence of volume flow rate for Glycol-DI mixture at T_{in} = 60°C: wall heat flux $\dot{q}_{wall}$ versus wall super heat ΔT_{sat} for matte and polished silicon surface.

The influence of the volume flow rate and the surface properties of silicon on the boiling curves is shown in Figure 6. It can be observed for the use of Glycol-DI water mixture that the surface properties (matte or polished) show between each of the analyzed volume flow rates almost the same behavior.

After the onset of nucleate boiling (ONB) at a wall super heat of around ΔT_{sat} = 30 K, the boiling curves tend to run together onto a single curve with increasing wall heat flux independently from the volume flow rate, as has already been observed with pure DI water, indicating the dominance of nucleate boiling over convective heat transfer.

Figure 7 shows the influence of the volume flow rate on the heat transfer coefficient (HTC) behavior for the TTV with matte and polished silicon surface. The results show that with increasing volume flow rate, the cooling behavior stays longer in the single-phase region. But once the two-phase regime is established, all curves consolidate to a single curve, regardless of the volume flow rate.

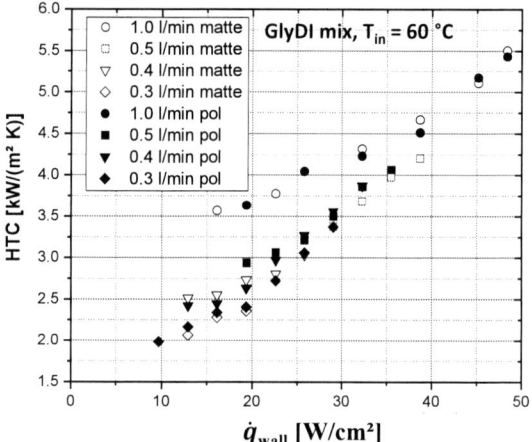

Fig. 7 Influence of volume flow rate for Glycol-DI mixture at T_{in} = 60°C: heat transfer coefficient HTC versus wall heat flux $\dot{q}_{wall}$ for matte and polished silicon surface.

Figure 8 and 9 compares the performance of pure DI water with that of a Glycol-DI mixture at a volume flow rate $\dot{v}$ = 0.5 l/min and an inlet temperature of T_{in} = 60 °C.

[Kandlikar 2002] reports that Glycol-DI water mixtures reduce the surface tension of the fluid compared to pure DI water. This facilitates the formation of nucleation bubbles, which leads to an earlier start of flow boiling although Glycol-DI water mixtures have a higher saturation temperature compared to pure DI water. However, from Figure 8 can be observed that there is an increase of the wall super heat (~ 20 K) for the Glycol-DI water mixture compared to pure DI water.

This is due to the fact that the reduced surface tension of the Glycol-DI water mixture facilitates the formation of nucleation bubbles, so that boiling starts earlier, but the effect of the higher saturation temperature is more dominant. The observed stronger increase in wall super heat with increasing heat flow for the Glycol-DI mixture compared to DI water can be attributed to the reduction in the thermal conductivity and specific heat capacity of the Glycol-DI mixture compared to pure DI water. These properties negatively influence the heat dissipation and ultimately lead to the observed higher wall temperature to ensure the heat flow [Kim 2019].

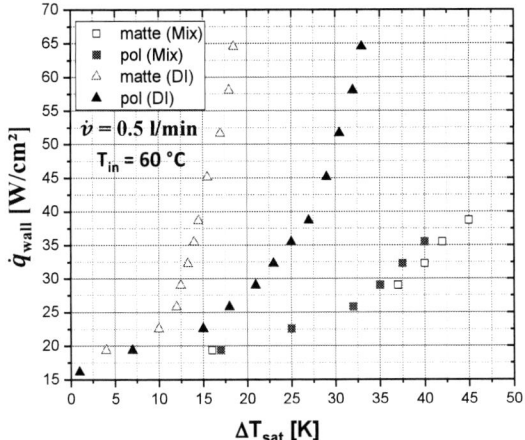

Fig. 8 Influence of matte/polished silicon surface and Glycol-DI water mixture (Mix) / pure DI water (DI): wall heat flux $\dot{q}_{wall}$ versus wall super heat ΔT_{sat} for volume flow rate $\dot{v} = 0.5$ l/min at $T_{in} = 60°C$.

So, the results fit together when considering the transition from the nucleation phase (at low wall super heat) to established flow boiling convection (at higher wall super heat). Glycol-DI-water mixtures show advantages in the initiation of boiling, while at the same time they require a higher wall temperature for higher heat flux densities due to their physical properties.

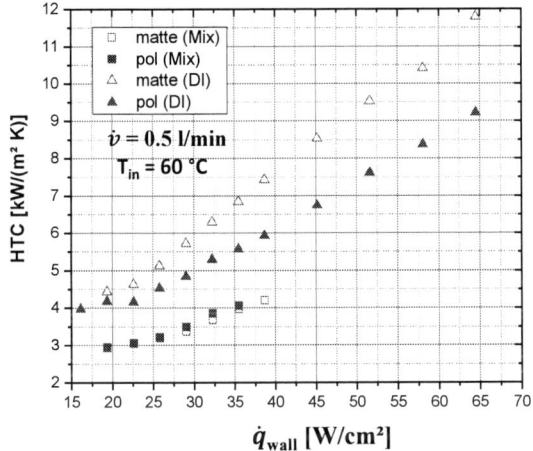

Fig. 9 Influence of matte/polished silicon surface and Glycol-DI water mixture (Mix) / pure DI water (DI): Heat transfer coefficient (HTC) versus wall super heat ΔT_{sat} for volume flow rate $\dot{v} = 0.5$ l/min at $T_{in} = 60°C$.

As noted from Figure 9, pure DI water performs best, especially on matte surfaces, as the higher thermal conductivity and lower viscosity of the fluid favor heat transfer, which is also consistent with the observations of [Bergles 1964].

Bubble growth for Glycol-DI water mixture on polished silicon surface is slower and the bubbles may grow larger before they dissipate, limiting heat transfer compared to pure DI water. While the matte silicon surface facilitates bubble formation and detachment, the lower wetting and higher viscosity of the mixture makes bubble release less efficient than DI water. From Figure 9 can also observed that Glycol-

DI water mixture shows on polished surfaces, contrary to the theory, a slightly higher heat transfer efficiency compared to matte ones.

This can may be attributed to a combination of improved wetting properties, more stable bubble formation and more uniform heat transfer of the Glycol-DI-water mixture on polished surfaces. Although matte surfaces theoretically provide more nucleation sites, practical factors such as bubble stability, localized heat build-up and chemical interactions on polished surfaces can increase performance in certain scenarios. The slightly improved performance of the polished surfaces may be due to measurement uncertainty. Factors such as surface preparation, fluid composition or minimal differences in the test conditions can also a reason for the deviation from the theoretical expectation.

IV. CONCLUSIONS

In this paper, the use of Glycol-DI-water mixture instead of DI water as a coolant for the introduced HPC application-oriented two-phase flow boiling cooler concept was experimentally investigated with regard to heat transfer performance. In addition, the influence of the surface properties of matte and polished silicon surface on the heat transfer coefficient performance was investigated.

A heat transfer coefficient of maximum HTC ~ 4.4 kW/(m²K) at a volume flow rate of $\dot{v} = 0.5$ l/min ($T_{in} = 60°C$) with a maximum wall heat flux of $\dot{q}_{wall}$ ~ 38 W/cm² and a wall temperature of T_{wall} ~ 165°C could be achieved with the Glycol-DI-water mixture as coolant.

In comparison to this, with pure DI water as cooling fluid a maximum $HTC = 11.8$ kW/(m²K) ($\dot{q}_{wall} = 65$ W/cm²) was achieved for matte silicon surface at a volume flow rate $\dot{v} = 0.5$ l/min ($T_{in} = 60°C$). The maximum chip temperature was measured to $T_{wall} = 118$ °C. For polished silicon surface a maximum $HTC = 9.2$ kW/(m²K) ($\dot{q}_{wall} = 65$ W/cm²) was achieved at a volume flow rate $\dot{v} = 0.5$ l/min ($T_{in} = 60°C$). The maximum chip temperature was measured to $T_{wall} = 130$ °C.

The influence of a matte and polished silicon surface property on the heat transfer performance in relation to the use of Glycol-DI-water mixture showed no significant difference. This could be due to a combination of optimised wetting properties, more stable bubble formation and more uniform heat transfer of the Glycol-DI-water mixture on polished surfaces. While matte surfaces theoretically offer more nucleation sites, practical factors such as bubble stability, localised heat concentration and chemical interactions on polished surfaces can lead to better performance in certain cases.

In total, the choice between Glycol-DI water mixtures and pure DI water depends on the specific operating conditions. Glycol-DI water mixtures are suitable for applications requiring improved bubble formation and lower freezing points, while DI water offers advantages for high heat flux densities and low flow resistance. The surface finish strongly influences both fluids: polished surfaces favor a lower viscosity, while matte surfaces benefit from improved wetting properties.

The two-phase flow boiling cooler used is a prototype that was developed for optical observations. The experiments have shown that the cooler can be scaled down further in the future, depending on the application.

The next step is to investigate a fluid that has a lower saturation temperature than pure DI water, is non-explosive and has no negative impact on global warming. The aim is to achieve chip temperatures below 100°C with wall heat fluxes above 250 W/cm² at a volume flow rate of $\dot{v} = 1.0$ l/min and an inlet temperature of $T_{in} = 60$°C.

REFERENCES

[Schacht 2024] R. Schacht, J. B. Majed, T. Grün, D. May, M. A. Ras and B. Wunderle, "Closed-loop flow boiling cooler test stand for investigations on future power package designs," 2024 30th International Workshop on Thermal Investigations of ICs and Systems (THERMINIC), Toulouse, France, 2024, pp. 1-8, doi: 10.1109/THERMINIC62015.2024.10732089.

[Wakil 2006] Jamil Wakil, "Thermal performance impacts of heat spreading lids on flip chip packages: With and without heat sinks", Microelectronics Reliability,Volume 46, Issues 2–4, 2006, Pages 380-385, ISSN 0026-2714, https://doi.org/10.1016/j.microrel.2005.01.007.

[Schaeffer 2020] G. Schaeffer, "Thermal Management in Electric Vehicles", Energy Procedia, 2020.

[Kim J. 2019] J. Kim et al., "Advanced Cooling Techniques for Electric Vehicles," Applied Thermal Engineering, 2019.

[Lee 2017] M. Lee et al., "Dielectric Coolants for Electric Vehicle Systems," Journal of Power Sources, 2017.

[Zhao 2022] R. Zhao, "Optimizing Thermal Performance in Glycol-Based Systems," Energy Conversion and Management, 2022.

[Bauer 2019] C. Bauer, "Long-Term Stability of Automotive Coolants," International Journal of Thermal Sciences, 2019.

[Kandlikar 2002] Kandlikar, S. G., "Heat Transfer Characteristics in Boiling with Enhanced Surfaces," Heat Transfer Engineering, 2002.

[Li 2007] Li, C., & Peterson, G. P., "Effect of Surface Wettability on Boiling Heat Transfer," International Journal of Heat and Mass Transfer, 2007.

[You 1999] You, S. M., Kim, J. H., & Kim, K. H., "Effect of Surface Roughness on Pool Boiling Heat Transfer," International Journal of Heat and Mass Transfer, 1999.

[Park 2011] Park, S. H., & Thome, J. R., "Bubble Dynamics on Rough and Smooth Surfaces in Flow Boiling," Experimental Thermal and Fluid Science, 2011.

[Cornwell 1994] Cornwell, K., & Houston, S., "Nucleate Pool Boiling on Micromachined Surfaces," Journal of Heat Transfer, 1994.

[Stephan 2003] Stephan, P., "Fundamentals of Pool and Flow Boiling Heat Transfer," Heat Transfer Handbook, 2003.

[Kreith 2001] Kreith, F., & Bohn, M. S., "Principles of Heat Transfer," Cengage Learning, 2001.

[Bergles 1964] Bergles, A. E., & Rohsenow, W. M., "The Influence of Surface Roughness on Nucleate Boiling," ASME Journal of Heat Transfer, 1964.

[prokuehl 2003] pro KÜHLSOLE GmbH, "GLYKOSOL N Datenblatt", Stand 01.01.2003, https://www.glykolundsole.de/Downloaddateien/GLYKOSOL%20N%20klein%20deutsch.pdf

[Kim 2019] Kim, J., et al. "Bubble Dynamics and Heat Transfer in Glycol-Water Mixtures: Implications for Flow Boiling", International Journal of Heat and Mass Transfer, 2019.

[Sternberg 2023] M. Sternberg et al., "Thermal Test Vehicle for HPC – System Level Approach for Investigation of the Thermal Heat Path Signature with the Property of Spatial Resolution," 2023 29th International Workshop on Thermal Investigations of ICs and Systems (THERMINIC), Budapest, Hungary, 2023, pp. 1-7, doi: 10.1109/THERMINIC60375.2023.10325887.

A TinyML Model Driven Approach To Dynamic Fan Power Control For Energy Efficiency In Edge Devices

Ashok Kumar Sankaran, Mukul Golash, Damaruganath Pinjala, Mohan K, Manikandan R
Cisco Systems Inc, Bangalore, India
ashsanka@cisco.com, mgolash@cisco.com, dpinjala@cisco.com, mohk@cisco.com, mar2@cisco.com

Abstract — This study addresses the significant challenge of fan power consumption in enhancing the throughput of edge devices. With fan power being a primary constraint, an efficient fan speed controller becomes essential for optimizing energy use without compromising system performance. We propose a TinyML-based fan speed control model that dynamically adjusts fan speed based on real-time sensor data, such as the temperature of critical components and current traffic conditions. This model is designed to maintain temperatures within predefined thresholds, thereby ensuring system stability and efficiency.

Our experimental setup involved evaluating the TinyML model on an edge device, simulating maximum traffic conditions to test its efficacy under load. Experiments show that this method reduces fan power consumption by up to 33% across various ambient temperatures, while maintaining system stability. Compared to traditional methods, it offers a nearly 30% improvement in energy efficiency, with only a 7.5% increase in component temperatures, making it an effective solution for modern edge computing environments.

Overall, the TinyML model-driven fan speed control presents a compelling solution for reducing energy consumption in edge devices, providing significant power savings and maintaining system stability under various operating conditions. Also being lightweight with limited computing requirements, this approach is an attractive option for optimizing performance and energy efficiency in modern systems.

Key Words — TinyML, Fan power consumption, Fan speed control, Energy efficiency, Edge devices, Dynamic fan adjustment, Power savings, Real-time data

NOMENCLATURE

TinyML: Tiny Machine Learning

NN: Neural Network

SVM: Support Vector Machine

NPU: Network Processing Unit

PWM: Pulse Width Modulation

SMART: Systematic Management of Adaptive Response Technology

I. INTRODUCTION

In recent years, the telecommunications industry has witnessed a substantial increase in the demand for network bandwidth, driven by the proliferation of data-intensive applications and services. Concurrently, the ongoing miniaturization of transistors has enabled higher integration densities in telecom systems. While this advancement in integration density enhances system capabilities, it also leads to increased power density and heat generation. These thermal challenges can significantly impact system reliability, reduce the lifespan of components, and impair operational speed. [1]

Despite the impact of fan speed on power consumption, noise levels, and reliability, fan speed is often not managed in an intelligent or efficient manner. Traditional fan controllers tend to be reactive, relying on a static correlation between limited temperature sensor value and fan speed. This method typically results in simply increasing airflow in response to rising system temperatures, without optimizing for energy efficiency or other factors.

To address these issues, dynamic management of fan power control has emerged as a viable solution. By intelligently adjusting fan speeds, systems can efficiently dissipate heat, thereby maintaining optimal operating conditions. This approach not only preserves the integrity of telecom systems but also enhances energy efficiency [2].

A promising technique for achieving dynamic fan control is the use of neural network-based predictive approaches. Leveraging machine learning models, these systems can predict the appropriate fan speed for the current instance of sensor readings and adjust fan speeds accordingly. This predictive capability ensures that telecom systems maintain temperatures within desired thresholds while minimizing power consumption. By proactively managing fan speeds, these systems ensure reliable operation even under fluctuating workloads and environmental conditions [3].

Furthermore, the integration of TinyML, a branch of machine learning focused on deploying efficient models on edge devices with limited computational resources, presents new opportunities for telecom systems. TinyML enables the implementation of sophisticated neural network models directly on embedded systems, allowing for real-time, low-latency fan control without the need for extensive computational power [4].

Overall, the application of neural network-based predictive approaches for fan power management in telecom systems represents a forward-thinking solution that aligns with the industry's demand for higher performance and energy efficiency. Through intelligent control mechanisms, telecom systems can achieve greater reliability, extend component lifespans, and optimize operational speed, ultimately meeting the growing demands of modern telecommunications infrastructure.

This paper introduces a Neural Network-based model designed to minimize fan power consumption by dynamically controlling fan speed, while ensuring that temperatures remain close to expected levels.

- Our experimental results demonstrate that our approach significantly outperforms traditional fan controllers in terms of energy efficiency, without compromising the expected temperature thresholds.
- Additionally, our proposed model leverages TinyML, allowing the thermal-aware algorithm to run efficiently on existing hardware without requiring additional components. This makes it a cost-effective solution for telecom systems, as it enables real-time fan speed control and energy management directly on embedded devices with minimal computational overhead.

This paper is organized as follows: Section II introduces the problem statement and justification for our proposed solution. Section III details the experiment set up analysis and outcome, with final conclusions drawn in Section IV.

II. BACKGROUND

In this section, we present the problem statement and the proposed solutions in detail.

A. Problem statement

Traditional fan control strategies often use fixed temperature thresholds or basic algorithms to manage fan speeds. These methods can be inefficient, leading to overcooling or insufficient cooling as they don't adapt to changing conditions. The constant or stepwise speed adjustments can result in high energy consumption and accelerated fan wear, affecting their lifespan. Additionally, these strategies overlook workload and environmental variations, potentially compromising cooling efficiency and increasing the risk of thermal stress and equipment failure.

This study focuses on an edge device comprising various components such as Networking Processing Units (NPUs), routing processors, and other critical processing elements. Each component is equipped with its dedicated sensors. The system is cooled using a forced-convection air-cooled mechanism. The cooling fan circulates ambient air through the system and the heat sink of critical components. Thermal sensors provide critical feedback signals to closed-loop controllers tasked with adjusting the fan speed.

The challenge addressed in this paper is the dynamic control of fan speed to minimize power consumption while ensuring that temperatures do not exceed specified thresholds. The thermal performance of a heat sink can be adjusted (lowering thermal resistance) by varying the fan speed; however, achieving lower thermal resistance through increased fan speed leads to higher power consumption. This is particularly challenging due to the cubic relationship between fan speed and power consumption, which can significantly escalate cooling costs [5]. Moreover, increasing the fan speed can lead to higher noise levels, potentially affecting system reliability. Conversely, reducing the fan speed offers the benefit of extending the fan's lifespan. For instance, decreasing the fan speed by 35% can double its operational lifetime [6].

Key Aspects of the Problem:
- **Fan Power Consumption Optimization:** The objective is to reduce power consumption associated with fan operation without compromising on cooling effectiveness. This involves optimizing fan speed to achieve the desired thermal performance with minimal energy usage.

- **Lightweight TinyML Model:** Traditional machine learning models, while effective, can be resource-intensive and may not be feasible for deployment on constrained devices within telecom systems. In contrast, TinyML models offer a lightweight, efficient alternative that can run on-device with minimal computational overhead. This study emphasizes the advantages of TinyML for real-time fan speed prediction, leveraging its low power and resource requirements.

- **Limitations of Dynamic Fan Control:** Frequent adjustments to fan speed can lead to decreased fan lifespan. To mitigate this, the proposed model predicts fan speed in class intervals, reducing the frequency of control changes and extending the fan's operational life. This strategic approach ensures that fan adjustments are made judiciously, balancing temperature regulation with mechanical longevity.

By integrating these considerations, the proposed solution seeks to enhance the edge device thermal management strategy, achieving a synergy between energy efficiency and component reliability.

B. Solution proposed

The dynamic control of fan speed in edge devices presents a unique challenge that involves optimizing power consumption while maintaining effective thermal management. Our solution addresses this by leveraging advanced algorithms and machine learning techniques, focusing on three main aspects: fan power optimization, the implementation of a lightweight TinyML model, and enhanced fan control.

Fan Power Optimization:

To achieve optimal fan power consumption, we utilize a data-driven approach powered by the SMART (Systematic Management of Adaptive Response Technology) fan algorithm. This algorithm is central to our solution, as it facilitates the collection of real-time data from multiple sensors within the edge device. These sensors monitor critical components such as Network Processing Units (NPUs) and routing processors, providing vital information about the device's thermal state.

The SMART fan utility helps thorough data collection process to train its neural network effectively. This meticulous approach ensures that the collected data covers the full range of operational parameters of the equipment, thereby improving the system's accuracy and reliability. For example, the data includes a broad spectrum of operating temperatures, enabling the system to perform optimally across different thermal environments. Furthermore, it considers various load scenarios, including diverse traffic conditions and patterns, to ensure the equipment can meet a wide array of operational demands. By incorporating such extensive data, the SMART fan utility is better equipped to predict and adapt to changing conditions, enhancing its overall efficiency and performance.

Lightweight TinyML Model:

To implement fan speed predictions effectively within resource-constrained environments of edge devices, we employ a lightweight TinyML model. Traditional machine learning models, although effective, are often resource-intensive and unsuitable for real-time applications on constrained devices. In contrast, the TinyML model is specifically quantized to operate efficiently with minimal computational and power requirements.

The TinyML model provides a robust solution by enabling on-device inference, allowing the device to make real-time decisions regarding fan speed adjustments. This approach not only reduces the latency associated with cloud-based processing but also minimizes the energy overhead, aligning with our goal of optimizing fan power consumption. By integrating TinyML, we achieve a balance between maintaining device performance and ensuring energy efficiency.

Enhanced Fan Control:

Temperature fluctuations, caused by discrete fan speed adjustments, can lead to instability in fan operation. To address this, we introduce an enhanced fan control strategy that categorizes Pulse Width Modulation (PWM) values into distinct classes. This classification forms the basis for developing a predictive model that manages fan speed changes more effectively.

By grouping PWM values, the classification model can anticipate optimal fan speeds, smoothing out transitions and reducing unnecessary toggling between consecutive speed settings. This approach not only stabilizes fan speed control but also contributes to the overall efficiency and longevity of the cooling system. By minimizing frequent speed adjustments, we reduce wear and tear on the fan, further extending its operational life while ensuring consistent thermal management.

This comprehensive solution leverages the SMART fan algorithm, TinyML models, and classification-based fan control to address the challenges of dynamic fan speed management. By optimizing fan power consumption, implementing lightweight models, and enhancing control mechanisms, the proposed solution ensures effective cooling, energy efficiency, and component reliability within edge devices.

III. EXPERIMENT

A. Machine learning model preparation

Traditional fan controllers typically rely on a fixed mapping between the temperatures of critical components and fan speed, aiming to maintain these temperatures below a certain threshold. However, this approach often overlooks fan power consumption. To address this, we propose an intelligent, power saving fan speed model. Our method aims to select an appropriate fan speed that keeps the temperature of critical components close to the desired level while minimizing power consumption.

The presented technique involves multiple stages. Initially, an appropriate dataset is generated using the SMART fan utility, capturing the effects of various fan speeds and the temperatures of critical components. This dataset is then used to train different models, including SVM classifiers, random forest classifiers, and sequential neural networks. A detailed parametric study was conducted with each classifier to determine the optimal regularization parameters. The outcome as follows

- Best SVM parameters: {'C': 10, 'gamma': 'auto', 'kernel': 'rbf'}
- Best SVM accuracy: 0.9456957819894776
- Best Random Forest parameters: {'max_depth': None, 'min_samples_split': 2, 'n_estimators': 150}
- Best Random Forest accuracy: 0.9354129528240129
- Best Neural Network parameters: {'model__optimizer': 'adam', 'model__num_neurons': 64, 'model__activation': 'relu', 'epochs': 200, 'batch_size': 64}

- Best Neural Network accuracy: 0.9526640853599909

The trained neural network model is designed to select optimal fan speeds, with its architecture clearly defined in terms of neurons, weights, and activation functions. The fan model includes multiple input parameters and one output parameter, relying on a dataset that reflects the impact of different fan speeds on component temperatures.

SMART fan algorithm collected data from sensors monitoring NPUs and routing processors to dynamically determine optimal fan speeds. This data was normalized across classes for training the TinyML model. This data-driven approach allows the neural network to effectively learn and predict optimal fan speeds, improving cooling efficiency and ensuring component longevity.

The simple distribution of the data shown in the below picture.

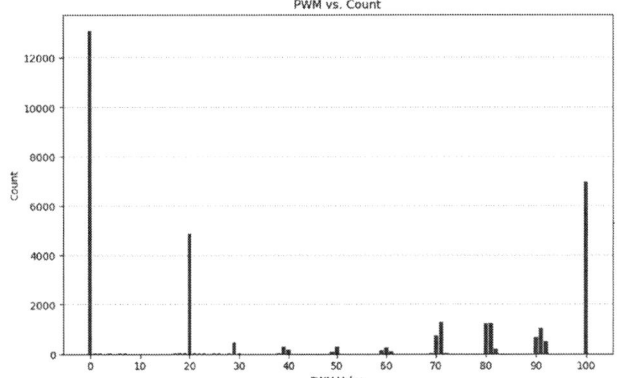

Figure 1. The distribution of data count with PWM

Due to minor fluctuations in temperature caused by the discretization of fan speeds, there can be instances where fan speed toggles between two consecutive discrete values. To address this issue, the fan PWM values are grouped into distinct classes. This grouping aids in developing a classification model that can more effectively predict optimal fan speeds. By categorizing the PWM values, the model can better manage the subtle variations in temperature, resulting in smoother transitions and more stable fan speed control. This approach minimizes unnecessary fan speed changes, enhancing the overall efficiency and longevity of the cooling system. The training data was grouped into following class intervals.

Class	PWM Range	Count
0	0	4000
1	1 - 20	4333
2	21 - 29	491
3	30 - 39	299
4	40 - 50	504
5	51 - 60	357
6	61 - 70	696
7	71 - 79	1061
8	80 - 81	1789
9	82 - 90	778
10	91 - 99	1450
11	100	4000

Table 1. Balanced class distribution of data

There are 24 input variables in the system.

15

S. No	Sensor	Pearson Correlation Coefficients	Decision tree Regression MSE	Decision tree Regression R² value
1	Sensor 1	0.929	152.01	0.91
2	Sensor 2	0.904	156.73	0.91
3	Sensor 3	0.908	193.6	0.89
4	Sensor 4	0.910	199.2	0.89
5	Sensor 5	0.910	190.45	0.89
6	Sensor 6	0.909	191.39	0.89
7	Sensor 7	0.909	191.52	0.89
8	Sensor 8	0.907	193	0.89
9	Sensor 9	0.882	195.65	0.89
10	Sensor 10	0.833	316.5	0.82
11	Sensor 11	0.908	176.84	0.9
12	Sensor 12	0.825	314.02	0.82
13	Sensor 13	0.831	310.2	0.82
14	Sensor 14	0.837	302.93	0.83
15	Sensor 15	0.837	306.8	0.83
16	Sensor 16	0.904	161.33	0.91
17	Sensor 17	0.191	1393.53	0.21
18	Sensor 18	0.737	623.25	0.65
19	Sensor 19	0.161	1404.21	0.21
20	Sensor 20	0.125	1313.96	0.26
21	Sensor 21	0.639	974.21	0.45
22	Sensor 22	0.297	1382.56	0.22
23	Sensor 23	0.909	180.38	0.9
24	Sensor 24	0.923	120.1	0.93

Table 2. Correlation between Input and output variables

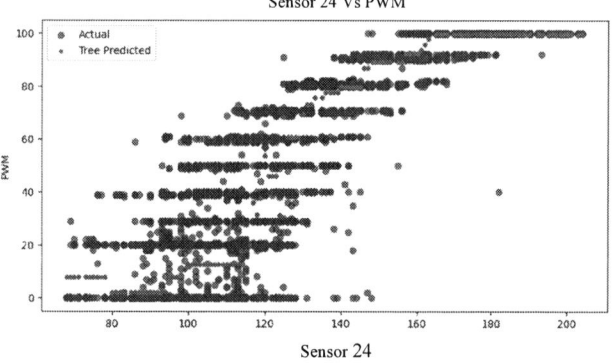

Figure 2. Correlation between Sensor 24 and PWM

To select input variables for predictive model, we can analyze the data provided using two key metrics: Pearson Correlation and Decision Tree Regression performance (MSE and R² values). Here's how to interpret these metrics and make informed decisions on which sensors to include as input variables:

Key Metrics:

Pearson Correlation: This measures the linear relationship between each sensor's readings and the target variable (e.g., PWM or fan speed). A value close to 1 indicates a strong positive correlation, while a value close to 0 suggests little to no linear relationship.

Decision Tree Regression Metrics:

- Mean Squared Error (MSE): This indicates the average squared difference between the predicted and actual values. Lower MSE values signify better predictive accuracy.
- R² Value: This measures the proportion of the variance in the target variable that is predictable from the input features. A value closer to 1 indicates a better fit.

Selecting Input Variables:

- Strong Candidates: Sensors with Pearson correlation and R² above 0.9, such as Sensor 1, 2, 16, 23, and 24.
- Moderate Candidates: Sensors with correlation from 0.7 to 0.9 and R² around 0.8 to 0.9, like Sensor 8 and 11.
- Weak Candidates: Sensors with correlation below 0.3 and R² below 0.5, e.g., Sensor 17, 19, and 20, unless critical.
- Reduce Redundancy: Opt for the most reliable sensors with similar data.
- Priority Sensors: Include Sensor 1, 2, 5, 8, 11, 15, 16, 23, and 24.

Algorithm 1: DNN Model Training
1: Import necessary components from TensorFlow.
2: Create a Sequential model.
3: Add a Dense layer with specified units, activation function, and input dimension, inner layers and final Dense layer with units equal to the number of classes and a softmax activation function.
4: Compile Model: Set optimizer, loss function and define performance metrics.
5: Train Model: Use fit with training data and labels. Set number of epochs, batch size, and validation split.
6: Track Training History: Retrieve training metrics from the model's history object.
7: Iterate or Adjust: Continue training or adjust parameters based on performance evaluation.

This algorithm outlines the sequential steps to set up, compile, and train a deep neural network model using regularization to enhance the training process by adjusting the learning rate and preventing overfitting.

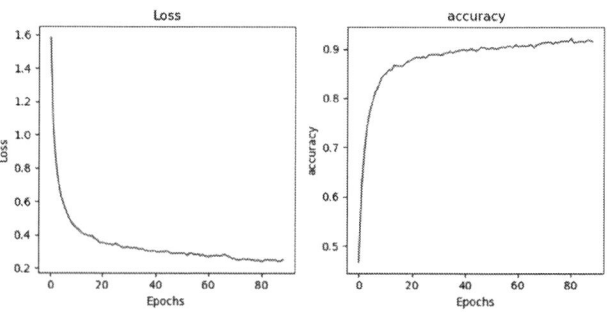

Figure 3. Residual plots

We developed a Fan Speed Predictor model using a dataset, which was pre-processed and balanced for equitable training. Key features were selected based on Pearson Correlation and decision tree analysis. A Deep Neural Network (DNN) was employed for its ability to capture complex patterns, resulting in a strong predictive performance with an error rate of less than 2%, showcasing effective feature selection and model design.

C. TinyML model preparation

In preparing the TinyML model, we adapted our machine learning strategies to fit the constraints and requirements of edge devices, focusing on minimal resource usage while maintaining high accuracy. Building on the foundational work with the Fan Speed Predictor and LiteRT model in few KBs ensuring lightweight processing.

The pre-processing stage involved further optimization techniques, such as quantization and pruning, to reduce the model size and computational demands without significant loss of accuracy. Key features identified through Pearson Correlation and decision tree analysis in previous models were retained to ensure the model's effectiveness in capturing essential data patterns.

We employed a simplified neural network architecture, tailored for low-power environments, which could still leverage the predictive capabilities demonstrated by our earlier models. By focusing on lightweight operations and efficient memory usage, the TinyML model was designed to provide real-time fan speed predictions, adapting to dynamic temperature changes on edge devices. This approach allows for smart, autonomous operation, enhancing cooling efficiency and promoting longevity in a variety of embedded systems.

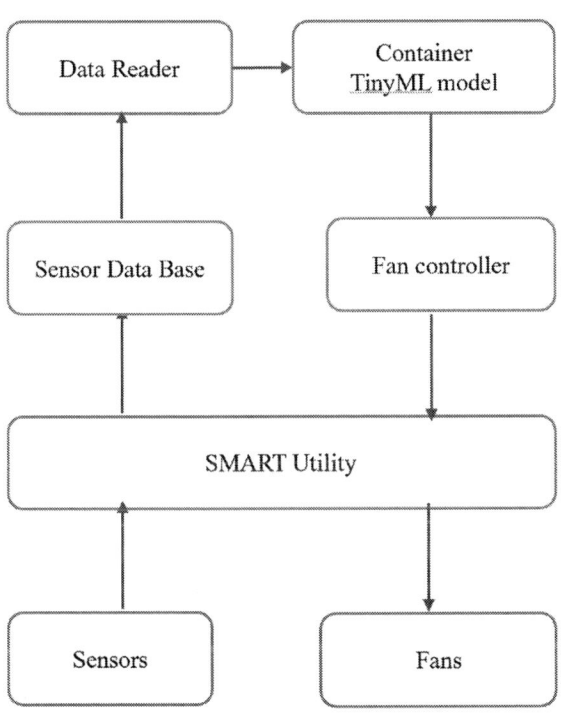

Figure 4: TinyML implementation in device

The SMART Utility platform is designed to optimize environmental conditions through a system that integrates sensors, data processing, and machine learning. Here's a concise overview of its components and functionality:

a

Components:
- Sensors: Collect real-time environmental data such as temperature, humidity, and air quality.
- SMART Utility: Collects and manage run-time data from sensors and helps program fan power.
- Sensor Database: Stores sensor data for input to model and historical tracking.
- Data Reader: Extracts data from the database and prepares it for machine learning processing.
- Containerized TinyML Model: A lightweight machine learning model that analyses run-time sensor data to predict optimum fan speed.
- Fan Controller: Adjusts fan operations based on the TinyML model's output.
- Fans: Actuators that control environmental conditions by adjusting ventilation or cooling.

Data Flow:
- Sensors → SMART Utility: Sensors send collected data to the SMART Utility platform.
- SMART Utility → Sensor Database: Data is stored for further analysis.
- Sensor Database → Data Reader → TinyML Model: Data is prepared and processed by the TinyML model.
- TinyML Model → Fan Controller: The model provides control signals to the fan controller.
- Fan Controller → Fans: Fans are adjusted to optimize environmental conditions.

This system automates the monitoring and control of environmental conditions using machine learning on edge devices, enabling efficient fan control and improved energy efficiency.

D. Experimental results

The implementation of the TinyML model for fan speed control demonstrated notable improvements in efficiency and adaptability. Leveraging its compact architecture, the model efficiently operated within the constraints of edge devices, with only KBs of memory. During the experiments, the TinyML model dynamically adjusted fan speeds based on real-time temperature readings and traffic conditions, optimizing cooling performance.

Simulated maximum traffic conditions were applied to test the model's responsiveness under high load scenarios. The TinyML approach showed a marked reduction in power consumption when compared to traditional static fan controllers, which rely on preset temperature thresholds to adjust fan speeds linearly. The model's adaptive capabilities allowed it to maintain optimal temperatures while minimizing energy use, as reflected in the comparative data presented in Table 3.

Ambient temp	Fan power (W)		% Saving
	TinyML	Existing	
28	11.5	16.20	29
38	21.5	31.4	32
46	31.2	46.90	33
55	69.3	87.2	21
65	93.7	93.7	0

Table 3. Fan power comparison

The experiments confirmed the TinyML model's effectiveness in diverse scenarios, as various target temperatures were tested to ensure comprehensive evaluation. The results underscored the model's potential to enhance cooling efficiency and extend component longevity, making it a viable solution for advanced thermal management in edge computing environments.

IV. CONCLUSION

Fan power consumption remains a significant constraint in enhancing the throughput and energy efficiency of edge devices. An effective fan speed controller is crucial for optimizing fan power usage. In this study, we introduced a TinyML-based fan speed control model aimed at reducing fan power consumption while preventing temperature threshold violations in edge devices. The TinyML model adjusts future fan speeds based on the current temperatures of critical components, adhering to temperature thresholds. We validated this approach on an edge device. Experimental results from practical benchmarks demonstrate that the TinyML model outperforms traditional fan controllers in power savings, with only a slight increase in average temperature that stays within acceptable limits. Our proposed technique improves mean fan power consumption by nearly 30% compared to traditional controllers, with an approximate 7.5% increase in temperatures of critical components.

ACKNOWLEDGMENT

I thank Prof. Murali P, BITS Pilani, for his essential guidance. I am also grateful to Ramesh Kumar Velugoti and Prashanth Pavithran for their support and resources.

REFERENCES

[1] G. Refai-Ahmed, I. Barber, A. Torza and B. Philofsky, "A holistic view of chip-level thermal architecture from heterogeneous stacked dice to system level in telecoms applications," *2015 International 3D Systems Integration Conference (3DIC)*, Sendai, Japan, 2015, pp. TS10.1.1-TS10.1.4, doi: 10.1109/3DIC.2015.7334615.

[2] Iranfar, F. Terraneo, G. Csordas, M. Zapater, W. Fornaciari and D. Atienza, "Dynamic Thermal Management with Proactive Fan Speed Control Through Reinforcement Learning," *2020 Design, Automation & Test in Europe Conference & Exhibition (DATE)*, Grenoble, France, 2020, pp. 418-423, doi: 10.23919/DATE48585.2020.9116510.

[3] V. S. Simon, A. Siddarth and D. Agonafer, "Artificial Neural Network Based Prediction of Control Strategies for Multiple Air-Cooling Units in a Raised-floor Data Center," *2020 19th IEEE Intersociety Conference on Thermal and Thermomechanical Phenomena in Electronic Systems (ITherm)*, Orlando, FL, USA, 2020, pp. 334-340, doi: 10.1109/ITherm45881.2020.9190431.

[4] Alajlan NN, Ibrahim DM. TinyML: Enabling of Inference Deep Learning Models on Ultra-Low-Power IoT Edge Devices for AI Applications. Micromachines (Basel). 2022 May 29;13(6):851. doi: 10.3390/mi13060851. PMID: 35744466; PMCID: PMC9227753.

[5] M.K. Patterson, "The effect of data center temperature on energy efficiency," In Thermal and Thermomechanical Phenomena in Electronic Systems, 2008. ITHERM 2008. 11th Intersociety Conference on, pp. 1167-1174, 2008.

[6] G. Paparrizos, "An Integrated fan speed control solution can lower system costs, reduce acoustic noise, power consumption and enhance system reliability," Technical report, Microchip Technology Inc, 2003.

[7] I. Goodfellow, Y. Bengio, A. Courville, and Y. Bengio, Deep learning. 2016.

[8] K. Willems, "Keras Tutorial: Deep Learning in Python," *https://www.datacamp.com/*, 2019

[9] Aurélien Géron - Hands-On Machine Learning with Scikit-Learn, Keras, and TensorFlow_ Concepts, Tools, and Techniques to Build Intelligent Systems

[10] Seth Weidman - Deep Learning from Scratch Building with Python from First Principles

[11] Eugene Charniak - Introduction to Deep Learning (2019, The MIT Press)

[12] Pete Warden, Daniel Situnayake, TinyML: Machine Learning with TensorFlow Lite on Arduino and Ultra-low-power Microcontrollers.

A Practical Metric for Cold Plate Thermal Performance in Two-Phase Direct-to-Chip Cooling

Qingyang Wang, Serdar Ozguc, Richard W. Bonner III
Accelsius
Austin, TX, USA
qwang@accelsius.com

Abstract

The increasing chip-level and rack-level power densities in data centers have necessitated liquid cooling solutions, in which pumped two-phase (2P) direct-to-chip (DTC) cooling shows promising performance and attracts great attention. The cold plate used in a 2P DTC system is a critical component dictating the cooling performance of the system. The thermal performance of a cold plate for single-phase DTC cooling is commonly characterized by the case-to-fluid thermal resistance with the inlet temperature as the characteristic fluid temperature. For 2P DTC, a case-to-fluid thermal resistance R_{cf} has been used to describe cold plate thermal performance with the characteristic fluid temperature taken as the saturation temperature, or the weighted average fluid temperature considering both subcooled sensible heat and latent heat. In this work, we developed thermohydraulic analysis and showed that R_{cf} can fail to represent 2P cold plate thermal performance and result in faulty conclusions in certain conditions, where a lower thermal resistance value could associate with higher case temperature, given server-level and rack-level conditions unchanged. We proposed a case-to-outlet thermal resistance R_{co}, which incorporates the temperature rise caused by the 2P pressure drop. R_{co} is more practically accessible and physically accurate, making it a good performance metric for 2P cold plates in 2P DTC systems for data centers.

Keywords

two-phase cooling, direct-to-chip, cold plate, thermal resistance, data center

Nomenclature

A_{chip}	area of the chip, [m^2]
Bo	boiling number
$c_{p,l}$	liquid specific heat capacity, [J/kg K]
d_h	hydraulic diameter, [m]
F_K	a constant in the Kandlikar correlation
f	friction factor
G	mass flux, [kg/m^2 s]
H_{ch}	channel height, [m]
h	heat transfer coefficient, [W/m^2 K]
h_{fg}	latent heat of vaporization, [J/kg]
k	thermal conductivity, [W/m K]
L	length, [m]
m	fin parameter, [m^{-1}]
$\dot{m}$	mass flow rate, [kg/s]
N_{ch}	number of channels
Nu	Nusselt number
P	pressure, [Pa]
Q	heat/power, [W]
q''_{fp}	footprint heat flux, [W/m^2]
q''_w	wall heat flux, [W/m^2]
R	thermal resistance, [K/W]
R_{cf}	case-to-fluid thermal resistance, [K/W]
R_{co}	case-to-outlet thermal resistance, [K/W]
Re	Reynolds number
T	temperature, [°C]
T_{case}	case temperature, [°C]
T_f	fluid average temperature, [°C]
t_{base}	base plate thickness, [m]
u	velocity, [m/s]
W_{ch}	channel width, [m]
W_{fin}	fin width, [m]
x	vapor quality

Greek symbols

ΔP	pressure drop, [Pa]
η	fin efficiency
μ	viscosity, [Pa·s]
ρ	density, [kg/m^3]

Subscripts

ac	acceleration component
ave	average between inlet and outlet
boil	boiling heat transfer process
CBD	convective boiling dominant
Cu	copper
ch	channel
fr	friction component
in	inlet of the cold plate
l	liquid phase
lat	latent heat part
le	liquid only
m	mean value for two-phase mixture
NBD	nucleate boiling dominant
nom	nominal
out	outlet of the cold plate
sat	saturation condition
sen	sensible heat part
tot	total heat
v	vapor phase

Acronyms

1P	single-phase
2P	two-phase
AI	artificial intelligence
CDU	coolant distribution unit
CP	cold plate
CPU	central processing unit
DTC	direct-to-chip

GPU   graphics processing unit
GWP   global warming potential
HTC   heat transfer coefficient
QD   quick-disconnect
TDP   thermal design power
TIM   thermal interface material

1. Introduction

The rapidly developing artificial intelligence (AI) technology has posed serious challenges to the thermal management of data centers, where traditional air convective cooling is no longer able to cope with the increasing thermal design power (TDP) from advanced processors and AI accelerators in densely populated server racks. Liquid cooling solutions are considered inevitable to keep up with the soaring cooling demand, among which two-phase (2P) direct-to-chip (DTC) cooling shows great promise. 2P DTC cooling offers high heat transfer potential using the latent heat of a dielectric refrigerant, requiring lower flow rate while ensuring no disastrous damage to IT equipment in case of leakage.

2P cooling has been extensively studied in the academic community. The forms of liquid-vapor phase change employed include pool boiling [1], flow boiling [2], thin film evaporation [3], spray cooling [4], and jet impingement [5]. Significant enhancements in heat transfer have been demonstrated using both water and dielectric fluids, with demonstrated heat flux dissipation approaching or exceeding 1 kW/cm² [6-8]. Due to its high performance, 2P heat transfer has been employed by thermal engineers for data centers. Compared with 2P immersion, 2P DTC allows retrofitting of existing data centers while still offering the high performance of 2P heat transfer.

A number of publications already exist for 2P DTC cooling for data centers. Heydari et al. [9] evaluated cold plates and cooling loops using experiments and models and provided suggestions on design considerations for 2P DTC cooling. Heydari et al. [10] analyzed the performance of different refrigerants for 2P DTC cooling and discussed the effects of different operating conditions. Wang et al. [11] established an experimental system for server-level 2P DTC characterization and demonstrated low thermal resistance up to a TDP of 1000 W using R1233zd(E). Ozguc et al. [12] modeled the rack-level flow distribution and demonstrated successful 2P flow control through the implementation of flow restrictors. Narayanan et al. [13] demonstrated high-performance cooling using 2P DTC with R1233zd(E) for a high-power GPU thermal test vehicle with a TDP of up to 2.2 kW. Wang et al. [14] conducted experiments at the cold plate-level and compared the performance of R1233zd(E) and R515B for 2P DTC cooling on different thermal test vehicles. Kulkarni et al. [15] and Wang et al. [16] introduced the concept of using a universal cold plate for both single-phase (1P) and 2P DTC cooling and demonstrated better thermal performance of 2P than 1P, suggesting that a faster and cheaper adoption of 2P DTC technology with existing 1P DTC infrastructure is possible.

For 1P DTC cooling, the case-to-fluid thermal resistance of a cold plate is widely used as the performance metric to evaluate cold plates, where the characteristic fluid temperature is taken as the cold plate inlet temperature. On the contrary, since 2P DTC for data center cooling is still in its infancy, there is no unanimously accepted performance metrics for 2P cold

plates in data center thermal management. Many prior works used the saturation temperature to calculate the case-to-fluid thermal resistance [11, 13-16]. In this work, we provide analysis of some exemplary 2P cold plates, and show that the case-to-fluid thermal resistance defined using the saturation temperature (or the average temperature considering both 1P and 2P) could fail to represent 2P cold plate performance under certain conditions, with a lower resistance potentially leading to worse performance. We propose a case-to-outlet thermal resistance to serve as the performance metric for 2P cold plates, which incorporates the pressure drop of 2P flow in the cold plate. This metric correctly represents cold plate performance under given server-level and system-level conditions, and is easily accessible during cold plate testing. This allows data center engineers to quickly and accurately evaluate 2P cold plates during product development and product comparison, especially for people without a 2P thermal background, and helps to bridge the gap between 2P heat transfer research and data center cooling applications.

2. 2P DTC Fundamentals

Recently, Kulkarni et al. gave a comprehensive introduction and discussion of the 2P DTC cooling systems for data center thermal management [17]. The main components of a 2P coolant distribution unit (CDU) include the pump, the condenser, the reservoir, and the tubes, hoses, and fittings (including the quick-disconnect (QD) fittings) connecting all the components. Compared with a 1P DTC CDU, a refrigerant reservoir is added into a 2P DTC system to accommodate the volume expansion during heat load variations due to vastly different densities between liquid and vapor. The CDU can be in-rack or in-row, providing cooling for a single rack or for multiple racks, respectively. The CDU can also use liquid, air, or another circulated refrigerant as the coolant for the condenser. The different forms of 2P CDUs do not affect the analysis of this work, and an in-rack refrigerant to liquid CDU is employed in this work as an example.

The basic flow diagram of an in-rack 2P DTC system for data center cooling is shown in Figure 1. During operation, both vapor and liquid phases coexist in the reservoir and the refrigerant remains saturated. The liquid refrigerant is pumped from the saturated reservoir into the liquid manifold, distributed into the servers populated in the rack, and delivered to the cold plates attached to the high-power processors (CPUs/GPUs/AI accelerators). The refrigerant dissipates the heat and vaporizes, and exits the cold plates in the form of

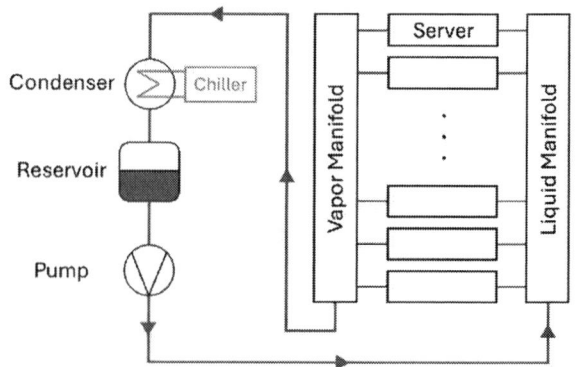

Figure 1. *Flow diagram of a 2P DTC system.*

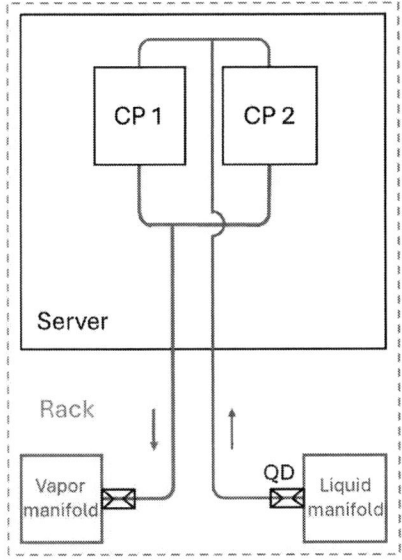

Figure 2. Top view schematic of a two-processor server cooled in a 2P DTC system.

liquid/vapor mixture. The saturated mixture enters the vapor manifold, and then gets condensed back into liquid and returns to the reservoir.

Figure 2 shows a top-view schematic of a server cooled in a 2P DTC system. For simplicity, the server is assumed to have two high-power processors. Within the server, the two cold plates (CPs) are hydraulically connected in parallel. The hoses delivering liquid into the server and taking saturated 2P mixture out of the server are interfaced with the liquid and vapor manifolds through QD couplings, which ensures spill-free and leak-free connection/disconnection of the flow paths without having to drain the refrigerant from any components.

3. Existing Performance Metric

Cold plates are one of the most important components dictating the thermal performance of a DTC system, either 1P or 2P. The thermal resistance of a cold plate is a representation of its cooling performance and is thus widely used as the performance metric for cold plates. The case-to-fluid thermal resistance is commonly used to describe cold plate performance, defined by the difference between the case temperature of the chip package and the characteristic fluid temperature divided by the dissipated total heat (the processor power). The case temperature can be taken as the maximum temperature across the case surface, the temperature at the center point of the case surface, or the average temperature across the case surface, depending on different applications and requirements.

The case-to-fluid thermal resistance, based on its definition, is a comprehensive lumped total resistance including contributions from thermal interface materials (TIMs) and base plate conduction. For 1P DTC systems, the characteristic fluid temperature is taken as the inlet coolant temperature T_{in}. However, in a 2P cold plate, liquid-vapor phase change process occurs under the constant saturation temperature T_{sat}. The inlet fluid is usually subcooled with $T_{in} < T_{sat}$. As 1P contribution in 2P cold plates is usually small, it is not thermally reasonable to use T_{in} in thermal resistance calculation for 2P.

The heat transfer in a 2P cold plate can be divided into two processes: 1) subcooled liquid absorbs heat as 1P liquid to raise its temperature from T_{in} to T_{sat}; 2) saturated liquid absorbs heat and transitions into vapor under T_{sat}. Consequently, the characteristic fluid temperature in the cold plate can be calculated as the energy-weighted average of these two processes:

$$T_f = \frac{Q_{sen}}{Q_{tot}} \frac{T_{in} + T_{sat}}{2} + \frac{Q_{lat}}{Q_{tot}} T_{sat} \quad (1)$$

where $Q_{tot} = Q_{sen} + Q_{lat}$. For a given refrigerant mass flow rate $\dot{m}$, the 1P sensible heat contribution Q_{sen} is

$$Q_{sen} = \dot{m} c_{p,l}(T_{sat} - T_{in}) \quad (2)$$

and the latent heat contribution Q_{lat} is

$$Q_{lat} = \dot{m} x_{out} h_{fg} \quad (3)$$

where x_{out} is the outlet vapor quality of the saturated mixture. A large vapor quality close to 1 ensures minimized pumping power, but could result in low thermal performance and potential dry-out. On the other hand, as 2P boiling usually has much higher heat transfer coefficient (HTC) than 1P convection, a small vapor quality indicates excessive flow rate and unwanted large 1P contribution. The designed vapor quality for a 2P cold plate is usually ~0.7 to avoid dry-out and flooding simultaneously.

With the characteristic fluid temperature T_f defined, the case-to-fluid thermal resistance for 2P cold plates is obtained as

$$R_{cf} = \frac{T_{case} - T_f}{Q_{tot}} \quad (4)$$

In the designed working conditions when 2P process dominates over 1P, T_f is very close to T_{sat} (usually within 1 °C), and given the much larger temperature difference between T_{case} and T_{sat} (>20 °C), it is convenient to use T_{sat} to replace T_f in Eq. (4) without introducing any significant error.

4. Model

4.1. Model formulation

Here we establish a model with classical correlations to analyze the performance of a set of microchannel-based 2P cold plates. To compare the thermal performance of different cold plates in a 2P DTC system, the chip-level, server-level, and rack-level conditions should all be maintained the same. The chip-level heating power and the form factor of the heating surface are fixed. The cold plates have the same mass flow rate. Given fixed CDU design, server plumbing, and operating conditions, the facility water temperature and flow rate determine the temperature/pressure of the saturated liquid in the reservoir as well as the pressure drop from the outlet of the cold plates to the reservoir. Consequently, the cold plate inlet temperature T_{in} (equal to the reservoir temperature) and outlet temperature/pressure for the saturated 2P mixture T_{out} and P_{out} are all fixed.

The 2P microchannel cold plates modeled here are used to cool a processor with a TDP of 2000 W. The processor has a form factor of 50×70 mm^2 and generates uniform heat flux on the case surface. The microchannel projected area matches the heated area, with the channel length L_{ch} matching the long edge of the chip surface. Figure 3 shows a drawing of the microchannel structures with the important geometrical

Figure 3. *Geometrical parameters of the microchannel cold plate in the model.*

parameters labeled. The channels are closed at the top and adjacent channels are sealed. The base plate thickness t_{base} is taken as 2.5 mm with a copper thermal conductivity of 390 W/mK, and the TIM thermal resistance R_{TIM} is taken as 10 mm²K/W. The inlet fluid is subcooled at 35 °C, representing a simplified case when the facility cooling water outlet temperature is 35 °C and the CDU heat exchanger performs very efficiently. The outlet fluid is set as saturated 2P mixture at 45 °C, which is consistent with our rack-level experiments.

Refrigerant R1233zd(E) is employed as the working fluid. The thermophysical properties of the fluid are all evaluated at the outlet temperature of 45 °C and are assumed to be temperature independent. The mass flow rate of the cold plate is obtained by prescribing a nominal exit vapor quality x_{nom} of 0.7 for the given processor total power,

$$\dot{m} = \frac{Q_{tot}}{h_{fg} x_{nom}} \quad (5)$$

The nominal vapor quality ignores the 1P contribution. Given the channel geometrical parameters (Figure 3), the number of channels N_{ch} can be calculated based on the width of the processor. Uniform flow distribution is assumed such that the mass flow rate inside each channel is the same and can be calculated by $\dot{m}_{ch} = \dot{m}/N_{ch}$.

By assuming uniform heat flux on the processor surface, there are two segments along the flow length: a first 1P convection segment where subcooled liquid is heated to saturation pressure, and a second 2P flow boiling segment where saturated liquid vaporizes and the fluid exits the channels as a saturated liquid/vapor mixture. In the 2P segment, the flow is considered homogeneous and treated as 1P flow with 2P average properties. the 1P flow length L_{1P} is obtained as

$$L_{1P} = \frac{Q_{1P}}{Q_{tot}} L_{ch} \quad (6)$$

where Q_{1P} is the sensible heat,

$$Q_{1P} = \dot{m} c_{p,l} \left(T_{2P,in} - T_{in} \right) \quad (7)$$

$T_{2P,in}$ is the inlet temperature of the 2P segment, which is dependent on the pressure drop of the 2P segment. Consequently, an initial guess of $T_{2P,in} = T_{out}$ is taken to proceed the modeling with an iterated calculation until $T_{2P,in}$ is converged.

4.2. Channel pressure drop

The pressure drop along the channel is calculated by summing the 1P pressure drop and the 2P pressure drop. The 1P pressure drop is frictional, and given the small channel size, the flow is laminar, and the pressure drop can be calculated by the Darcy-Weisbach equation as

$$\Delta P_{1P} = \frac{L_{1P} f_{1P} \rho_l u_{1P}^2}{2 d_h} \quad (8)$$

where $u_{1P} = G_{ch}/\rho_l$ is the 1P flow velocity in the channel, $G_{ch} = \frac{\dot{m}_{ch}}{W_{ch} H_{ch}}$ is the channel mass flux. d_h is the channel hydraulic diameter, $f_{1P} = 64/Re_{1P}$ is the 1P laminar friction factor, $Re_{1P} = G_{ch} d_h/\mu_l$ is the 1P Reynolds number. It is noted that the constant value of 64 for fRe is valid for laminar flow in circular tubes, and value would be different for the rectangular channel shape here. Nonetheless, equations for circular tubes are implemented for simplicity, since 1P contribution is generally much smaller than 2P.

The 2P pressure drop in the channel includes friction and acceleration contributions. The friction pressure drop is obtained by

$$\Delta P_{2P,fr} = \frac{L_{2P} f_{2P} \rho_m u_{2P}^2}{2 d_h} \quad (9)$$

where $L_{2P} = L_{ch} - L_{1P}$ is the 2P flow length. For homogeneous 2P flow, the mixture density is calculated by

$$\rho_m = \left(\frac{x_{ave}}{\rho_v} + \frac{1 - x_{ave}}{\rho_l} \right)^{-1} \quad (10)$$

The average vapor quality in the 2P segment is $x_{ave} = x_{out}/2$, where the outlet vapor quality

$$x_{out} = \frac{Q_{2P}}{\dot{m} h_{fg}} = \frac{Q_{tot} - Q_{1P}}{\dot{m} h_{fg}} \quad (11)$$

The average 2P velocity is obtained by

$$u_{2P} = \frac{\dot{m}_{ch}}{\rho_m W_{ch} H_{ch}} \quad (12)$$

and the 2P friction factor f_{2P} is calculated depending on the flow type:

$$f_{2P} = \begin{cases} \dfrac{64}{Re_{2P}}, & \text{if } Re_{2P} \leq 2300 \\ (0.790 \ln Re_{2P} - 1.64)^{-2}, & \text{if } Re_{2P} > 2300 \end{cases} \quad (13)$$

where the average 2P Reynolds number is calculated by

$$Re_{2P} = \frac{G_{ch} d_h}{\mu_m} \quad (14)$$

in which the average mixture viscosity μ_m is obtained by

$$\mu_m = \left(\frac{x_{ave}}{\mu_v} + \frac{1 - x_{ave}}{\mu_l} \right)^{-1} \quad (15)$$

In most cases, $Re_{2P} < 2300$ and the flow is laminar due to the small channel size.

The 2P acceleration pressure drop in the channel can be calculated by

$$\Delta P_{2P,ac} = G_{ch}^2 \left(\frac{1}{\rho_{out}} - \frac{1}{\rho_l} \right) \quad (16)$$

where the outlet mixture density is calculated by

$$\rho_{out} = \left(\frac{x_{out}}{\rho_v} + \frac{1 - x_{out}}{\rho_l} \right)^{-1} \quad (17)$$

The outlet temperature is given, and the outlet fluid is saturated at T_{out} and P_{out}. Consequently, the inlet temperature of the 2P segment $T_{2P,in}$ can be obtained as the saturation temperature for the inlet pressure of the 2P segment $P_{2P,in}$, where

$$P_{2P,in} = P_{out} + \Delta P_{2P,fr} + \Delta P_{2P,ac} \tag{18}$$

and the obtained $T_{2P,in}$ is used to continue another iteration of calculation until convergence.

4.3. Channel heat transfer

Similar to pressure drop, the heat transfer characteristics in the channels are also based on the two-segment model. In the 1P segment, the convective HTC inside the channel is obtained by

$$h_{1P} = \frac{Nu_{1P}k_l}{d_h} \tag{19}$$

The 1P Nusselt number Nu_{1P} is taken as 4.36 for simplicity, which is the case for fully developed laminar flow inside a circular tube with constant wall heat flux.

In the 2P segment, the HTC can be estimated using the Kandlikar correlation [18]:

$$h_{2P} = \max(h_{NBD}, h_{CBD}) \tag{20}$$

where the nucleate boiling dominant and convective boiling dominant HTCs are calculated by

$$h_{NBD} = 0.6683 \left(\frac{\rho_l}{\rho_v}\right)^{0.1} x_{ave}^{0.16}(1 - x_{ave})^{0.64}h_{le} \\ + 1058.0 Bo^{0.7}F_K(1 - x_{ave})^{0.8}h_{le} \tag{21}$$

$$h_{CBD} = 1.1360 \left(\frac{\rho_l}{\rho_v}\right)^{0.45} x_{ave}^{0.72}(1 - x_{ave})^{0.08}h_{le} \\ + 667.2 Bo^{0.7}F_K(1 - x_{ave})^{0.8}h_{le} \tag{22}$$

Due to the laminar flow regime, the liquid only HTC h_{le} is equal to h_{1P}. The value of fluid-dependent parameter F_K is not available for R1233zd(E), and is taken as 1 in this work [19]. The boiling number Bo is defined by

$$Bo = \frac{q_w''}{G_{ch}h_{fg}} \tag{23}$$

Taking into consideration the fin efficiency, the wall heat flux q_w'' is calculated as

$$q_w'' = q_{fp}'' \frac{W_{ch} + W_{fin}}{W_{ch} + 2\eta H_{ch}} \tag{24}$$

where the footprint heat flux is defined by

$$q_{fp}'' = \frac{Q_{tot}}{A_{chip}} \tag{25}$$

with A_{chip} being the area of the processor (heated area). The fin efficiency is calculated by

$$\eta = \frac{\tanh(mH_{ch})}{mH_{ch}} \tag{26}$$

$$m = \sqrt{\frac{2h_{2P}}{k_{Cu}W_{fin}}} \tag{27}$$

where k_{Cu} is copper thermal conductivity (390 W/mK). The fin efficiency η is dependent on the 2P HTC, and an iteration is needed to converge the value.

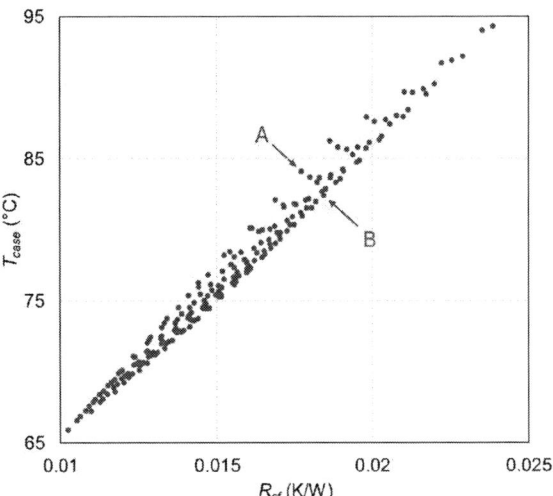

Figure 4. *Modeling results for different microchannel geometrical parameters.*

With the 1P and 2P HTCs obtained, the overall HTC along the channel wall is calculated as

$$h_{ch} = \left(\frac{Q_{1P}}{Q_{tot}}\frac{1}{h_{1P}} + \frac{Q_{2P}}{Q_{tot}}\frac{1}{h_{2P}}\right)^{-1} \tag{28}$$

The average case temperature is then obtained by

$$T_{case} = T_f + \frac{q_w''}{h_{ch}} + q_{fp}''\left(\frac{t_{base}}{k_{Cu}} + R_{TIM}\right) \tag{29}$$

where the characteristic fluid temperature T_f is given by

$$T_f = \frac{Q_{1P}}{Q_{tot}}\frac{T_{in} + T_{2P,in}}{2} + \frac{Q_{2P}}{Q_{tot}}\frac{T_{2P,in} + T_{out}}{2} \tag{30}$$

in which the T_{sat} in Eq. (1) is replaced by the average saturation temperature along the 2P segment.

5. Results and Discussion

Calculations are performed under different channel geometrical parameters. The fin width (0.15~0.25 mm), channel width (0.15~0.25 mm), and channel height (1~2 mm) are all swept for their respective ranges of values. Figure 4 shows the data points plotted as the case temperature against R_{cf}. In general, the case temperature increases with increasing R_{cf}, as expected. However, the trend is not monotonic, and for some cases, a lower R_{cf} would associate with a higher case temperature. For example, with $(W_{fin}, W_{ch}, H_{ch})$ = (0.23, 0.15, 1) mm for Point A in Figure 4, R_{cf} is 0.0177 K/W and T_{case} is 84.1 °C; whereas for the Point B, with geometries of (0.21, 0.25, 1.4) mm, R_{cf} is 0.0184 K/W and T_{case} is 82.4 °C. This indicates that R_{cf} could potentially give faulty conclusions when comparing two 2P cold plates under the exact same system-level conditions (with the same flow rate, the same subcooled inlet temperature, and the same outlet temperature). That is mainly because the fluid temperature T_f varies with different channel geometries, because the HTC and the pressure drop both increase with decreasing channel size. Consequently, in some cases, a smaller channel with a higher HTC could also have a higher saturation temperature in the

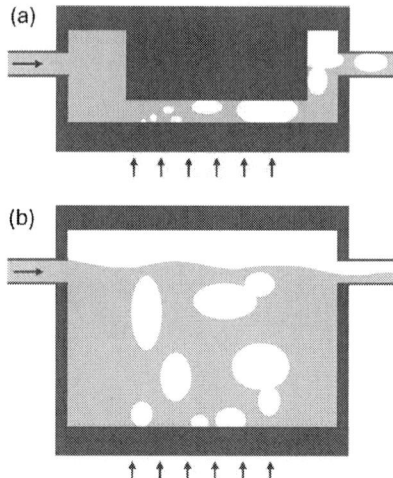

Figure 5. *(a) A microchannel flow boiling based cold plate with high pressure drop across the channels and the outlet manifold. (b) A pool boiling based cold plate with negligible internal pressure drop.*

channel given a constant outlet temperature, resulting in higher case temperature.

It is worth noting that the model is not aimed to be absolutely accurate to predict thermal performance of a cold plate. Instead, it is developed primarily to show that potential error exists in using R_{cf} to measure 2P cold plate performance. Therefore, the model is not numerically structured, simplified assumptions and correlations are used, and flow boiling instabilities are not considered. Moreover, the model neglects the pressure drop from outlet of the channels to the exit tube through an outlet plenum or manifold, which could be significant and result in more complexity in the pressure distribution. It is difficult to quantitatively model the pressure drop of the 2P flow with unknown flow pattern through a plenum/manifold with multi-dimensional shape. Nevertheless, it can be expected qualitatively that this pressure drop brings more possibility for R_{cf} defined by Eq. (4) to fail in describing thermal performance. For example, if two cold plates have the exact same boiling/convection HTC (and hence same R_{cf}), the one with higher outlet manifold pressure drop would have a higher characteristic fluid temperature in the channel, and consequently result in a higher processor case temperature. In an extreme case, as schematically shown in Figure 5, a microchannel based cold plate with very high pressure drops both along the channel and across the outlet (e.g., with a small neckdown to connect to a certain fitting) can have higher HTC and lower R_{cf} than a pool-boiling based cold plate with almost no pressure drop across itself, but can also have much higher fluid saturation temperature due to the pressure drop, and therefore potentially result in higher case temperature.

The modeling results above have shown that in a 2P system, the temperature distribution is not only dependent on temperature itself, but also on pressure. The case temperature is of most practical interest for thermal management purposes, and the outlet temperature is dictated by the server- and system-level operating conditions. Therefore, we propose a case-to-outlet thermal resistance R_{co}, defined by

$$R_{co} = \frac{T_{case} - T_{out}}{Q_{tot}} \quad (31)$$

Figure 6 shows the schematic drawing of the important temperature locations in a 2P cold plate, and the thermal resistance network. Similar to R_{cf}, R_{co} also includes contributions from TIM R_{TIM}, conduction across the base plate R_{cond}, and convective/boiling resistance within the cold plate R_{boil} (which may also include the subcooled 1P convection). Additionally, R_{co} also includes a thermal resistance R_{dp} resulted from fluid temperature reduction due to 2P pressure drop from the heat transfer surface to the outlet. Consequently, any rise of fluid characteristic temperature within the cold plate due to internal pressure drop of the 2P flow can be included within the R_{co} value. As discussed before, in a 2P DTC cooling system, T_{out} is determined by the system-level conditions, since it equals the reservoir fluid temperature plus the temperature drop both across the condenser and along the vapor return lines. Therefore, a higher R_{co} will always lead to a higher case temperature given fixed server- and rack-level conditions, making it a good performance indicator for 2P cold plates.

During the research and development phase of 2P cold plates, as well as evaluation and comparison of 2P cold plate products, experimental testing and performance characterization are conducted under conditions benefiting the implementation of R_{co}. For practical cold plate characterization, the local fluid temperature within the cold plate is usually not available, and imbedding additional temperature sensors inside the cold plate could be intrusive to affect the performance, or compromise the mechanical integrity under high internal pressure. Common tests are conducted with a temperature sensor placed outside of the cold plate, usually somewhere along the outlet tubing. Therefore, some prior characterizations [11, 13-16] of 2P cold plates reported case-to-fluid resistance R_{cf} while the reported values are in fact R_{co}. That means that the parameter R_{co} is not only accurate in capturing the true thermal performance considering the pressure drop contributions neglected before, but also practical in experimental testing and characterization of a cold plate by requiring no internal thermal sensor.

It is noted that the error caused by using R_{cf} most likely occurs when the system uses low-pressure refrigerants (such as

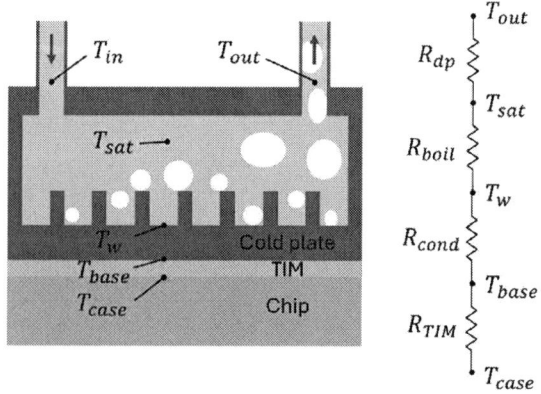

Figure 6. *Schematic drawing and thermal resistance network of a 2P DTC cold plate.*

R1233zd(E) used in this work) with high-power cooling, since the pressure drop is significant with small vapor density and large flow rate. When medium- to high-pressure refrigerants are used, the pressure drop inside cold plates can become minimal, so that $T_{out} \approx T_{sat}$ and R_{cf} value is approaching R_{co} (see Figure 6). Consequently, using R_{cf} might only result in negligible differences. For example, if we arbitrarily consider a 3 °C case temperature rise caused by R_{dp} to be a significant deterioration of performance, that translates to a 3 °C temperature drop from T_{sat} to T_{out}. Given a T_{out} of 45 °C, it corresponds to a ~3 psi pressure drop from the boiling site to the outlet for R1233zd(E), which is not impossible if the cold plate is not optimized (e.g. with long flow paths and narrow flow passages; complicated outlet manifold; neck-down at the outlet, etc.) and when high TDP requires large flow rate. However, for the same 3 °C temperature drop, the corresponding pressure drop for R515B (a medium-pressure refrigerant) becomes ~9 psi, which is highly unlikely for a reasonable cold plate design, especially since the larger vapor density of R515B already tends to produce smaller pressure drop given similar flow rate [14, 20]. Therefore, an estimation of the pressure drop within the cold plate based on refrigerant selection and working condition could be a quick and easy method to determine whether the proposed performance metric R_{co} is needed. When the pressure drop inside the cold plate can cause a rise of T_{case} larger than the threshold of tolerance, R_{co} will be needed to incorporate the pressure drop contribution and provide a fair evaluation of the performance. In other words, the necessity of using R_{co} is dependent on the refrigerant selection, designed flow rate, and the tolerance of temperature change due to cold plate pressure drop. Nonetheless, even when R_{co} does not result in material difference from R_{cf} and is not necessary, it is still easily accessible experimentally, since the outlet temperature is easier to obtain than the saturation temperature inside the cold plate.

It is also worth noting that the above discussion was based on a parallel configuration shown in Figure 2. When the cold plates are configured in series, the downstream cold plate receives a 2P mixture at its inlet, and the inlet manifold/plenum pressure drop could be significant, which increases the outlet temperature of the upstream cold plate given fixed rack-level conditions. In an exemplary case, if two downstream cold plates have different inlet 2P pressure drops, the same upstream cold plate will have different case temperatures. However, the situation described here does not contradict the analysis above. The pressure drop characteristics of the downstream cold plate should be viewed as a server-level feature when evaluating the upstream cold plate. Hence, the different case temperatures for the upstream cold plate should be attributed to the variation of server-level operating conditions instead of the cold plate itself.

6. Conclusions

With 2P cooling showing a promising future for data center thermal management, a performance metric for 2P cold plates is needed to compare cold plate performance and improve cold plate design [21]. In this work, we analyzed the thermal performance of microchannel 2P cold plates, and demonstrated that the case-to-fluid thermal resistance R_{cf} used in prior works based on saturation temperature can fail to represent the thermal performance of 2P cold plates. That is because the

saturation temperature at which the fluid boils can be higher than the outlet fluid temperature determined by the system-level conditions, due to 2P pressure drop induced temperature reduction. Therefore, for 2P DTC cooling systems, we introduced a practical performance metric for 2P cold plates as the case-to-outlet thermal resistance R_{co}, which includes the thermal resistance related to internal pressure drop of the 2P flow. A higher R_{co} would always result in higher case temperature given fixed server- and system-level conditions, making it an accurate performance indicator for 2P cold plates, which is also easily accessible during cold plate testing. The R_{co} metric provides data center engineers with a quick method to evaluate cold plates in 2P DTC cooling, and reminds cold plate designers to take into consideration the pressure drop effect on thermal performance.

References

1. G. Liang, I. Mudawar. "Review of pool boiling enhancement by surface modification." International Journal of Heat and Mass Transfer 128 (2019): 892-933.

2. S. G. Kandlikar. "History, advances, and challenges in liquid flow and flow boiling heat transfer in microchannels: a critical review." Journal of Heat Transfer 134.3 (2012): 034001.

3. J. L. Plawsky, A. G. Fedorov, S. V. Garimella, H. B. Ma, S. C. Maroo, L. Chen, Y. Nam. "Nano-and microstructures for thin-film evaporation—A review." Nanoscale and microscale thermophysical engineering 18.3 (2014): 251-269.

4. J. Kim. "Spray cooling heat transfer: The state of the art." International Journal of Heat and Fluid Flow 28.4 (2007): 753-767.

5. L. Qiu, S. Dubey, F. H. Choo, F. Duan. "Recent developments of jet impingement nucleate boiling." International Journal of Heat and Mass Transfer 89 (2015): 42-58.

6. Q. Wang, and R. Chen. "Ultrahigh flux thin film boiling heat transfer through nanoporous membranes." Nano letters 18.5 (2018): 3096-3103.

7. K. P. Drummond, D. Back, M. D. Sinanis, D. B. Janes, D. Peroulis, J. A. Weibel, S. V. Garimella. "A hierarchical manifold microchannel heat sink array for high-heat-flux two-phase cooling of electronics." International Journal of Heat and Mass Transfer 117 (2018): 319-330.

8. C. Woodcock, C. Ng'oma, M. Sweet, Y. Wang, Y. Peles, J. Plawsky. "Ultra-high heat flux dissipation with Piranha Pin Fins." International Journal of Heat and Mass Transfer 128 (2019): 504-515.

9. A. Heydari, Y. Manaserh, A. Abubakar, C. Caceres, H. Miyamura, A. Ortega, J. Rodriguez. "Direct-to-Chip Two-Phase Cooling for High Heat Flux Processors." International Electronic Packaging Technical Conference and Exhibition. Vol. 86557. American Society of Mechanical Engineers, 2022.

10. A. Heydari, O. Al-Zu'bi, Y. Manaserh, M. Mehrabi, F. Hosseini, B. Sammakia. "System-Level Assessment of Green Refrigerant Replacements for Direct-to-Chip Two-Phase Cooling." 2024 23rd IEEE Intersociety Conference on Thermal and Thermomechanical Phenomena in Electronic Systems (ITherm). IEEE, 2024.

11. Q. Wang, S. Ozguc, A. Narayanan, R. W. Bonner. "A Server-Level Test System for Direct-To-Chip Two-Phase Cooling of Data Centers Using a Low Global Warming Potential Fluid", 2024 23rd IEEE Intersociety Conference on Thermal and Thermomechanical Phenomena in Electronic Systems (ITherm). IEEE, 2024.

12. S. Ozguc, Q. Wang, A. Narayanan, R. W. Bonner. "Investigation of Flow Restrictors for Rack Level Two-Phase Cooling Under Nonuniform Heating." 2024 40th Semiconductor Thermal Measurement, Modeling & Management Symposium (SEMI-THERM). IEEE, 2024.

13. A. Narayanan, Q. Wang, S. Ozguc, R. W. Bonner. "Investigation of Server Level Direct-to-Chip Two-Phase Cooling Solution for High Power GPUs", International Electronic Packaging Technical Conference and Exhibition. Vol. 88469. American Society of Mechanical Engineers, 2024.

14. Q. Wang, A. Narayanan, S. Ozguc, J. D. Moore, R. W. Bonner. "Performance Comparison of R1233zd(E) and R515B for Two-Phase Direct-to-Chip Cooling", 2025 24th IEEE Intersociety Conference on Thermal and Thermomechanical Phenomena in Electronic Systems (ITherm). IEEE, 2025.

15. D. Kulkarni, J. Gulick, J. King, B. Jarrett, P. George, Y. Feldman. "Thermal Performance of Common Cold Plate for Pumped Single- and Two-Phase Direct Liquid Cooling for Next Generation High Power Server Processors", International Electronic Packaging Technical Conference and Exhibition. Vol. 88469. American Society of Mechanical Engineers, 2024.

16. Q. Wang, D. P. Kulkarni, R. W. Bonner, J. C. Gulick. "Universal Direct-to-Chip Cold Plates for Single- and Two-Phase Cooling", OCP Global Summit 2024 Future Technologies Symposium.

17. D. Kulkarni, R. Tipton, E. Leka, J. Bankston, Q. Wang. "2P Refrigerant-Based Direct Liquid Cooling (DLC) Technology for Next Generation AI Clusters with High TDP Accelerators", OCP White Paper.

18. S. G. Kandlikar. "Heat transfer mechanisms during flow boiling in microchannels." Journal of Heat Transfer 126.1 (2004): 8-16.

19. V. P. Carey. *Liquid-vapor phase-change phenomena: an introduction to the thermophysics of vaporization and condensation processes in heat transfer equipment.* CRC Press; 2020 Feb 28.

20. R. Tipton, A. Wise, S. Mutabdzija, C. Shores, D. Kulkarni. "Maturation of Pumped Two-Phase Liquid Cooling to Commercial Scale-Up Deployment." International Electronic Packaging Technical Conference and Exhibition. Vol. 88469. American Society of Mechanical Engineers, 2024.

21. S. Ozguc, Q. Wang, A. Narayanan, J. D. Moore, R. W. Bonner. "Design Optimization of Manifold Integrated Skived Cold Plates for Two-Phase Flow-Boiling." 2025 41st Semiconductor Thermal Measurement, Modeling & Management Symposium (SEMI-THERM). IEEE, 2025.

This page intentionally left blank.

Validation of 1-D Thermo-Fluid Analysis Tool for Liquid Cooling Systems in Data Center Product: A Comparison Study with 3-D Simulations and Experimental Data

Jiwon Yu, Bharath Ram Ravi, Stephen Keefe

Celestica

2791 Telecom Pkwy, Richardson, TX 75082

jiwon.yu@celestica.com

Abstract

Advances in artificial intelligence (A.I.) technology has brought about the need for hardware systems for data centers with significantly higher performance than before. Although air cooling system is still necessary to cool various components, it is inevitable to implement liquid cooling system on key components with high heat density such as GPUs, and CPUs. In this study, commercial 1-D nodal analysis tool is evaluated by comparing the results with those from 3-D CFD simulation and experiments. Although both simulation results showed reasonable prediction over the system, 1-D nodal analysis cannot capture detail features of the system and sometimes may lead over or under design. Also, 3-D CFD simulation is still indispensable to evaluate the performance of the core parts including cold plate. Nevertheless, 1-D nodal analysis tool provides a scalability with suitable accuracy and affordable time to build and analyze larger systems over rack level, starting from detailed simulation on the crucial parts using 3-D CFD analysis tools, which stands for the two types of analysis are complementary to each other.

Keywords

Liquid cooling, Servers, One dimensional analysis

Nomenclature

$\dot{q}$	Heat dissipation (W)
ΔP	Pressure drop (Pa)
Q	Volumetric flow rate (lpm)
T_{in}	Mean temperature of the coolant at inlet (°C)
T_{out}	Mean temperature of the coolant at outlet (°C)
T_{liquid}	Average temperature of T_{in} and T_{out} (°C)
θ	Thermal resistance (°C/W)
$A1 - A8$	DIMM set A
$B1 - B8$	DIMM set B

1. Introduction

The rapid advances in artificial intelligence (A.I.) technology and the Internet of Things (IoT) has driven unprecedented growth in high-performance data centers. As processing demands increase, the power consumption and corresponding heat generation of server components such as CPUs and GPUs have escalated significantly, reaching levels where traditional air-cooling methods struggle to provide adequate thermal management [1]. Consequently, liquid cooling systems or hybrid liquid-air cooled systems have emerged as a critical solution due to their superior thermal conductivity and specific heat capacity [2]. Despite its efficiency, liquid cooling system design poses challenges due to its complexity. Computational tools like three-dimensional (3-D) computational fluid dynamics (CFD) simulations provide detailed thermal and flow predictions but are computationally expensive, especially for large-scale data center configurations [3]. In designing liquid cooled systems for electronics, the ability to rapidly access options and to be able to rely on the results is necessary. One-dimensional (1-D) nodal analysis tools have gained popularity due to their ability to model entire systems quickly and with reasonable accuracy [4].

In this study, a commercial 1-D thermo-fluid analysis tool is evaluated by comparing the results with those from 3-D CFD simulations to experimental results. Both simulation tools predicted reasonable results but, 1-D analysis cannot capture detail features of the system and sometimes may lead to over or under design. In this regard, 3-D CFD simulation is still indispensable to evaluate the performance of the cold plate. 1-D analysis tools provide a scalability with suitable accuracy along with rapid build time to analyze larger systems at the rack level. A reasonable method is to start from detailed simulation of the non-standard parts using 3-D CFD to characterize those parts and that information is fed into the 1-D tool. The 1-D tool then can be used to rapidly parameterize the system as a whole. The two analysis tools thus complement each other and the complementary use of both approaches ensures balanced design accuracy and computational efficiency.

2. System overview

The system under investigation is a 2U, half width server with a hybrid liquid-air cooling system which contains two Intel Xeon Sapphire Rapids CPU chips and sixteen Samsung 64GB DIMM modules. Table 1 shows the list of liquid cooled components and their thermal design power. CPU temperatures were measured by machining the CPU lid and attaching a thermocouple. DIMM temperatures were read using internal temperature sensors. The test was conducted at a 25°C ambient air environment while inlet coolant temperature was kept at 40°C.

Table 1 Liquid cooled components and thermal design power

Component	$\dot{q}$ (W)	Quantity	Total $\dot{q}$ (W)
Intel Xeon Sapphire Rapids CPU	270	2	540
Sumsung 64GB DIMM	10	16	160

Figure 1 The server with a liquid-air cooling system

Figure 1 illustrates the system under test with the liquid cooling system and **Error! Reference source not found.** provides the reference designators of components in the liquid cooling loop of the system and demonstrates how the two different branches split and merge along with the flow directions. Each branch contains a cold plate designed to cool high-power CPU at 270 W in addition to ten cold plates array which are dedicated to cool eight DIMM modules. Copper is used for the CPU cold plate base and fins while a cover is made of aluminum. DIMM cold plates are the array of the narrow channels made of stainless steel and the connecting hoses are made of PTFE. The overall system pressure drop and coolant temperature were measured at the system's inlet and outlet.

PG-25, a mixture of water and polypropylene glycol with the ratio of 75% to 25%, is used as the coolant in this study. The coolant gains heat from the CPUs and DIMMs when it passes through the cooling loop and the hot coolant releases the heat via a three-kilowatt liquid to air external coolant distribution unit (CDU) which maintains the coolant temperature at an optimum inlet temperature of 40°C. In addition to the liquid cooling loop for the components with high-power density, three 40 mm dual rotor fans were used to generate the airflow required to cool other low power components such as DC-DC converters which are not included in the liquid cooling loop.

.

3. Analysis details

3.1. 1-D analysis

The liquid cooling loop consists of various components such as tubes with differing cross-sectional areas, elbows, tee junctions, manifolds, and barbed fittings. One dimensional (1-D) nodal network analysis tool, use a modeling technique known as flow network modeling to model these components. The governing flow and thermal equations are simplified and solved under the assumption that property variations occur in only one dimension. Flow losses in bends, expansions and contraction joints are estimated using existing correlations.

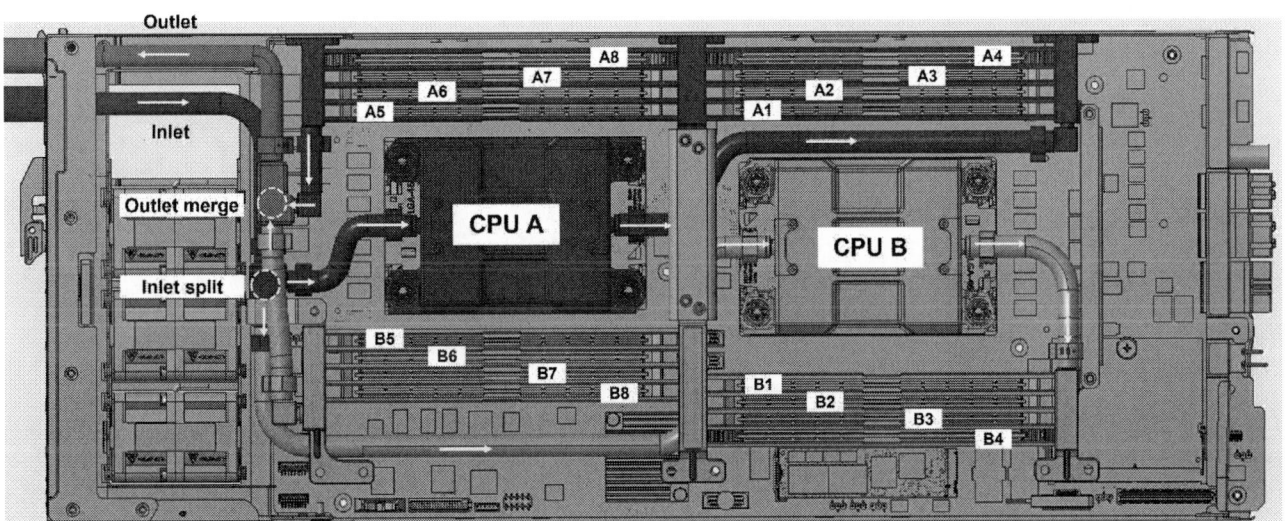

Figure 2 Description of the liquid cooling system

Studies have demonstrated that complex flow systems with multiple pathways can be accurately represented as 1-D networks through the integration of components such as tee junctions, elbows, and reducers. The authors compared the performance of two different 1-D tools in a previous study modelling the same system and discussed in detail the modelling methodology and shared results [5].

In this study, a similar methodology is employed to model the cooling system, incorporating cold plates and manifolds.

Tools designed for electronics cooling commonly feature a built-in cold plate component in their libraries. This functionality allows users to input cold plate data such as thermal resistance and pressure drop. On the other hand, this feature means that 1-D nodal analysis needs information from the test or 3-D CFD simulation in order to model non-standard component such as cold plates. For the 1-D analysis in this study, data on the thermal resistance and pressure loss of the CPU cold plate was applied directly by the cold plate manufacturer. Figure 3 illustrates the portion of the model featuring a built-in cold plate component. Figure 4 depicts the section of the model where the manifold is represented using Y-junctions, elbow segments, and fittings.

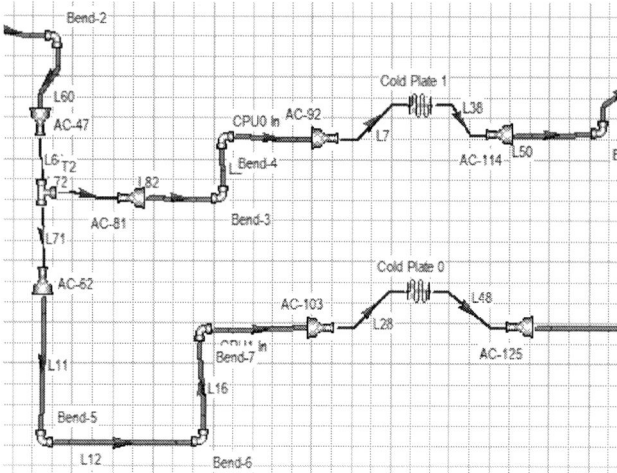

Figure 3 1-D simulation tool showing CPU cold plate feature

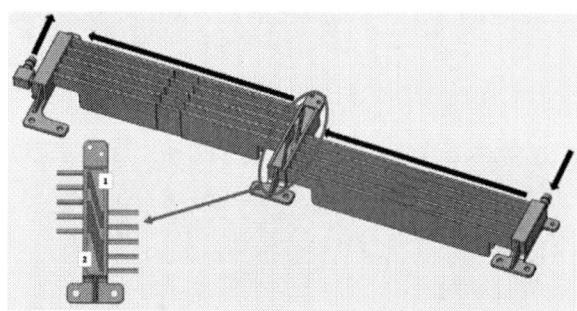

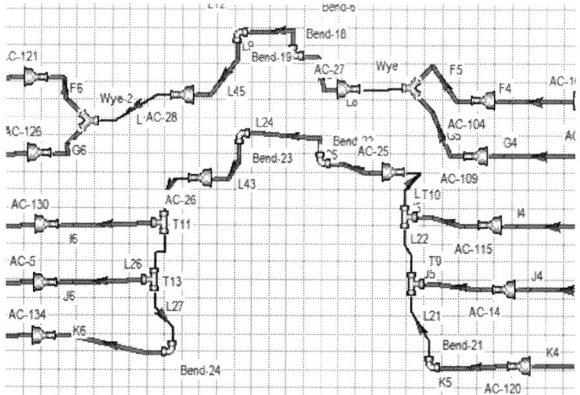

Figure 4 1-D simulation tool showing manifold between DIMM cold plates

3.2. 3-D CFD analysis

3-D thermal model was built by importing geometry from CAD as shown in Figure 5. Thermal model provided by the vendor was used for modeling CPUs. Since the aim of this

study is to investigate and compare the results of the liquid cooling loop obtained from the experiment, 1-D analysis, and 3-D CFD simulation, components outside the liquid cooling loop were not modeled. The three 40 mm dual rotor fans were modeled to capture the convective heat transfer in the system by the airflow. Table 2 describes the thermal properties of the coolant, PG25. The coolant flow path is demonstrated in Figure 6. The ambient temperature was set to 25°C while the inlet liquid temperature was set to 40°C.

Table 2 Thermal properties of PG25

Property	Unit	Value
k	W/m·K	0.47
ρ	kg/m³	1,014
c_p	J/kg·K	3,955

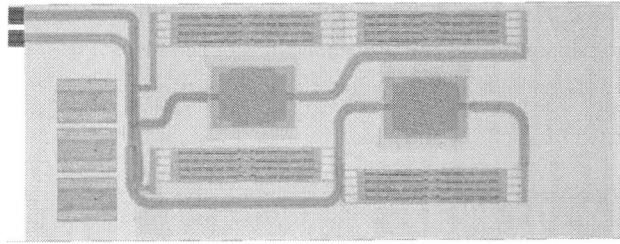

Figure 5 3-D thermal simulation model

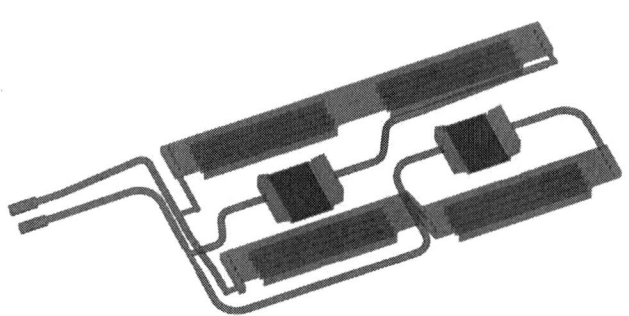

Figure 6 Liquid flow path in the system

4. Results and discussion

4.1. System pressure drop

The overall hydraulic performance of the liquid cooling system was evaluated, compared, and is illustrated in Figure 7. Both 1-D nodal analysis and 3-D CFD simulations provided good prediction of entire pressure drop over the cooling loop compared with the experimental data. A relatively large difference over 10% was found between 1-D nodal analysis result and the experimental result at high flow rate (2 lpm) while that of 3-D CFD simulation demonstrated a good agreement with the experiment. The assumptions in the 1-D nodal analysis solver regarding laminar to turbulent flow transition and the correlations used for pressure loss prediction might cause this discrepancy. In this regard, some 1-D nodal analysis tools provide the option to select the laminar to turbulent transition Reynolds number.

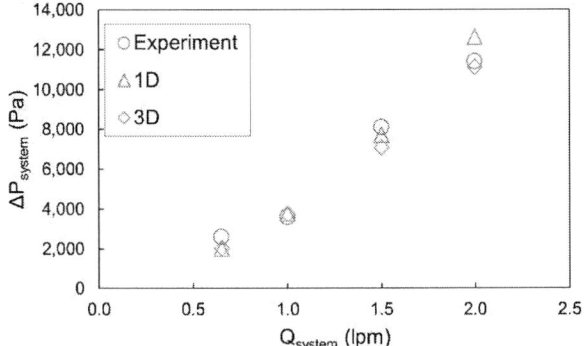

Figure 7 System pressure drop as a function of system flow rate

Flow rate and corresponding pressure drop across different sections at the flow rate of 1.5 lpm is shown in Table 3. This data was not measured in the experiments. Both 1-D nodal analysis and 3-D CFD simulation results demonstrated higher flow rate from branch B than that from branch A, with a deviation of less than 3%. Both CPU cold plates contribute 30~40% of the total pressure drop due to their dense fin design, which restricts the flow. It is also demonstrated in Table 3 that both 1-D nodal analysis and 3-D CFD simulation results show the pressure drop through DIMM set B is approximately 50% higher than that through DIMM set A. This is because of the connection manifold between DIMM B1~B4 and DIMM B5~B8, which has an inclined flow path as shown in Figure 2, causing additional pressure loss, whereas the manifold between DIMM A1~A4 and DIMM A5~A6 is straight.

According to Table 3, significantly large portion of the entire pressure drop comes from the pipes and fittings which is 36% and 40% of the total when evaluated by 1-D nodal analysis and 3-D CFD simulation, respectively. This result addresses the importance of the investigation and optimization of the pressure losses through the pipes and fittings during the early stage of the design phase. Using 1-D nodal analysis tools allow engineers to quickly obtain the maximum and minimum velocities as well as the locations of them. Local velocities are important design factors since it is a main contributor to pressure loss through a system. Building 3-D CFD simulation models at rack level and beyond requires significant effort and time. In contrast scaling up with the 1-D nodal analysis approach is much easier and fast. In this regard, it is beneficial to use 1-D nodal analysis tool for these kind of evaluations since 1-D nodal analysis tool provides the results very fast with reasonable accuracy even though 3-D CFD simulation predicts hydraulic and thermal behavior of the system more accurately.

Table 3 Flow rate and pressure drop across each section at the flow rate of 1.5 lpm

	Q (lpm)		ΔP (Pa)	
	1-D	3-D	1-D	3-D
System	1.50	1.50	7,700	7,098
CPU A	0.72	0.70	1,010	1,326
CPU B	0.78	0.80	1,189	1,500
DIMM A	0.72	0.70	1,050	541
DIMM B	0.78	0.80	1,690	874
Pipes and fittings	-	-	2,761	2,857

4.2. CPU temperature

The case temperatures of CPU from 1-D nodal analysis are calculated using the equation (1) ~ (3). The thermal resistance of the cold plate ($\theta_{CPUColdPlate}$) is obtained from vendor's testing data. Thermal grease was used as the TIM between the case of CPU and the cold plate pedestal with 42 psi loading pressure. The thermal resistance of TIM ($\theta_{TIM,CPU}$) was estimated to be 0.16°C·cm²/W taking into account the mounting pressure and surface tolerances based on the previous experience in using this TIM.

$$\dot{q} = \rho \times c_p \times w \times (T_{out} - T_{in}) \qquad (1)$$

$$T_{c\_CPUColdPlate} = \dot{q} \times \theta_{CPUColdPlate} + T_{in} \qquad (2)$$

$$T_{c\_CPU} = \dot{q} \times \theta_{TIM,CPU} + T_{c\_CPUColdPlate} \qquad (3)$$

The case temperatures of CPU which were measured or evaluated by experiments or 1-D/3-D simulations, respectively, are described in Figure 8. Both 1-D analysis and 3-D CFD simulations demonstrated the inversely proportional trend for CPU case temperature to the system flow rate. However, the absolute temperature values from 1-D analysis are off by up to 2.4°C compared to the test data, while those from the 3-D CFD simulation quite well match with the test data with less than 1°C discrepancy. This indicates that higher accuracy can be achieved from 3-D CFD simulations than 1-D nodal analysis.

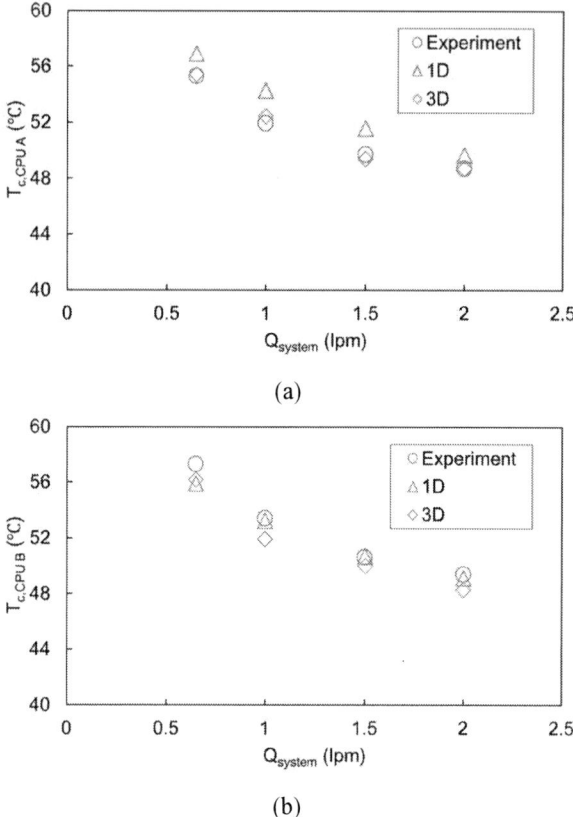

(a)

(b)

Figure 8 Comparison of CPU temperatures at different system flow rates between experiment, 1-D nodal analysis, and 3-D CFD simulation for (a) CPU A and (b) CPU B

4.3. DIMM temperature

The case temperatures of DIMM are calculated using equation (4) for 1-D analysis results. In this equation, it is assumed that the liquid temperature, T_{liquid}, is the average of

inlet and outlet liquid temperatures through the DIMM cold plate since the cold plate case temperature cannot be obtained from the 1-D nodal analysis tool and the power for the DIMMs are fairly low, causing negligible temperature rise across cold plate. In addition, the thermal resistance between DIMM case and the cold plate, $\theta_{TIM,DIMM}$, was estimated to be 0.0808 °C/W.

$$T_{c\,DIMM} = \dot{q} \times \theta_{TIM,DIMM} + T_{liquid} \qquad (4)$$

Figure 9 and Figure 10 show the comparison plot for DIMM temperatures from the experiments, 1-D analysis, and 3-D CFD simulations. In these figures, experimental and 3-D CFD simulation results demonstrate interesting trend that the DIMM temperatures located on the sides (A1, A4, A5, A8, B1, B4, B5, and B8) are lower than those at the center (A2, A3, A6, A7, B2, B3, B6, and B7). However, this trend was not clearly visible in the 1-D analysis results. This discrepancy mainly caused by the convective heat transfer from airflow through the system. DIMMs located in the sides of the bank are more exposed to the airflow and more heat is dissipated to the air from these DIMMs than from the DIMMs at the center. While this effect is illustrated in the experiments and 3-D CFD simulations, 1-D analysis tool does not capture the convective heat transfer from the DIMM cold plate.

In addition, there are some deviations in the absolute values of the temperatures between experimental results and 1-D/3-D simulation results even though 3-D CFD simulation predicts the general trend. This discrepancy might be associated with the several reasons. First possible reason is the non-uniform contact between the cold plate and DIMMs due to the friction fit during the installation. Additionally, thermal dissipation power might not have been uniform and could have varied across the forty individual memory modules on each DIMM. However, the calculations for DIMM temperatures based on the results from 1-D analysis and 3-D CFD simulation assumes uniform power distribution.

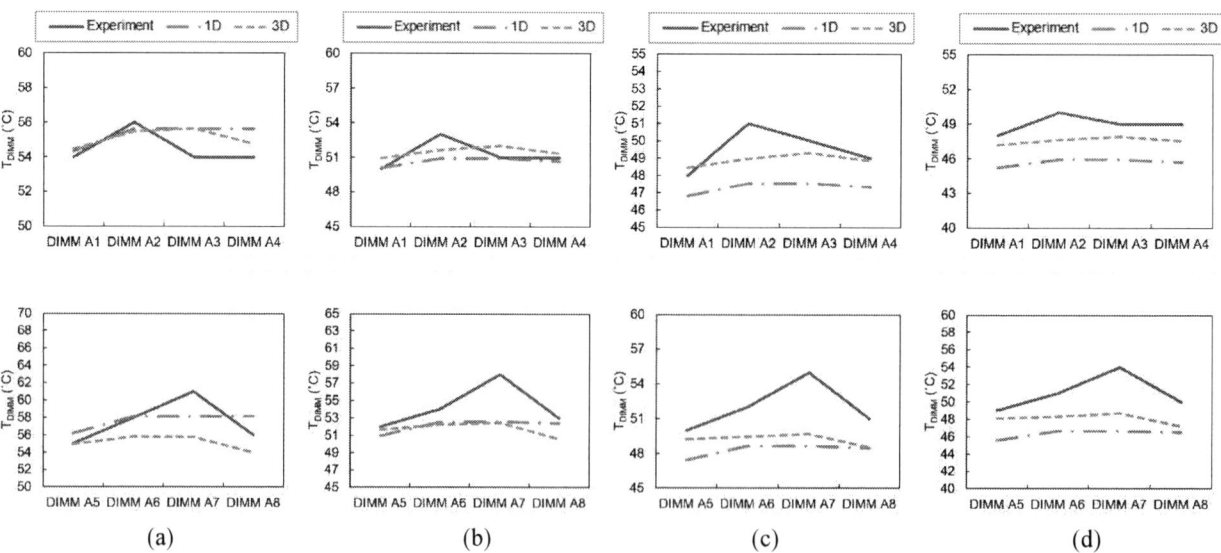

Figure 9 Comparison of DIMM A temperatures between experiment, 1-D analysis, and 3-D simulation at the system flow rate of (a) 0.65 lpm, (b) 1 lpm, (c) 1.5 lpm, and (d) 2 lpm

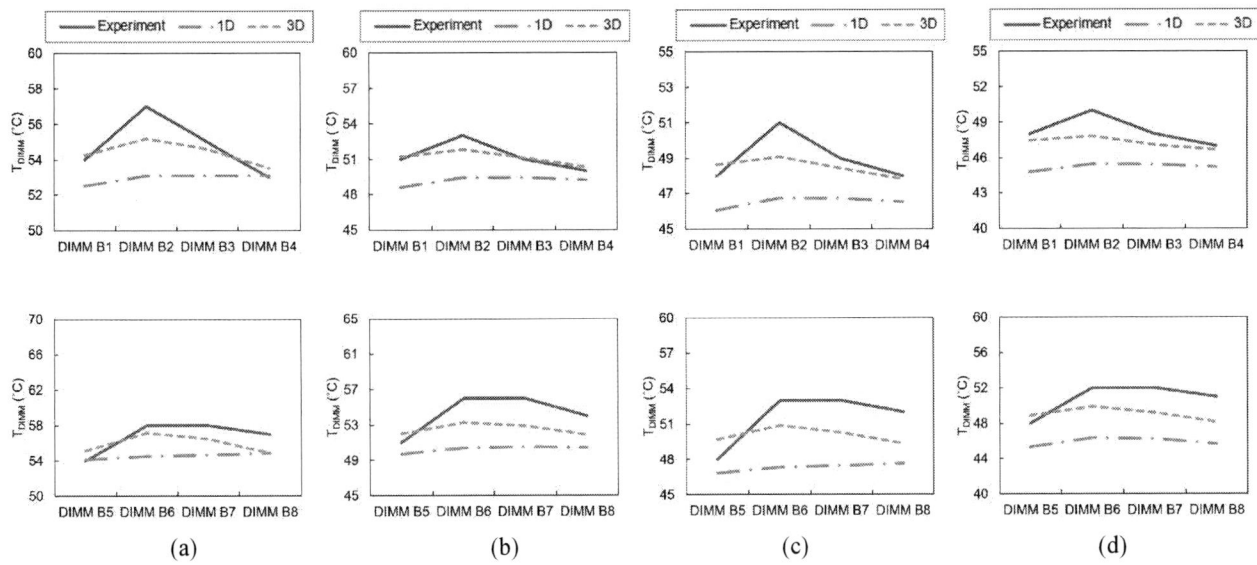

Figure 10 Comparison of DIMM B temperatures between experiment, 1-D analysis, and 3-D simulation at the system flow rate of (a) 0.65 lpm, (b) 1 lpm, (c) 1.5 lpm, and (d) 2 lpm

5. Conclusions

The traditional CFD analytical commercial tools have been used for liquid cooling analysis and design at the system level. However, it is cumbersome in modeling the entire liquid loop from the cold plate, through the manifolds to the coolant distribution unit (CDU). By applying the 1-D nodal network analysis, and CFD analysis at the system level, the current work demonstrated that they both have good results compared to empirical data. The CFD analysis results are closer to the empirical results.

This study provides confidence to users in applying 1-D nodal network analysis at the appropriate level, at sizing the overall liquid cooling system. Thermal engineers designing liquid cooling systems, should understand when it is appropriate to applying each type of analysis. The two types of analysis are complementary to each other.

Acknowledgments

We would like to thank the thermal team from Celestica, Shanghai for their support in this project.

References
1. Kheirabadi, A. C., & Groulx, D. "Cooling of server electronics: A design review of existing technology." Applied Thermal Engineering, 105 p622-638, 2016.
2. Schmidt, R. R. "Liquid Cooling Is Back." Electronics Cooling, 11(3), 2005.
3. Fernandes, J., et al. "Multi-design variable optimization for a fixed pumping power of a water-cooled cold plate for high-power electronics applications." 13th InterSociety Conference on Thermal and Thermomechanical Phenomena in Electronic Systems. IEEE, 2012.
4. Kelkar, K. M., & Patankar, S. V. "Analysis and design of liquid-cooling systems using flow network modeling (FNM)." International Electronic Packaging Technical Conference and Exhibition, 2003.
5. Bharath Ram Ravi, Jiwon Yu, Stephen Keefe, Herman Chu, and Chase Wealther. "Review of Commercial One-Dimensional Thermo-Fluid Solvers for Liquid Cooled Server System." In International Electronic Packaging Technical Conference and Exhibition, vol. 88469, p. V001T02A002. American Society of Mechanical Engineers, 2024.

Design Optimization of Manifold Integrated Skived Cold Plates for Two-Phase Flow-Boiling

Serdar Ozguc, Qingyang Wang, Akshith Narayanan, Jacob Moore, Richard W. Bonner III

Accelsius, Austin, TX, USA

sozguc@accelsius.com

Abstract

Two-phase cooling shows promise for data center applications due to high heat transfer coefficients and heat capacities associated with the boiling phenomena as well as the dielectric nature of the available two-phase coolants. Skived copper cold plates are commonly implemented in data center liquid cooling due to their low cost and ability to achieve fin and channel sizes on the order of 100s of μm. The demand for dissipation of increasing heat fluxes on the next generation of CPU and GPU chips necessitates the need for case-specific design of the cold plate parameters such as channel width, fin thickness, and channel height. To this end, this work performs numerical design optimization of skived cold plates for a set of operating conditions representative of the currently adopted chips in data center applications. Furthermore, a nylon insert to the cold plate package is designed as a manifold to the skived channels to reduce the pressure drop. An empirical convective heat transfer model is developed and calibrated using experimental data collected from two skived cold plates tested under various heat loads and mass flow rates. Developed correlation for the flow-boiling heat transfer coefficient matches with the measured data with less than 20% error. Optimization of the fin thickness and channel width is performed for a range of fin heights and with three different lower bounds on the fin thickness to represent manufacturing constraints. The overall thermal resistance of the optimized cold plates is up to 30% lower than the unoptimized cold plates used during model calibration. Reducing the lower bound on fin thickness enhances the thermal performance, however, reducing the fin thickness limit from 100 to 50 μm provides only a 2% improvement because the overall package resistance is dominated by the thermal interface and base thickness resistances. While thermal performance can be enhanced by incorporating more complex heat transfer features, such as optimized fin and channel shapes and topologies, greater reduction in thermal resistance can be achieved by improving the thermal interface and reducing the copper base thickness.

Nomenclature

D_H hydraulic diameter, m
f_D Darcy friction factor, -
G mass flux, kg/m^2-s
$\bar{h}$ average heat transfer coefficient, W/m^2-K
H_b base thickness, m
H_f fin height, m
h_{fg} latent heat, J/kg
k_s solid thermal conductivity, W/m-K
L fin length, m
$\dot{m}$ mass flow rate, kg/s
N number of fins, -
P_{sat} saturation pressure, Pa
Q heat input, W
q'' heat flux, W/m^2

R_{COND} base conduction resistance, K/W
R_{FB} flow boiling resistance, K/W
Re Reynolds number, -
R_{TIM} TIM resistance, K/W
T_{cp} cold plate surface temperature, K
t_f fin thickness, m
T_{sat} saturation temperature, K
T_w channel bottom wall temperature, K
W cold plate width, m
w_c channel width, m
x_{exit} exit vapor quality, -
y position along flow length, m

Greek Symbols

ΔP_{2P} two-phase pressure drop, Pa
η fin efficiency, -
μ_f liquid viscosity, Pa-s
μ_g vapor viscosity, Pa-s
μ_m mixture viscosity, Pa-s
ρ_f liquid density, kg/m^3
ρ_g vapor density, kg/m^3
ρ_m mixture density, kg/m^3

Acronyms

CAD   computer aided design
PEEK   polyetheretherketone
TIM   thermal interface material
TTV   thermal test vehicle

Keywords

flow boiling, direct-to-chip cooling, data center cooling, skived cold plates

1. Introduction

The thermal design power of CPU and GPU chips used in data centers are increasing rapidly with the artificial intelligence fueled demand for higher computational power. Currently available Nvidia H100 chips generate up to 700 W of heat [1] in a 814 mm^2 footprint [2]. Densely populated AI rack solutions reach up to 80 kW of power [3]. Two-phase liquid cooling is a potential solution to the thermal management demands of the next generation of data centers due to high heat transfer coefficients and heat capacities associated with boiling as well as the dielectric nature of the available two-phase coolants. Thermal resistance of the cold plates in a two-phase cooling system impacts the available heat dissipation at the maximum operating temperature, making it crucial to design efficient cold plates that enhance heat removal.

Numerical optimization of heat sinks and cold plates have been shown to provide significant performance improvements relative to intuition-based designs [4, 5, 6]. However, commonly adopted gradient-based algorithms rely on numerical models of the governing physics to calculate the gradients of the cost function with respect to the design variables. This poses a challenge for optimization of

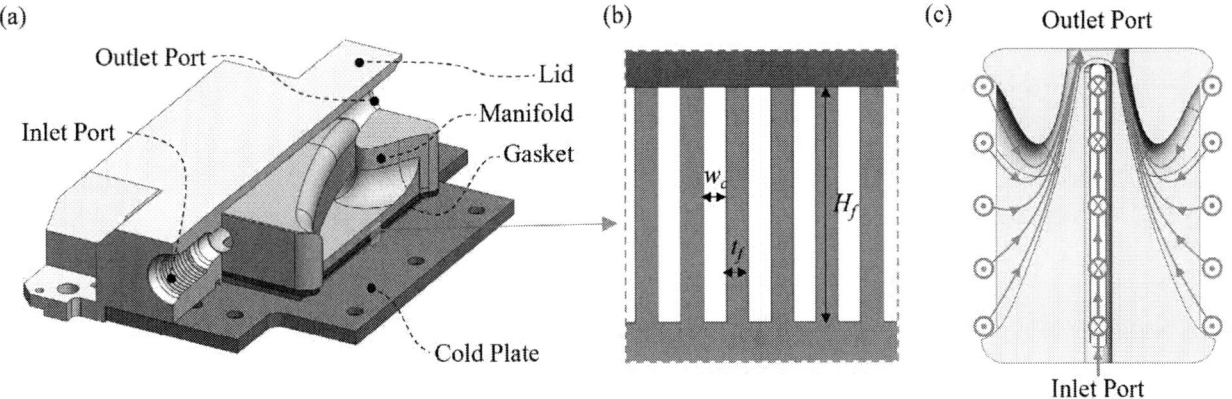

Figure 1. (a) Drawing of the TTV assembly with the aluminum lid cut-sectioned in half. (b) Side-view drawing of the skived fins with geometric parameters labeled. (c) Top-down cut-section view of the manifold with flow paths drawn (x symbols representing into the page, • out of the page).

components under flow-boiling because the available modeling approaches are highly specific to geometry and operating conditions. Flow-boiling models can be calibrated using experimental data prior to optimization to generate meaningful results [7].

Manufacturing capabilities need to be captured in the formulation of the design variables and the constraints. Skived copper cold plates are commonly implemented in data center liquid cooling due to their low cost and ability to achieve fin and channel sizes on the order of 100s of microns, which is critical to the cooling performance. However, design optimization of skived fin cold plates for two-phase cooling in data center applications is limited in literature. To this end, this study aims to generate optimized skived cold plates under flow-boiling with minimized thermal resistance. An insert to the lid is designed as a manifold to the skived channels to reduce the pressure drop across the cold plate, resembling a two-layer manifold heat sink commonly adopted in literature [8, 9, 10]. A numerical flow-boiling model is used for size optimization of the skived fins. To produce a reference dataset to calibrate the numerical model, the thermal resistances of different skived cold plates with various fin and channel sizes are experimentally measured under flow-boiling. The performances of the optimized cold plates are then investigated.

2. Methodology

Optimization results are dependent on the operating conditions and geometric constraints, which are summarized as follows. A cold plate with a finned footprint area of 60.2 mm × 41 mm is investigated. Uniform heating of 1 kW is applied along the bottom surface. The chosen working fluid is refrigerant R515B [11], which enters the cold plate, flows through the channels between the skived fins, and boils at 30°C while absorbing the heat as it changes phase. The target exit vapor quality is $x_{exit} = 0.7$, which corresponds to a mass flow rate of $\dot{m} = 9.02$ g/s.

2.1. Experimental Methods

Experimental data are needed to calibrate the numerical models used during optimization. To this end, a thermal test vehicle (TTV) is designed to emulate the boundary conditions of the investigated cold plate. Figure 1a shows a drawing of the TTV assembly with the aluminum lid cut-sectioned in half to show the internal components. The skived copper cold plate is bolted to the lid. Ports placed on the opposite sides of the lid act as the inlet and outlet respectively for refrigerant to flow through the cold plate. A copper block, not shown in the drawing, is compressed against the bottom surface of the cold plate with a thermal interface material (TIM) in between. Four cylindrical cartridge heaters inserted into the copper block generate heat, which transfers from the heater block to the cold plate through the TIM. The copper block is seated inside an insulating PEEK block to reduce heat losses to ambient. Thermocouples are inserted into the copper block and at the bottom surface of the cold plate, each centered within the heated area. Figure 1b shows a side-view drawing of the skived fins with the geometric parameters labeled where w_c is the channel width, t_f is the fin thickness, and H_f is the fin height.

A nylon manifold is placed inside the lid. Figure 1c illustrates a top-down cut-section of the manifold, with arrows depicting the flow path and symbols representing flow into and out of the page. Refrigerant enters through the inlet port and flows toward the outlet port, where a thin wall obstructs the direct path. This forces the refrigerant downward to the cold plate, where it splits into two outward directions within the channels. The channels augment the area over which heat transfer occurs and create a forced flow condition for the now-boiling refrigerant. The resulting liquid-vapor mixture exits the channels and flows back inward to the outlet port via the manifold. The manifold's geometry effectively divides the flow, reducing pressure drop by shortening the flow path. A gasket cut to match same profile of the bottom surface of the manifold is placed between the manifold and the fins to ensure the flow is directed through the channels rather than over them.

A two-phase flow loop is built to produce and control the operating conditions inside the TTV. Figure 2 shows a schematic diagram of the flow loop. Liquid refrigerant inside a reservoir is pumped through the loop with a gear pump (Micropump L28640). Refrigerant is pushed through an orifice before entering the TTV to suppress flow-boiling instabilities by inducing a large pressure drop upstream. Then, refrigerant flows through the TTV where it boils, and the resulting liquid-

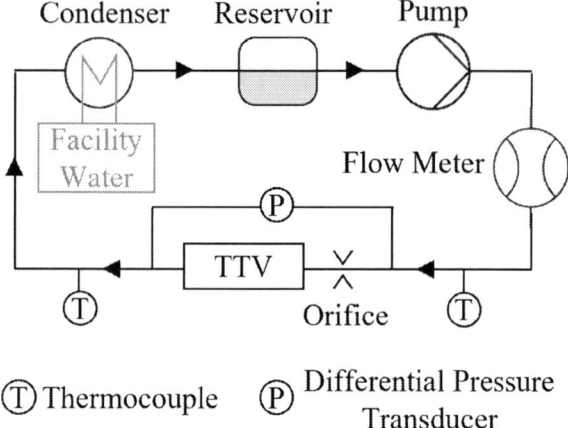

Figure 2. Schematic diagram of the flow loop.

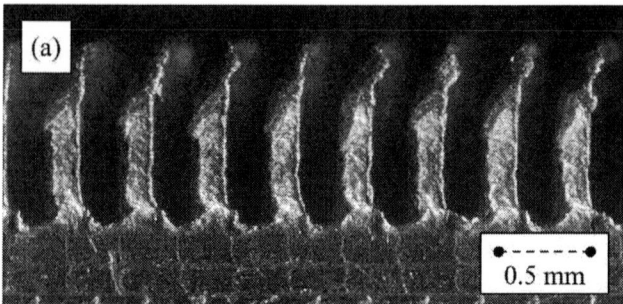

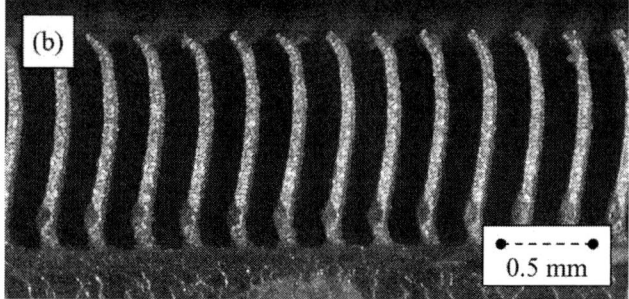

Figure 3. Side-view optical images of the skived fins of (a) cold plate #1 and (b) cold plate #2.

Table 1. Dimensions of the two cold plates tested.

#	t_f [mm]	w_c [mm]	H_f [mm]	H_b [mm]
1	0.20	0.20	0.94	4.54
2	0.10	0.15	0.13	4.08

vapor mixture condenses inside a brazed plate heat exchanger condenser. The heat is removed from the system by the facility water in a secondary loop. The condensed refrigerant flows back into the reservoir. The reservoir acts as a buffer for the changes in vapor volume in the system. Temperatures are measured before and after the TTV using thermocouple probes inserted into the flow. The pressure drop across the TTV is measured using a differential pressure transducer (Omega PX409-015DWU10V). An ultrasonic flow meter (Keyence FD-XS8) placed between the pump and the TTV measures the volumetric flow rate of the liquid refrigerant.

The flow loop and TTV enable the control of flow rate with the gear pump, heat input with a thyristor-based power controller (Gefran GFX4-IR) connected to the cartridge heaters, and boiling temperature by adjusting the facility water flow rate. Subcooling at the inlet of the TTV is not controlled but rather is an artifact of the vapor-line pressure drop in the loop. Subcooling was ensured to be below 5°C throughout all testing. Thermocouples and pressure transducer signals are measured and recorded using a data acquisition system (Keysight DAQ970A with DAQM901A). All data collection is performed at 1 Hz and averaged over a one-minute period to smooth out short-term fluctuations.

Two skived cold plates with different fin sizes are experimentally tested to collect model calibration data. Figure 3 shows side-view microscope images of the skived fins of (a) cold plate #1 and (b) cold plate #2. As a result of the skiving process, the fins are not purely rectangular as depicted in the CAD image in Figure 1a; instead, they are thicker near the base and thinner near the top. Cold plate #1 has fins with pointy tips, and cold plate #2 has curved fins. To reduce complexity of the model, numerical modeling is conducted based on the assumption of rectangular fins and channels, where the fin thickness and channel width are measured from the center along the height and averaged across all the fins shown in Figures 3a and 3b. Table 1 summarizes the geometric dimensions where H_b is the solid base thickness between the fins and the heated surface.

Calibration data are collected over a range of operating conditions. The cold plates are tested at various heat inputs (Q) between 0-1 kW. Exit vapor qualities (x_{exit}) between 0-1 are achieved by varying refrigerant flow rate for a given heat input. Boiling temperature is kept between 25-30°C, and subcooling is kept below 5°C for all tests. A total of 37 tests are conducted where heater temperature (T_h), cold plate bottom surface temperature (T_{cp}), inlet refrigerant temperature (T_{in}), outlet refrigerant temperature (T_{out}), and pressure drop across the cold plate (ΔP) are measured.

2.2. Numerical Methods

The overarching objective of the optimization is to lower the junction temperature of high heat flux chips in data centers. To this end, the overall thermal resistance of the cold plate package (R) is minimized. Thermal resistance is divided into three subcomponents: thermal resistance across the TIM (R_{TIM}), conduction resistance across the copper base thickness between the TIM and the fins (R_{COND}), and the thermal resistance through the fins and between the fins and the boiling refrigerant (R_{FB}). R_{TIM} is constant for a given heated surface area and therefore is not included in the optimization objective. Similarly, R_{COND} is assumed constant because heat input is uniform and convective resistance across the finned surface is expected to be uniform under boiling (i.e., heat spreading is negligible). Therefore, the objective is formulated as follows.

$$\min_{w_c, t_f} R_{FB}(w_c, t_f) \qquad [1]$$

$$R_{FB} = \frac{T_w - T_{sat}}{Q} \qquad [2]$$

where T_w is the bottom wall temperature of the channels, and T_{sat} is the saturation temperature of the refrigerant at the outlet of the cold plate. The design variables w_c and t_f are to be optimized to minimize the flow-boiling thermal resistance R_{FB}.

Flow-boiling thermal resistance is estimated using an average heat transfer coefficient for flow boiling ($\bar{h}$) and the extended surface analysis from Incropera et al. [12] wherein the effective surface area provided by the fins is a fraction of the total surface are defined by the fin efficiency (η).

$$R_{FB} = \frac{1}{\bar{h}NL(w_c + 2H_f\eta)} + \frac{dT_{sat}}{dP_{sat}}\frac{\Delta P_{2P}}{Q} \qquad [3]$$

$$\eta = \frac{\tanh\left(H_f\sqrt{2\bar{h}/k_s t_f}\right)}{H_f\sqrt{2\bar{h}/k_s t_f}} \qquad [4]$$

where N is the number of fins across the width ($W = 60.2$ mm), L is the length of the fins along the cold plate ($L = 41$ mm), and k_s is the thermal conductivity of the copper (388 W/m-K). The last term in Equation 3 represents the resistance arising from the increase in saturation temperature caused by the frictional pressure drop resulting from the two-phase flow within the channels (ΔP_{2P}). Pressure drop due to acceleration is calculated to be negligible for the investigated cold plate geometries, heat inputs, and mass flow rates. The term dT_{sat}/dP_{sat} is the gradient of saturation temperature with respect to saturation pressure and is equal to 0.41 K/psi at the saturation temperature of 30°C used throughout this study.

The pressure drop inside the channels is estimated by using the following assumptions. The flow is laminar and fully developed. Two-phase mixture is homogeneous, i.e., liquid and vapor phases are at equal speed. A mixture modeling approach [12] is used with the Darcy-Weisbach equation as follows.

$$\frac{dP_{2P}}{dy} = \frac{(f_D Re)\mu_m \dot{m}}{2\rho_m N w_c H_f D_H^2} \qquad [5]$$

where y is the position along the flow length inside a channel, f_D is the Darcy friction factor, Re is the Reynolds number, μ_m is the mixture viscosity, $\dot{m}$ is the mass flow rate across the cold plate, ρ_m is the mixture density, and D_h is the hydraulic diameter. The optimized channels are expected to have high aspect ratio; therefore, value of 96 is used for $f_D Re$ [13] and $D_h = 2w_c$, which corresponds to flow between parallel plates. The mixture viscosity is estimated using the correlation by McAdams et al. [14] and the mixture density is calculated as follows.

$$\frac{1}{\rho_m} = \frac{1-x}{\rho_f} + \frac{x}{\rho_g} \qquad [6]$$

$$\frac{1}{\mu_m} = \frac{1-x}{\mu_f} + \frac{x}{\mu_g} \qquad [7]$$

where x is the local vapor quality, and subscripts f and g represent liquid and vapor phases respectively. Heat dissipation by the refrigerant is assumed to be uniform along the flow length. Therefore, the vapor quality increases linearly along the flow position and Equation 5 is integrated across the channel as follows.

$$\Delta P_{2P} = \frac{L(f_D Re)\dot{m}}{4Nw_c H_f D_H^2 x_{exit}}\int_0^{x_{exit}} \frac{\mu_m}{\rho_m}dx \qquad [8]$$

where x_{exit} is the vapor quality at the outlet of the cold plate.

The heat transfer coefficient in Equation 3 is to be calibrated using the experimental measurements taken from the two skived cold plates tested. Therefore, a general formulation is used for the average heat transfer coefficient as follows.

$$\bar{h} = C_1 q''^{C_2} G^{C_3} w_c^{C_4} x_{exit}^{C_5}(1 - x_{exit})^{c_6} \qquad [9]$$

$$q'' = \frac{Q}{NL(w_c + 2H_f\eta)} \qquad [10]$$

where q'' is the heat flux at the fin walls, G is the mass flux in the channels, and C_1-C_6 are the model coefficients which are determined by curve fitting Equation 9 to the heat transfer coefficients calculated from the experimental measurements. The definition of heat flux provided in Equation 10 contains the term η (Equation 4) which is a function of $\bar{h}$. Therefore, when the thermal resistance is calculated during curve fitting and design optimization, Equation 9 is solved iteratively with an initial guess of $\eta = 1$ until convergence.

Calibration of the model coefficients requires calculation of the heat transfer coefficient from the experimental measurements which can be achieved using Equations 2-4, 9, and 10. However, T_w in Equation 2 (the bottom wall temperature of the channels) is not directly measured during testing. Instead, T_{cp}, cold plate bottom surface temperature, is measured and T_w is estimated assuming 1D conduction across the solid cold plate base as follows.

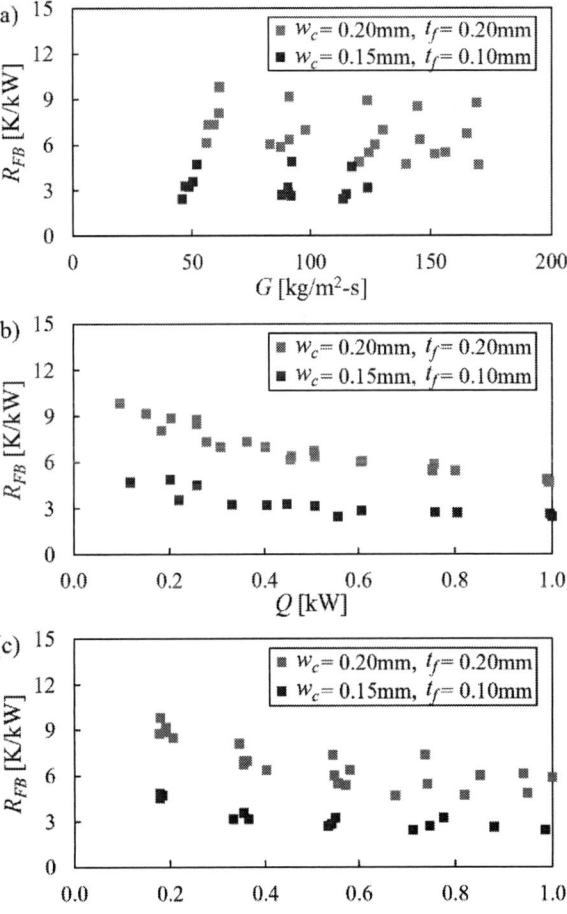

Figure 4. Thermal resistance with respect to mass flux, heat input, and exit quality for the two cold plates tested.

$$T_w = T_{cp} - \frac{QH_b}{k_s LW} \qquad [11]$$

The optimization problem formulated in Equation 1 is solved using the interior-point algorithm. Constant-value properties of R515B evaluated at 30°C are used throughout the calibration and optimization, which are shown in Table 2.

Table 2. Properties of R515B at 30°C [15].

ρ_f [kg/m³]	ρ_g [kg/m³]	μ_f [μPa-s]	μ_g [μPa-s]	h_{fg} [kJ/kg]
1163.9	31.3	181.3	12.5	158.2

3. Results

Two skived cold plates were tested under various flow rates and heat inputs for a total of 37 experiments. Figure 4 shows the flow boiling thermal resistance plots with respect to mass flux, heat input, and exit vapor quality. Thermal resistance decreases with higher heat input and exit vapor quality. However, the three variables are not independent; exit vapor quality is calculated from the mass flux and heat input. Therefore, the reduction in thermal resistance at higher exit vapor qualities may be attributed to the increased heat loads required to achieve those higher qualities or vice versa. Thermal resistance shows a significant dependence on the cold plate dimensions, as seen by the noticeable offset in resistances between the two cold plates.

Heat transfer coefficient correlation shown in Equation 9 is fit to the experimental data by optimizing the six coefficients to minimize the mean-square-error between measurements and predictions. Coefficients C_3 and C_6 resulted in small values and therefore are dropped to further simply the equation without significant loss in prediction accuracy. The equation below shows the resulting heat transfer coefficient correlation after calibration.

$$\bar{h} = 90.46 q''^{0.27} w_c^{-1.32} x_{exit}^{0.11} \qquad [12]$$

where all variables are defined in SI units except w_c which is in millimeters. Figure 5 compares the measured heat transfer coefficient with the predictions from Equation [12]. The predictions closely match the measurements, within 20% error bounds. However, it is important to note that this correlation is specific to the boundary and operating conditions investigated in this study.

The minimum fin thickness achievable by the skiving process is limited. Therefore, optimization defined in Equation 1 is performed with a lower bound on the fin thickness. Three different lower bounds of 0.2, 0.1, and 0.05 mm are used to capture the effect of the lower bound on the optimal performance. The commercially advertised limit is represented by 0.2 mm, minimum fin thickness achieved by fine-tuning the skiving process by the authors is 0.1 mm, and 0.05 mm is investigated to understand the performance implications of producing thinner fins. Figure 6a shows the flow boiling thermal resistance of the cold plates optimized at varying fin heights (H_f = 0.2-2.5 mm) and for three different lower bounds on fin thickness. Thermal resistance is decreasing with increasing fin height as more surface area becomes available and pressure drop is reduced by the larger available flow area. The fin efficiency defined in Equation 4 decreases with fin height. Therefore, thermal resistance asymptotes at higher fin heights, providing diminishing returns. The optimized fin thickness is calculated to always match the lower bound, indicating that a fin thickness smaller than the imposed lower bound is preferred for optimal thermal performance. The difference in thermal resistance between the three lower bounds is small at higher fin heights, reaching R_{FB} = 1.13, 0.89, and 0.74 K/kW for $t_f \geq$ 0.2, 0.1, and 0.5 mm, respectively, at a fin

(a)

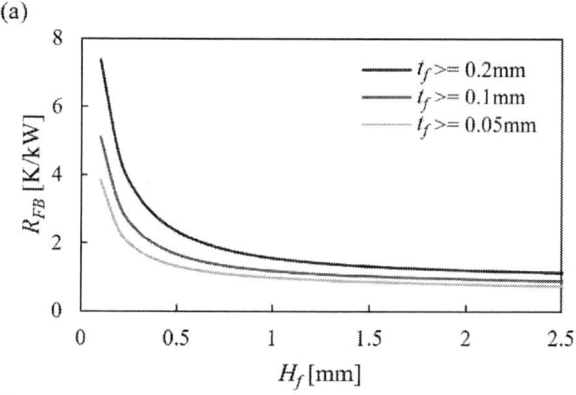

(b)

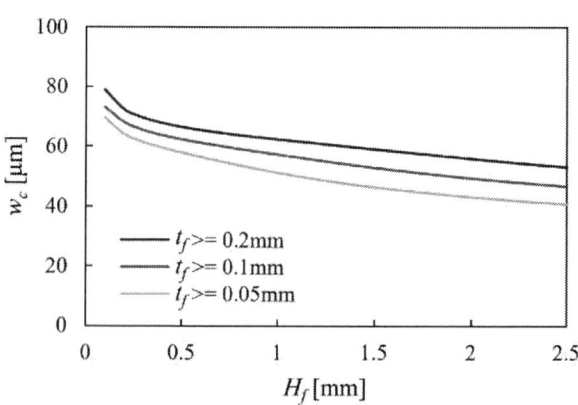

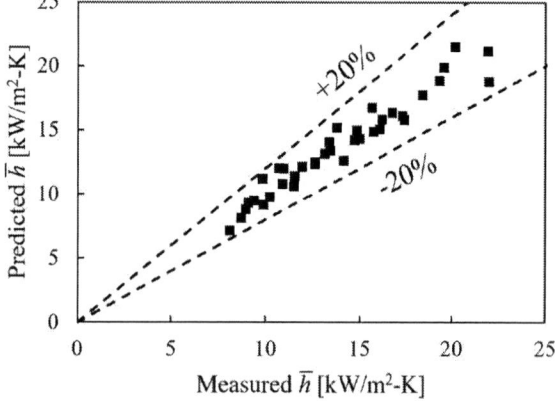

Figure 5. Measured average heat transfer coefficient vs prediction by the calibrated model.

Figure 6. (a) Thermal resistance and (b) channel width of the cold plates optimized at various fin heights and fin thickness bounds.

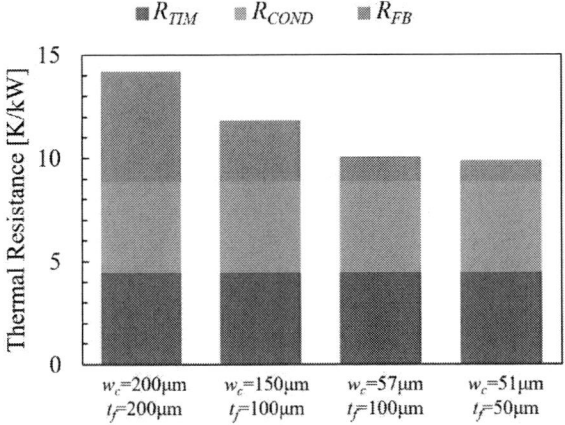

Figure 7. Overall thermal resistance stack-up for four cold plates with different channel widths and fin thicknesses.

height of 2.5 mm. Therefore, the benefit of using thinner fins diminishes with increasing fin height. Figure 6b shows the channel widths for the optimized thermal resistances shown in Figure 6a. Optimized channel width is less than 0.1 mm throughout and decreases with increasing fin height. Also, the channel width is smaller when a lower bound on fin thickness is used indicating that a smaller channel is preferred for thinner fins.

Thermal resistance of the overall packaging is investigated to evaluate the performance enhancement provided by the optimization. Figure 7 shows thermal resistance stack-up for four different cold plates, each with a fin height of 1 mm. Cold plates with 200 μm and 150 μm channel width represent the two cold plates fabricated and tested. The other two bars belong to the cold plates optimized using a fin height of 1 mm with the two of the lower bounds on fin thickness (0.1 and 0.05 mm). Flow boiling thermal resistance, R_{FB}, is calculated using the calibrated model. Base conduction resistance, R_{COND}, is estimated to be 4.43 K/kW assuming 1D conduction across a 4 mm copper thickness. Thermal resistance across the TIM, R_{TIM}, is measured to be 4.44 K/kW during experimentation of the two cold plates. R_{COND} and R_{TIM} are estimated to be equal between all cold plates because the same TIM and base thickness are used, and heat spreading is assumed negligible. Predictions of the flow boiling thermal resistances of the optimized cold plates are significantly lower than those of the two cold plates tested (60-80% reduction). Likewise, the overall thermal resistances of the optimized cold plates are significantly lower than those of the two cold plates tested, with a 30% reduction between the worst and best performers. The flow-boiling-related component of the thermal resistance reduces by 15% between the two optimized cold plates due to the difference in fin thickness. However, this difference shrinks to only 2% when the overall resistance of the package is considered.

4. Conclusions

Optimization of skived cold plates is investigated for direct-to-chip two-phase cooling for data centers using refrigerant R515B. A numerical model of flow-boiling in skived cold plates is developed and calibrated using experimental measurements. Optimization of the channel width and fin thickness is performed. The predicted thermal resistances of the optimized cold plates are then compared to the experimentally measured resistances of the unoptimized cold plates. Conclusions are summarized as follows.

- Average heat transfer coefficient for flow-boiling in skived cold plates is strongly dependent on the heat flux, exit vapor quality, and channel width. Mass flux of refrigerant does not have a significant impact on the heat transfer coefficient for the operating conditions and the cold plate dimensions investigated.

- Optimized fin thicknesses are always equal to the enforced lower bounds, imposed by manufacturability rules of thumb. Thermal resistance is lower when the lower bound is reduced, however, the performance enhancement for using a smaller fin thickness limit diminishes at higher fin heights.

- Smaller channels provide higher heat transfer coefficients, however, the increase in pressure drop increases the boiling temperature and thermal resistance. Therefore, optimized channel width is smaller when the fin thickness is reduced or fin height is increased, both of which provide more flow area and enable the smaller channels without causing a detrimental increase in pressure drop.

- Optimization of the channel width and fin thickness offers significant reduction to R_{FB}, the flow-boiling resistance. The optimized cold plate with 50 μm fin thickness and 51 μm channel width has 80% lower flow boiling resistance compared to the fabricated 200 μm fin thickness and 200 μm channel width. However, the improvement reduces to 30% when the overall thermal resistance consisting of the TIM and the copper base is considered.

- Reducing the fin thickness of optimized cold plates from 100 to 50 μm provides a 15% reduction in flow boiling thermal resistance. The improvement reduces to 2% when overall package resistance is considered because the resistance of the TIM and the cold plate base thickness dominate. For the optimized cold plates, flow boiling thermal resistance accounts for as small as 10% of the cold plate assembly resistance. Therefore, further improvements to the cold plate performance for two-phase direct-to-chip cooling in data centers will come from reducing the TIM and base conduction resistances.

References

[1] "NVIDIA H100 Tensor Core GPU," Nvidia. Accessed: Sep. 26, 2024. [Online]. Available: https://resources.nvidia.com/en-us-tensor-core/nvidia-tensor-core-gpu-datasheet

[2] "NVIDIA H100 PCIe 80 GB." TechPowerUp. Accessed: Sep. 26, 2024. [Online]. Available: https://www.techpowerup.com/gpu-specs/h100-pcie-80-gb.c3899#:~:text=Since%20H100%20PCIe%2080%20GB,mm%C2%B2%20and%2080%2C000%20million%20transistors

[3] "Large Language Models (LLM) to the AI Edge," Super Micro. Accessed: Sep. 26, 2024. [Online]. Available: https://www.supermicro.com/manuals/brochure/Brochure_AI_GPU_Workloads.pdf

[4] K. L. Kirsch and K. A. Thole, "Numerical Optimization, Characterization, and Experimental Investigation of Additively Manufactured Communicating

Microchannels," *J. Turbomach.*, vol. 140, no. 11, p. 111003, Nov. 2018, doi: 10.1115/1.4041494.

[5] J. H. Ryu, D. H. Choi, and S. J. Kim, "Three-dimensional numerical optimization of a manifold microchannel heat sink," *Int. J. Heat Mass Transf.*, vol. 46, no. 9, pp. 1553–1562, Apr. 2003, doi: 10.1016/S0017-9310(02)00443-X.

[6] S. Ozguc, L. Pan, and J. A. Weibel, "Optimization of permeable membrane microchannel heat sinks for additive manufacturing," *Appl. Therm. Eng.*, vol. 198, p. 117490, Nov. 2021, doi: 10.1016/j.applthermaleng.2021.117490.

[7] S. Ozguc, L. Pan, and J. A. Weibel, "An approach for topology optimization of heat sinks for two-phase flow boiling: Part 1 – Model formulation and numerical implementation," *Appl. Therm. Eng.*, vol. 249, p. 123337, Jul. 2024, doi: 10.1016/j.applthermaleng.2024.123337.

[8] D. Copeland, H. Takahira, W. Nakayama, and B.-C. Pak, "Manifold microchannel heat sinks: theory and experiments," *Adv. Electron. Packag.*, vol. 10, no. 2, pp. 829–834, 1995.

[9] M. A. Arie, A. H. Shooshtari, S. V. Dessiatoun, E. Al-Hajri, and M. M. Ohadi, "Numerical modeling and thermal optimization of a single-phase flow manifold-microchannel plate heat exchanger," *Int. J. Heat Mass Transf.*, vol. 81, pp. 478–489, Feb. 2015, doi: 10.1016/j.ijheatmasstransfer.2014.10.022.

[10] I. L. Collins, J. A. Weibel, L. Pan, and S. V. Garimella, "A permeable-membrane microchannel heat sink made by additive manufacturing," *Int. J. Heat Mass Transf.*, vol. 131, pp. 1174–1183, Mar. 2019, doi: 10.1016/j.ijheatmasstransfer.2018.11.126.

[11] "Solstice N15 (R-515B) Technical Data Sheet," Honeywell. Accessed: Jan. 01, 2025. [Online]. Available: https://www.honeywell-refrigerants.com/europe/wp-content/uploads/2022/01/Solstice-N15-TDS_EN.pdf

[12] W. Chao-Yang and C. Beckermann, "A two-phase mixture model of liquid-gas flow and heat transfer in capillary porous media—I. Formulation," *Int. J. Heat Mass Transf.*, vol. 36, no. 11, pp. 2747–2758, Jul. 1993, doi: 10.1016/0017-9310(93)90094-M.

[13] *F. P. Incropera, D. P. DeWitt, T. L. Bergman, A. S. Lavine, "Fundamentals of Heat and Mass Transfer," Wiley, 2011.*

[14] W. H. McAdams, W. K. Woods, and L. C. Heroman, "Vaporization Inside Horizontal Tubes—II Benzene-Oil Mixtures," *J. Fluids Eng.*, vol. 64, no. 3, pp. 193–199, Apr. 1942, doi: 10.1115/1.4019013.

[15] M. Huber, A. Harvey, E. Lemmon, G. Hardin, I. Bell, and M. McLinden, "NIST Reference Fluid Thermodynamic and Transport Properties Database (REFPROP) Version 10 - SRD 23." National Institute of Standards and Technology, 2018. doi: 10.18434/T4/1502528.

Superposition Method of Predicting Junction Temperature of Multi Chip Packages - Important Thermal Parameters

M Kamrul Russel, Adeel Ahmad, Muhammad Ahmad, Otto Joe
Advanced Mico Devices Inc., Santa Clara, CA 95054
kamrul.russel@amd.com

Abstract

For years, semiconductor industry has observed the trend of increasing number of transistors on a single chip. With the advent of miniaturization transistors' size also tend to shrink and use of different technology nodes on a single package has become advantageous. At the same time the number of transistors on a single chip has increased. Use of multi-chip, including but not limited to computing chips, IO chips, AI chips, memory chips - on a single package has become more useful in terms of performance, functionality etc. Combining all these factors may lead to higher heat flux - thermal management of which can be challenging and may demand innovations in cooling technologies. One key parameter for designing the thermal management solution is the thermal resistance from the heat generating elements to the common casing of the multi-chip package (MCP) or to the surrounding ambient. In a complex MCP each chip may have its own design temperature beyond which the chip might fail or experience lower than expected life – and hence degrade the overall performance of the MCP. Each chip may undergo different power map and power levels – generating its unique junction temperature. To cater for various end users, it is advantageous during the design phase of such MCP to know the junction temperature of all junctions to understand the thermal interaction, limitations and benefits of assigning asymmetric and arbitrary power distribution. To accurately determine the junction temperature with an arbitrary power map, superposition method has been studied and validated in the past with great level of accuracy. In this method the impact of one chip/heat source on other chips/heat sources and on itself is determined and presented as an influence matrix, which is later used to determine the chips' junction temperature when power applied to each chip is known. The influence matrix can be generated from experiments. For MCP with very large number of chips/heat sources it would be more practical to generate this matrix using thermal simulation, while validating a few key cases against experiments. In this study the effects of asymmetry in applied power to the chips and the boundary condition impact on the accuracy of the well-established superposition method has been studied at steady state.

Keywords

Multi-chip packages, Linear superposition, Influence matrix

1. Introduction

Semiconductor industry had seen a rapid advancement in technology shift over the past years. To accommodate for the growing demand of high-performance multi-chip packages (MCP) have emerged as a critical solution to integrate multiple chips into a single package. MCP offers significant benefits in terms of functionality, space utilization, performance, data transfer and cost. The compact design of MCP to integrate multiple chips, however, poses a critical engineering challenge of thermal management. The challenge comes in two fronts: increased heat flux and maintaining each of the chips within its safe design temperature. Predicting and managing the junction temperature accurately of each chip in an MCP becomes essential to ensure reliability [1], longevity, production yield, optimized performance and to prevent thermal related mechanical failures. Typically, thermal performance of a single chip is characterized using a single junction-to-ambient thermal resistance, which is not applicable for MCPs as multiple chips are integrated within it. Thermal analysis of MCPs is inherently complex due to the complex interaction of multiple heat sources, heat dissipation paths and varying material properties. Non-uniform power consumption by various functional components of a chip within the MCP results in local hot spots and makes thermal analysis further challenging. Additionally, the end users might configure a given MCP to operate at various power levels. Modern day chip technologies and design decisions critically depend on prior knowledge of thermal capabilities and interaction between various chips of an MCP.

A methodology to accurately estimate the junction temperature for a set of given boundary conditions is essential for various practical power maps and can be utilized for optimum package layout during the design phase. One approach could be traditional thermal modeling methods, such as finite element analysis (FEA), which may become computationally intensive and time consuming if all potential field application cases are studied or during iterative design purposes. To address these challenges, the influence coefficient matrix based on superposition approach has been developed and studied for years as an efficient and accurate method to predict junction temperatures in multichip modules [2-10]. In this method junction temperature of a given chip or heat source is calculated by adding the contribution made by each device powered in the MCP. An influence matrix dictates the contribution of one chip on another. The value of the influence matrix on a given chip is greater when a contributing chip is closer. The influence of a chip on itself is the greatest.

For an MCP with multiple dies linear superposition is a classic approach to predict temperature field or junction temperature of an active layer for various power maps. Lall et.al.[2] studied an experimental technique to predict the junction temperature of an MCP of up to 4 chips of equal size for both natural and forced convection. The approach was based on average junction to ambient thermal resistance. An empirical equation for natural convection was proposed to predict the average junction to ambient thermal resistance and was in good agreement with experimental data with mostly uniform power levels. The error increased when the non-uniformity in the power levels on various chips was significant.

Forced convective experiments were performed in a wind tunnel at air speed of 400 to 1000 lfpm. The prediction from superposition method was in good agreement with the experimental data. Individual die temperature at various power configurations could be predicted using the data from a limited number of tests. Goh [3] used 5 active cells on a chip as heat sources and used ANSYS as an FEA tool to generate the influence matrix and to compare junction temperature results from simulation against the ones obtained from thermal model incorporating the influence matrix. Temperature prediction from the simulation was found to be in good agreement with the thermal model. Treurniet and Lammens [5] studied multichip LED modules with 4 LED chips for 4 different colors. They derived the influence matrix from experiments with arbitrary heat loads applied separately to each of the chips. Approximately13% error in the junction temperature was observed when the influence of the neighboring chip is ignored. Tat et.al. [6] used a 2 chip MCP in their experiments to obtain influence matrix using linear superposition method. Ansys 13.0 was used as simulation tools to compare against junction temperature from tests. They found a maximum of 7% error between test data and simulation results when single die is powered, while it was 3% when both dies were active. Chang and Guta [7] developed a cooling envelope of a 2 chip MCP that predicts the maximum cooling capacity for a given cooling solution. Kelly et. al. [9] in their study proposed a reverse inversion method to obtain influence matrix to predict accurate junction temperature. They found that superposition models can accurately predict junction temperature for arbitrary power maps if package average temperature is similar for both tests and prediction models. In their forced convective experiments, they had 13 chip MCP with asymmetric size and heat load. An error margin of 26% was observed when experimental temperature data were compared to the thermal model prediction that uses an influence matrix for which the average package temperature is significantly dissimilar than the experiment. This error was reduced to about 8% with a revised influence matrix derived from 13 random power sets.

The above examples involve a limited number of discrete heat sources, making it less time-intensive to obtain the influence matrix through experiments or simulations. However, for a complex MCP with multiple dissimilar chips, some containing sub-millimeter-sized functional elements, the effective number of heat source can reach several thousand. Obtaining an influence matrix through experiments or simulations for such MCPs across a wide range of power maps and cooling solutions or heat transfer coefficients becomes extremely time consuming, if not impractical. The objective of this study is to evaluate the applicability of the magnitude and asymmetry of heat dissipation from the chips, along with ambient temperature and heat transfer coefficient, which differ from those used to generate the influence matrix. A simple MCP consisting of two chips [6, 7] has been considered. The chips are positioned between a printed circuit board (PCB) and a copper lid, with thermal interface layers in between (Fig. 1). A heat transfer coefficient (h) is applied to the top surface of the lid to model the effect of a heat sink over the lid. This assumes convection heat transfer path through the heat sink to the ambient.

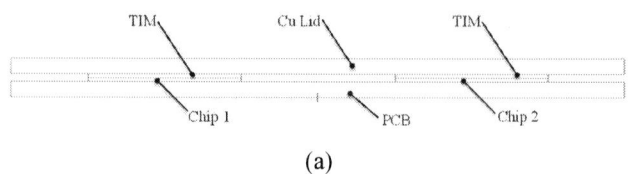

(a)

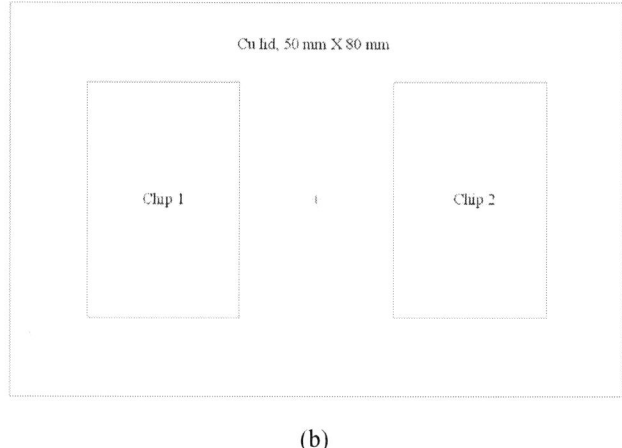

(b)

Figure 1: Schematic of the MCP (a) cross-sectional view (b) top view

2. Influence Matrix

Linear superposition approach to obtain an influence matrix was achieved following Chang, J. Y., Gupta [7], with 1W separately applied to each chip, and for ambient temperature (T_∞) of 30 °C and heat transfer coefficient, h of 1000 Wm^{-2}K^{-1}. The influence matrix (Ψ_{ij_ja}, °C/W), which includes the case to ambient thermal resistance (θ_{ca}), is obtained from simulation [3] and shown in Table 1. Ansys Icepak was used for all simulation purposes. Mesh refinement for any simulation was performed until junction temperature or other derived thermal values changed by less than 3%.

For the given lid (surface area, A_S of 50 mm X 80 mm) the thermal resistance from the top of the lid, i.e., case to the ambient ($\theta_{ca} = 1/(hA_S)$) is 0.25 (°C/W). The junction to case influence matrix (Ψ_{ij_jc}) is obtained by subtracting this case-to-ambient thermal resistance from Table 1 and presented in Table 2. Using the junction to case influence matrix (Ψ_{ij}) the junction temperature of i-th chip ($T_{j,i}$) can be obtained from equation (1)

Table 1: Junction to ambient influence matrix (Ψ_{ij_ja}) with heat transfer coefficient of 1000 Wm^{-2}K^{-1}

Influence matrix (°C/W)	Chip 1	Chip 2
Chip 1	0.42	0.14
Chip 2	0.14	0.42

Table 2: Junction to case influence matrix (Ψ_{ij})

Influence matrix (°C/W)	Chip 1	Chip 2
Chip 1	0.17	-0.11
Chip 2	-0.11	0.17

if the power dissipation from each chip (P_i), ambient temperature (T_∞) and case to ambient thermal resistance (θ_{ca}), i.e., the cooling solution is known.

$$T_{j,i} = (\Psi_{ij} + \theta_{ca}) P_i + T_\infty \qquad (1)$$

3. Parametric Studies and Discussion

In this paper effects of various key thermal parameters which differ from those used in obtaining an influence matrix has been studied. The simulation domain has been changed towards a more practical one for these studies and shown in Fig 2 and has top surface completely open as cold air inlet and a 5 mm high opening near the bottom edge of the domain as hot air exhaust. Intake air flow rate, Q of 1002 CFM was found to result in equivalent heat transfer coefficient of 1000 $Wm^{-2}K^{-1}$.

3.1. Effect of Symmetric Power Chips

In this section various uniform power dissipation level (P_i) from two chips has been studied at an ambient temperature of 30 °C with $P_1 = P_2$ and $h \approx 1000$ $Wm^{-2}K^{-1}$. Various cases studied are shown in Table 3.

Table 3: Uniform power levels studied

Case	Chip 1, W	Chip 2, W	Order, O
1	1	1	0
2	10	10	1
3	100	100	2

While generating the influence matrix 1 W heat dissipation was used for each chip, which is checked for the new simulation domain shown in Fig. 2. The order, O in Table 3 is the order of magnitude of heat dissipation from a chip compared to the magnitude used in obtaining influence matrix. The results for case 3 is shown in Fig. 3. For all 3 cases studied off-centric local hot spots on each chip were observed [3] which are closer to each other. Average junction temperature instead of local hot spot temperature is considered for all comparisons made against those obtained from linear superposition method using influence matrix (equation 1) in this study. The average junction temperature for case 3 from simulation for chip 1 and chip 2 are 89.82 °C while from equation 1 it is 86.4 °C. Linear

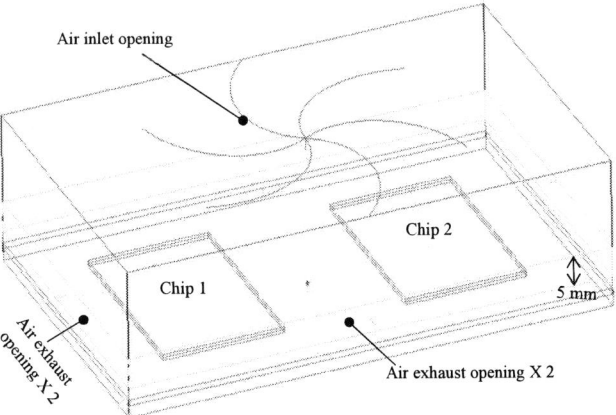

Figure 2: Simulation domain for forced convective cooling

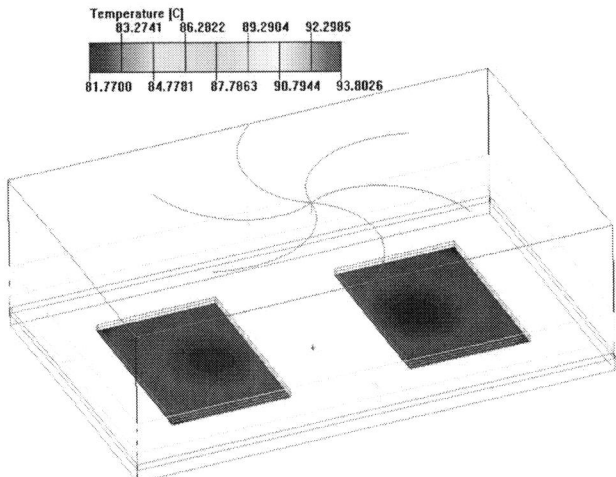

Figure 3: Simulation result for $P_1 = P_2 = 100W$, $T_\infty = 30$ °C and Q = 1002 CFM ($h \approx 1000$ $Wm^{-2}K^{-1}$)

superposition method underpredicts the junction temperature by ~ 3.81%. These values are ~ 0.1% and 1% for case 1 and case 2 respectively. For this study it is observed that heat dissipation from chips of up to order of magnitude of 2 with respect to the heat dissipation used in influence matrix can be considered acceptable.

3.2. Effect of Asymmetric Power Chips

In this section various asymmetric power dissipation levels (P_i) from two chips ($P_1 \neq P_2$) have been studied at an ambient temperature of 30 °C with $h \approx 1000$ $Wm^{-2}K^{-1}$. Various cases studied are shown in Table 4.

Table 4: Asymmetric power levels studied

Case	Chip 1, W	Chip 2, W	Order, O	O(chip2/chip1)
4	1	9	0	0
5	33	100	1	1
6	7	100	2	2

In this section, the order of the ratio of heat dissipation from one chip to the other is added to Table 4 – to indicate the level of asymmetry of heat loads. Results for case 5 is shown in Fig. 4. The average junction temperature for case 5 from simulation for chip 1 and chip 2 are 60.44 and 79.12 °C respectively while from equation 1 they are 58.14 and 76.87 °C respectively. Linear superposition method underpredicts the junction temperature by ~ 3.8%. These values are ~ 0.5% and 3.8% for case 4 and case 6 respectively. It is observed that the discrepancy between simulation and linear superposition method does not change for O (≥ 1) while this discrepancy can be considered acceptable for up to O (2).

3.3. Effect of Ambient Temperature

In this section the effect of ambient temperature is studied for power dissipation of (P_1, P_2) = (33, 100) W with $h \approx 1000$ $Wm^{-2}K^{-1}$. Ambient temperature studied are 0, 10, 20, 30, 40 and 50 °C. For the range of the ambient studied simulation was overpredicting junction temperature compared to superposition

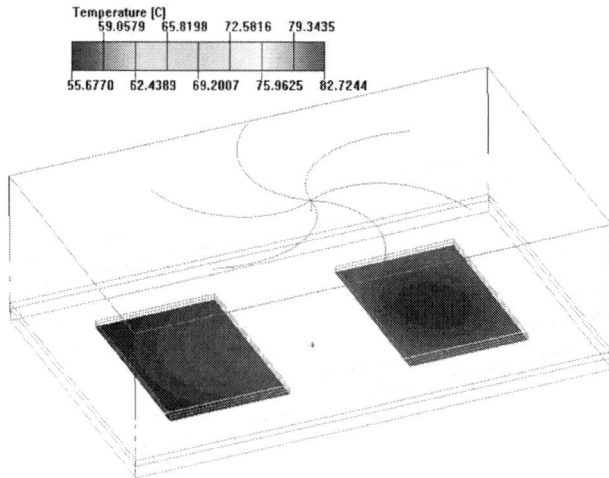

Figure 4: Simulation result for $(P_1, P_2) = (33, 100)$ W, $T_\infty = 30$ °C and Q = 1002 CFM (h ≈ 1000 Wm^{-2}K^{-1}).

method. There is an inherent discrepancy of maximum 3.8% due to the power dissipation from the chips of $(P_1, P_2) = (33, 100)$ W as discussed Section 3.2 for case 5. This is subtracted from the discrepancy obtained at various ambient temperatures between simulation and superposition method to achieve the maximum discrepancy caused by only ambient temperature. The results are shown in Fig. 5. The overall error remains within ~ 4%, which is acceptable for the range of ambient studied.

3.4. Effect of Case to Ambient Thermal Resistance

In this section the effect of case to ambient thermal resistance (θ_{ca}) is studied for power dissipation of $(P_1, P_2) = (33, 100)$ W at an ambient temperature of 30 °C. Case to ambient thermal resistance (θ_{ca}) is varied for the system by changing the air flow rate at the inlet of the computational domain in Fig. 1. For a given air flow rate θ_{ca} is calculated using the total applied power to the chips (e.g., 133 W), lid surface area (e.g., 50 mm X 80 mm), average temperature of the lid obtained from simulation and inlet air temperature (e.g., 30 °C). Air flow rate studied was from 50 to 4000 CFM, resulting in average heat transfer coefficient ranging from ~ 154 to 2700 Wm^{-2}K^{-1}. The case to ambient thermal resistance ranged from 0.09 to 1.6

°C/W. Junction temperature of the chips for each air flow rate were obtained from simulation and linear superposition model using equation (1). The relation between case to ambient thermal resistance and intake air flow rate is presented in Fig. 6. As expected, the case to ambient thermal resistance decreases with an increase in the intake air flow rate. The range of θ_{ca} studied covers a broader range of cooling techniques: low air flow-air cooling (e.g., $\theta_{ca} \gtrsim 0.5$ °C/W), forced air cooling (e.g., $0.1 \lesssim \theta_{ca} < 0.5$ °C/W) and liquid cooling (e.g., $\theta_{ca} < 0.1$ °C/W).

The discrepancy in junction temperature between simulation results and those obtained from superposition model, caused by only case to ambient thermal resistance, is obtained considering the effect of inherent maximum discrepancy due to power dissipation from chips – as discussed in section 3.3. These results are shown in Fig. 7. The error increases for a real case cooling solution if its θ_{ca} deviates away from the one used to generate influence matrix. If ≤ 4% is considered acceptable error, studies with $\theta_{ca} \gtrsim 0.15$ °C/W or intake air flow rate range of 50 to 2000 CFM fall withing acceptance. This implies that the methodology studied in this paper is applicable for low flow air cooling and forced air cooling. For liquid cooling applications, a separate influence matrix needs to be generated using a with θ_{ca} appropriate for liquid cooling.

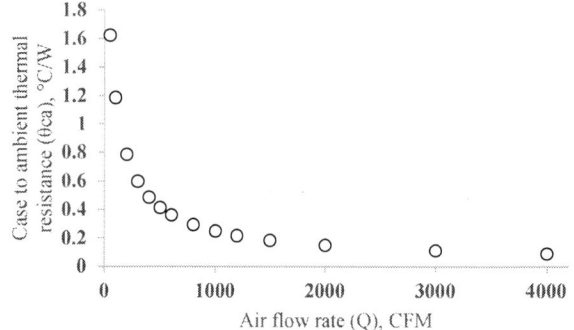

Figure 6: Relationship between intake air flow rate to case-to-ambient thermal resistance with $(P_1, P_2) = (33, 100)$ W and $T_\infty = 30$ °C

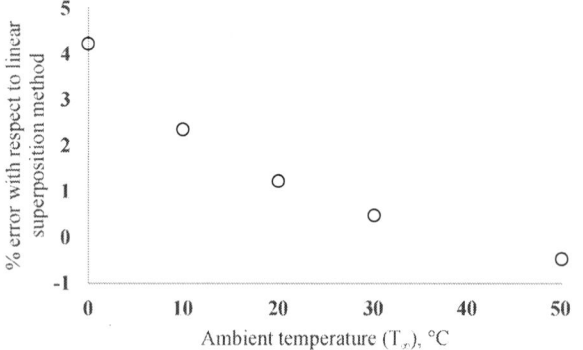

Figure 5: Effect of ambient temperature on discrepancy between simulation and superposition junction temperature with $(P_1, P_2) = (33, 100)$ W and h ≈ 1000 Wm^{-2}K^{-1}

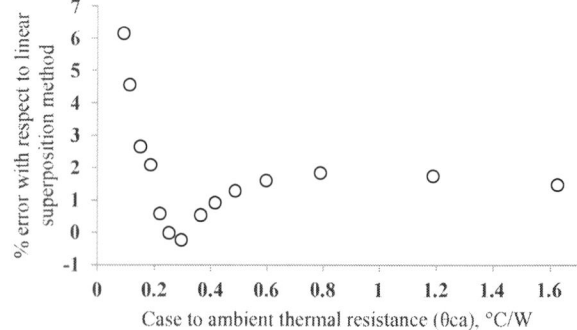

Figure 7: Effect of case-to-ambient thermal resistance on discrepancy between simulation and superposition junction temperature with $(P_1, P_2) = (33, 100)$ W and $T_\infty = 30$ °C

4. Conclusions

Thermal management of multi-chip packages (MCPs) presents a significant engineering challenge. Linear superposition method had been extensively used and validated through experiments demonstrating good accuracy. Given the wide range of possible combinations of applied power to the chips, relative power levels (i.e., the ratio of power applied to any two chips), ambient temperatures, and cooling solutions (case-to-ambient thermal resistance), selecting appropriate values for these parameters is crucial for obtaining the influence matrix. This paper investigates the impact of these parameters on the accuracy of the well-established superposition method. It was observed that a difference of two orders of magnitude between the applied power used to generate the influence matrix and the power applied during practical application resulted in only a 3.81% deviation in junction temperature, provided all boundary conditions were consistent. For asymmetric power distributions, the junction temperature accuracy remained within 3.8% for power asymmetry up to two orders of magnitude and within 0.5% for the same order of magnitude. The study also examined the applicability of ambient temperatures different from that used to generate the influence matrix. As the difference between the ambient temperature during matrix generation and the application increased, the deviation in junction temperature grew but remained below 4% in this study. Furthermore, the effect of boundary conditions under forced convection was analyzed. The thermal resistance from the case to the ambient was set to 0.25 °C/W during influence matrix generation. When thermal resistance varied between 0.1 and 1.6 °C/W, it was found that significantly smaller values (e.g., those typical of liquid cooling) led to a substantial deviation in junction temperature, reaching ~6%. For higher thermal resistance values e.g., 1.6 °C/W, the deviation was moderate, around 2%. The findings suggest that the superposition method is applicable across a range of case-to-ambient thermal resistances, provided the cooling techniques fall under the same category as the one used to generate the influence matrix. For different cooling techniques, such as liquid cooling, a different influence matrix is required. This study highlights both the limitations and benefits of using the superposition method with an influence matrix to predict junction temperatures in MCPs. The method demonstrated reasonable accuracy over a wide range of ambient temperatures and power distributions among chips. However, the accuracy is moderately affected by variations in case-to-ambient thermal resistance.

References

1. Czerny, B., Lederer, M., Nagl, B., Trnka, A., Khatibi, G, K., Thoben, M., "Thermo-mechanical analysis of bonding wires in IGBT modules under operating conditions", Microelectronics Reliability Vol. 52, No. 9-10, pp. 2353-2357, 2012.
2. Lall, B. S., Guenin, B. M., Molnar, R. J., "Methodology for thermal evaluation of multichip modules", Proceedings of IEEE/CPMT 11th Semiconductor Thermal Measurement and Management Symposium (SEMI-THERM), pp. 72-79, 1995.
3. Goh, T. J., "Thermal methodology for evaluating the performance of microelectronic devices with non-uniform power dissipation", 4th Electronics Packaging Technology Conference, pp. 312-317, 2002.
4. Fan, X., "Development, validation, and application of thermal modeling for a MCM power package", Nineteenth Annual IEEE Semiconductor Thermal Measurement and Management Symposium, pp. 144-150, 2003.
5. Treurniet, T., Lammens, V., "Thermal management in color variable multi-chip LED modules", Twenty-Second Annual IEEE Semiconductor Thermal Measurement And Management Symposium, pp. 173-177, 2006.
6. Tat, G. F., Lee, L. H. M. E., Peng, G. K., "Thermal measurement method for multi-chip packages", 13th International Conference on Electronic Packaging Technology & High Density Packaging, pp. 131-136, 2012.
7. Chang, J. Y., Gupta, A., "Estimating the thermal interaction between multiple side by side chips on a multi-chip package", Electronics Cooling 20. 2014.
8. Zhang, H. Y., Zhang, X. W., Lau, B. L., Lim, S., Ding, L., Yu, M. B., "Thermal characterization of both bare die and overmolded 2.5-D packages on through silicon interposers", IEEE Transactions on Components, Packaging and Manufacturing Technology Vol. 4, No. 5, pp. 807-816, 2014.
9. Kelly, M., Fosnot, P., Wei, J., Min, M., Galloway, J., "Multi-die packaging and thermal superposition modeling", 34th Thermal Measurement, Modeling & Management Symposium (SEMI-THERM), pp. 21-25, 2018.
10. Ghaisas, G., Krishnan, S., "Thermal influence coefficients-based electrothermal modeling approach for power electronics", IEEE Transactions on Components, Packaging and Manufacturing Technology Vol. 11, No. 8, pp. 1187-1196, 2021.

Experimental Study for Single-Phase High Heat Flux & High Coolant Temperature Spray Cooling

Mengyuan Yuan, Nan Chen, Zhipeng Zhao and Zhibin Yan
Advanced Liquid Cooling Technologies 4480 Liberty Hwy, Anderson, SC, USA 29621
+1 864-633-7174, mengyuan.yuan1@advancedliquidcooling.com

Extended Abstract

SUMMARY

Powerful processors consume significant power and produce a large amount of heat. For them to work properly, heat must be removed timely. Traditional air cooling is struggling to keep pace with the power increase in those processors and the waste heat from the processors needs a way for reuse to save energy. We developed a single-phase spraying cooling solution that is capable of cooling high heat flux with high coolant temperature, which can achieve worldwide four-season free cooling or be used for several heat reuse purposes. The single-phase spraying head is featured by 3D printing and is highly customizable for different server configurations.

1. INTRODUCTION

The fierce competition in AI computing power unequivocally poses significant challenges and opportunities in the realm of thermal management. Intel's scalable processors have reached a Thermal Design Power (TDP) of 385W [1], while Nvidia has unveiled its next-generation GPU surpassing a TDP of 1000W [2]. This escalation necessitates a more advanced and reliable cooling method, as traditional air-cooling solutions falter in meeting the increasing demands for processor cooling. Additionally, the concept of heat recovery has gained traction as an effective strategy to mitigate the looming energy crisis fueled by the AI surge. A notable potential for heat reuse highlighted by McKinsey [3] shows an anticipated increase in power requirements for data centers, from 17GW to 35GW from 2022 to 2030, in the US market alone. The cornerstone of achieving high heat flux and elevated coolant temperatures simultaneously lies in achieving High Heat Transfer Coefficients (HTC), described by Power Usage Efficiency (PUE) and Energy Reuse Factor (ERF) [4], respectively. Within the framework of Net Zero commitments, the ability to get low PUE and high ERF is a crucial element.

Current established liquid cooling technologies have proved their capabilities to deal with high-density heat rejection. Immersion cooling uses the specially designed heat sink and the dedicated flow injection from the boosting pump to elevate its heat rejection capability by the pumped forced convective flow [5]. Direct Liquid to Chip (DLC) as a hybrid cooling method, 70% of the heat is removed through the cold plate while the remaining 30% is still by forced air cooling [6]. Whether they can meet low PUE and high ERF has not been studied systematically, regarding the practical diversities of geometrical, thermal, and hydraulic boundaries. Concerning their physical essences, the following equation [7] provides an indicative evaluation of HTC to be of O(10) for natural convection.

$$h = \frac{k}{L}\left(0.68 + \frac{0.67 Ra_L^{1/4}}{\left(1 + \left(\frac{0.492k}{\mu c_p}\right)^{9/16}\right)^{4/9}}\right) \; if \; Ra_L < 10^9$$

The effectiveness of cold plate cooling hinges on design intricacies, such as the shapes of microchannels, flow configurations, and material choices. Zhang [8] achieved a thermal resistance of 0.40 K/W at a water flow rate of 2.73 L/min through both experimental and simulation studies.

Murshed [9] demonstrated that forced convection flow can achieve a Heat Transfer Coefficient (HTC) on the order of 1000 W/m²·K, which underpins our design philosophy. In practical terms, selecting an appropriate spray velocity involves balancing opposing factors such as the strength of forced convection, the static pressure on the cooled surface, and the volume of mist produced. For our experiments, we opted for a spray velocity range between 1 m/s and 3 m/s. We have designed a sprayer featured with twp slits for investigations, where the strategic positioning, patterning, and orientation of the slits help prevent dry spots. Besides, the sprayer is designed to fully enclose the heat sink, therefore reducing mist formation and restricting evaporation. Installing servers at a 70° angle ensures the formation of an oil film over components, effectively preventing liquid bypass. The test ensured that the processor's junction temperature remained below 90 °C, a critical factor for processor reliability, while the coolant temperature was kept above 50 °C. This temperature regulation supports year-round, dry, free cooling and opens avenues for efficient heat reuse. Our goal is to devise a cooling solution that is not only reliable but also capable of handling high heat loads with elevated coolant temperatures.

2. EXPERIMENTAL METHODS

An 80mm * 80mm copper plate installed on a server frame is used to simulate the heat source. Two 49mm * 43mm plate heaters, whose power can be adjusted by the input power voltages, are mounted on the back side of the copper plate. A 1U copper heat sink with a dimension of 110mm * 90mm is attached to the front of the copper plate while all other surfaces of the heating plate are insulated. The heat sink has a traditional air-cooling design. The heat sink fins have a thickness 0.5mm and a height of 15mm, and the pitch is 2mm. The base plate thickness is 5mm. An indium foil with a thickness of 0.1mm is installed between the heat sink and the plate to reduce the contact thermal resistance. A schematic of the spray testing rig is shown in Figure 1. Heat rejection in the test rig can be achieved by either a chiller or a dry cooler. The heat sink is mounted at an angle of 70° to the horizontal.

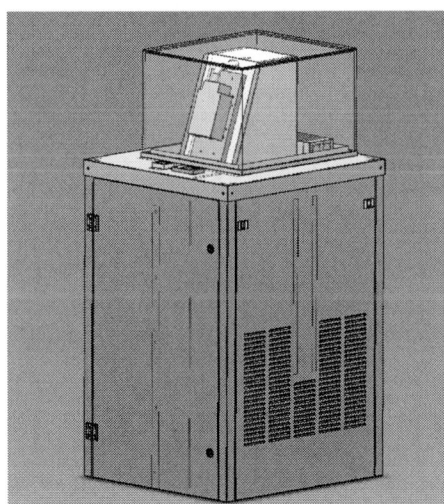 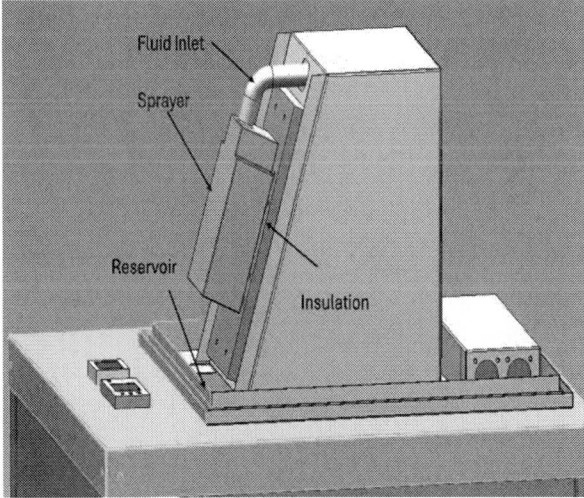

Figure 1. Schematic of Spray Test Rig

The sprayer is 3D printed and sits on the heat sink. 3D printing provides great flexibility to accommodate different heat sink shapes. In our experiment, the top and side surfaces are enclosed to contain oil mist and vapor while the bottom surface is open as the fluid exit. There are two slits on the top surface of the sprayer, generating consistent fluid curtains with the help of pressure in the plenum. Both slits have a width of 1mm, and the fluid velocity is 2.5m/s at the slits. Pictures of the copper heat sink and 3D-printed sprayer are shown in Figure 2. The spacing between the two slits and the orientation of the slits are optimized. The fluid curtains will impinge on the heat sink and then cascade down through the heat sink with the help of gravity, forming an even oil film over the heat sink. A temperature sensor is placed at the center of the heating plate to measure the maximum surface temperature point. A dielectric fluid is used as the coolant in our experiment and is directly sprayed on the heat sink. Gradually increase the input power and record the copper plate center temperature. We stop the experiments when the coolant temperature needs to drop below 55 °C to keep the junction temperature below 80 °C as the case temperature is limited to that for many chips.

A DC pump is used to circulate the dielectric oil, and a chiller is used to maintain the spraying oil at around 30 °C through a plate heat exchanger. In our experiments, heating power started from 800W and went up to 2300W with a 100W interval increase each time, then decreased from 2300W to 800 with a 100W interval to repeat the experiments. Oil and copper plate center temperatures are recorded accordingly.

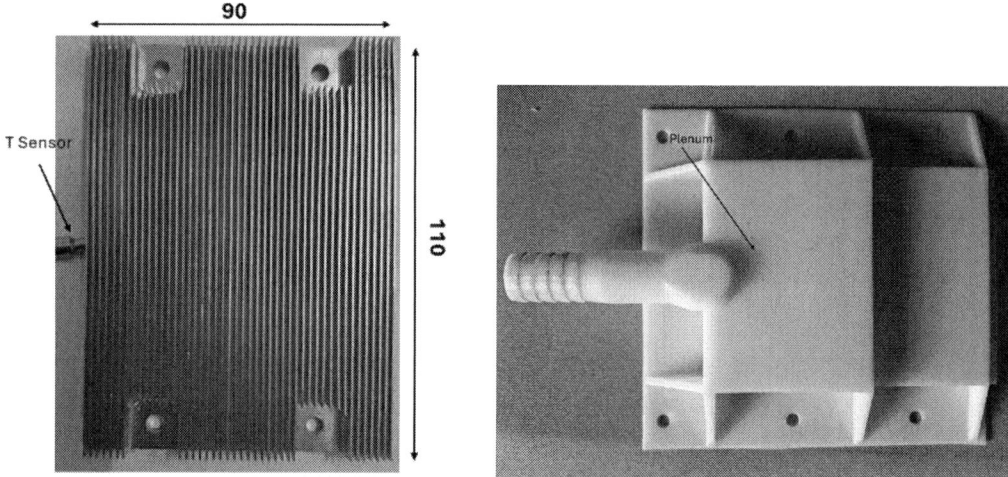

Figure 2. Copper Heat Sink and 3D Printed Sprayer

3. RESULTS

The overall thermal resistance is defined as:

$$R = \frac{\Delta T}{Q} \tag{1}$$

Where R is the thermal resistance in the system, ΔT is the temperature difference between the heating plate surface temperature and the coolant temperature, and Q is heat flow. The results are plotted in Figure 3 with red circles denoting data points. Linear fitting is used to calculate the thermal resistance. The linear fitting is forced to pass through the origin point, as ΔT should be zero when there is no heating power. The fitted dashed line is shown in Figure 3, and the fitted line equation is also listed. The linear fitting gives a value of 0.022 K/W for the thermal resistance. The red circles move from above the fitted line to below with increasing heating power. With oil temperature kept constant, larger heating powers result in higher heat sink temperatures. With higher heat sink temperatures, the boundary layer of the oil has a higher average temperature and, thus, a lower oil viscosity. A lower oil viscosity will in turn enhance the heat transfer over the heat sink, leading to lower thermal resistance.

Therefore, the fitted thermal resistance value is the mean value over the power range in our testing. At higher heating power cases, the actual thermal resistance will be slightly lower than the fitted value. The effective heat transfer coefficient is defined as:

$$h = \frac{1}{RA} \tag{2}$$

Where A is the area of the heat sink base. With a 0.022 K/W thermal resistance value, the effective heat transfer coefficient is 4591 $W/(m^2 \cdot K)$. This value is magnitude higher than the natural convective heat transfer coefficients.

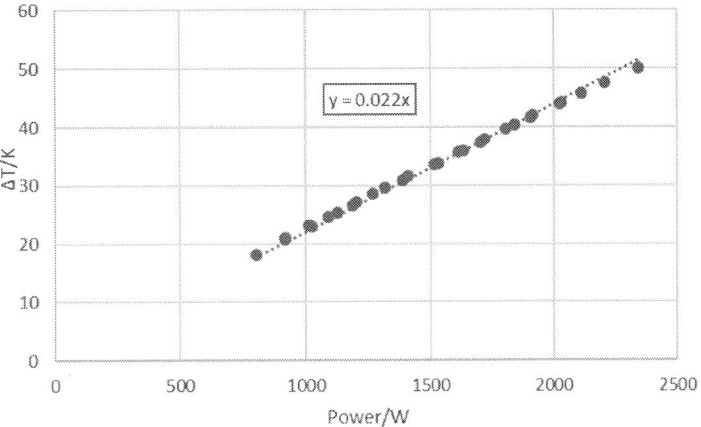

Figure 3. Temperature difference values at different heat powers

At 1100W heating power, the temperature difference is below 25 K. With the max case temperature of 80 °C, the coolant temperature can be above 55 °C, enabling free heat rejection to the ambient with dry coolers. At 2300W heating power, only room temperature water is needed in our system.

The oil pump has a power of 55W, therefore the power utilization efficiency (PUE) for our system can go down to 1.024. Further increasing the spraying velocity can further improve the cooling performance. However, the PUE value will increase, which means the power efficiency is deteriorating. There is a trade-off between the cooling performance and cooling system power consumption.

4. CONCLUSIONS

Our single-phase spraying test result shows a low thermal resistance and a high heat transfer coefficient in our system. With a 25 K temperature difference, we can properly cool a heat load of 1100W while maintaining the coolant temperature at 55 °C. This coolant temperature opens the possibility for heat reuse and a whole year of free cooling around the world. With high coolant temperatures, heat rejection can be achieved by dry coolers without chilled water even in hot and dry regions. Therefore, our single-phase liquid spray cooling can provide another possible and sustainable solution to the IT industry amid the explosion of AI. In our results, we used the air-cooling designed copper heat sink, but the TDP gets more than tripled by switching from air to spray cooling.

5. REFERENCES

1. https://www.intel.com/content/www/us/en/products/sku/237252/intel-xeon-platinum-8593q-processor-320m-cache-2-20-ghz/specifications.html
2. https://nvidianews.nvidia.com/news/nvidia-blackwell-platform-arrives-to-power-a-new-era-of-computing
3. https://www.mckinsey.com/industries/technology-media-and-telecommunications/our-insights/investing-

in-the-rising-data-center-economy

4. https://sales.sfs.fi/fi/index/tuotteet/ISO/ISO/ID9998/3/1020905.html.stx?_ga=2.165650047.1156658577.1712281785-972149575.1712281780

5. https://submer.com/business-cases/case-studies/forced-convection-heat-sink-with-intel/

6. https://www.grcooling.com/wp-content/uploads/2020/04/grc-blog-library-tech-comparison-%E2%80%94-cold-plate-vs-single-phase-immersion-cooling.pdf

7. F.P. Incropera, D.P. DeWitt, T.L. Bergman, and A.S. Lavine, Fundamentals of Heat and Mass Transfer, 6th ed., John Wiley & Sons, 2006.

8. Zhang, K., Ye, X., Hou, Z. et al. Study on heat transfer performance of cold plate with grid channel. Sci Rep 14, 4913 (2024).

9. S. M. S. Murshed, 'Introductory Chapter: Electronics Cooling — An Overview', Electronics Cooling. InTech, Jun. 15, 2016.

Thermal Characterization Testing with a Three-Resistor Finite Difference Model

Adolfo Lozano III, PhD, PE
Collins Aerospace, An RTX Business
3200 E Renner Road
Richardson, TX 75082
adolfo.lozano@rtx.com

Abstract

This manuscript outlines a mathematical methodology for developing a three-resistor, four-node finite difference numerical model representing a thermal characterization (or thermal survey) of a hardware system. The numerical model was calibrated to experimental data by determining the hardware system's characteristic thermal resistance-capacitance (RC) values for each resistor and node, respectively. Further, the construction of the numerical model was validated by a separate computational thermal model in Icepak (ANSYS) using the same RC value inputs. The numerical model outlined herein is straightforward to implement on a spreadsheet and can be used to quickly and easily predict the transient thermal response of a hardware system that has been characterized. Implementing the finite difference model presented herein can enable potentially significant time savings during environmental stress screening (ESS) procedures of hardware by developing time-efficient temperature cycling and burn-in profiles.

Keywords

Thermal characterization, thermal survey, transient thermal response, numerical model, finite difference, temperature cycling, environmental stress screening

Introduction

Thermal characterization testing, also known as a thermal survey, entails the experimental quantification of the transient thermodynamic properties of a hardware system in a known environment—specifically, the density, specific heat capacity, and convective heat transfer coefficient of the environment. These properties are related by the thermal time constant, τ, as shown in the equation below. The thermal time constant is represented as an RC value (thermal resistance and thermal capacitance) in an electrical circuit analogy.

$$\tau = \frac{\rho V c_P}{h A_S} = RC$$

Methodology

The primary objective of a thermal characterization effort is to identify and measure the physical locations with the least and greatest thermal inertia within the hardware. The location of least thermal inertia will correspond to the quickest transient thermal response to the environment, whereas the location of greatest thermal inertia will correspond to the slowest transient thermal response to the environment. Hence, experimentally quantifying RC values at key locations throughout the hardware system is the primary focus of a thermal characterization. A thermal characterization effort may be performed to experimentally validate a computational thermal model of the hardware system or to develop environmental

chamber temperature profiles during the temperature cycling or burn-in portions of the environmental stress screening (ESS) of hardware deliverable to the customer.

Accordingly, a hardware system can be represented as a one-dimensional, three-resistor, four-node series network as shown in Figure 1. The hardware system is assumed to consist of multiple internal assembly levels that nodal assignments are based on. Here, node N_A represents the ambient environment, node N_1 represents the location of least thermal inertia such as the exterior surface of the hardware system's highest-level assembly, node N_2 represents an intermediate interior location such as a lower-level assembly within the hardware, and node N_3 represents the interior location with the greatest thermal inertia such as the lowest-level assembly within the hardware. These four nodes fully bound the transient thermal response of the hardware. Each hardware node (N_1, N_2, and N_3) locally represents a lumped thermal capacitance of the solid and all nodes are assumed to be at uniform temperatures.

The three-resistor, four-node network model presented herein only accounts for internal heat conduction between hardware nodes (N_1, N_2, and N_3). Internal convection and radiation exchange within the hardware system are neglected in this model. This network model accounts for convection between the exterior of the hardware system and the ambient environment but neglects radiation.

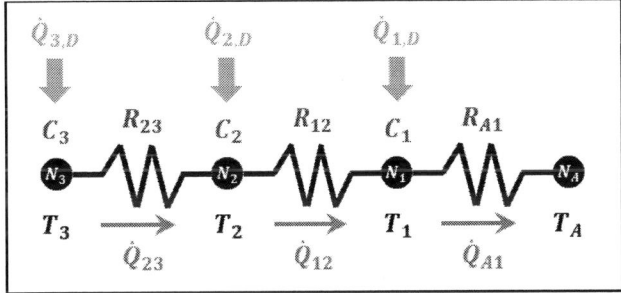

Figure 1: The finite difference numerical model quantified the transient thermal response of a three-resistor, four-node, lumped capacitance network model representing a hardware system.

Drawing from an undergraduate heat transfer course [1], the lumped capacitance model is validated by a Biot number much smaller than unity (Bi≪1). Here, the internal conductive temperature gradient within each node is neglected as long as the thermal inertia of the hardware is fully captured by the resistor-node network. Since hardware systems are typically housed by metal structure assemblies, this validation test with the Biot number typically passes reasonably well for all nodes.

Governing Equations

Two governing equations are relevant and necessary to solve this problem.

First, an energy balance (i.e., the first law of thermodynamics or conservation of energy) is applied to each nodal control volume representing the hardware (N_1, N_2, N_3) to account for heat input, output, storage, and generation [2].

$$\dot{E}_{IN} - \dot{E}_{OUT} = \dot{E}_{STOR} - \dot{E}_{GEN}$$

For all energy balances, the transient energy storage term is represented as:

$$\dot{E}_{STOR} = \rho V c_P \frac{\partial T}{\partial t} = C \frac{dT}{dt}$$

For nodes N_1, N_2, and N_3, the energy balance reduces to the following coupled, first-order, linear ordinary differential equations (ODEs):

$$C_1 \frac{dT_1}{dt} = \dot{Q}_{12} - \dot{Q}_{A1} + \dot{Q}_{1,D}$$

$$C_2 \frac{dT_2}{dt} = \dot{Q}_{23} - \dot{Q}_{12} + \dot{Q}_{2,D}$$

$$C_3 \frac{dT_3}{dt} = -\dot{Q}_{23} + \dot{Q}_{3,D}$$

Second, the thermal equivalent of Ohm's law is applicable to each resistance. This equation is used to quantify the heat flow across each resistance based on nodal temperatures. Ohm's law assumes a linear dependence on the temperature difference for heat transfer, which is applicable to conduction or convection. For resistances R_{A1}, R_{12}, and R_{23}, applying Ohm's law yields:

$$\dot{Q}_{A1} = \frac{(T_1(t) - T_A(t))}{R_{A1}}$$

$$\dot{Q}_{12} = \frac{(T_2(t) - T_1(t))}{R_{12}}$$

$$\dot{Q}_{23} = \frac{(T_3(t) - T_2(t))}{R_{23}}$$

Known quantities in this problem are the thermal resistance terms (R_{A1}, R_{12}, R_{23}), thermal capacitance terms (C_1, C_2, C_3), heat generation (or heat dissipation) terms ($\dot{Q}_{1,D}$, $\dot{Q}_{2,D}$, $\dot{Q}_{3,D}$), and the ambient temperature ($T_A(t)$). The first six parameters (R_{A1}, R_{12}, R_{23}, C_1, C_2, C_3) effectively represent the input variables for the numerical model, and because they are characteristic of the hardware, they are assumed to be time-independent constants (i.e., constant for all times). The heat dissipation terms ($\dot{Q}_{1,D}$, $\dot{Q}_{2,D}$, $\dot{Q}_{3,D}$) may be either zero ($\dot{Q}_{n,D} = 0$), time-independent constants ($\dot{Q}_{n,D} \neq 0$), or time-dependent constants ($\dot{Q}_{n,D}(t)$); these terms are explored later. Finally, the ambient temperature profile over time ($T_A(t)$) represents the boundary condition and is a known value at every time step (i.e., a time-dependent constant). Since node N_A represents the boundary condition, it does not require a thermal capacitance.

Analytical model

By considering the governing equations, the three-resistor, four-node network is characterized by a system of ODEs, which can be expressed in matrix form. Substituting the three resistance Ohm's law equations into the three nodal energy balances and simplifying yields:

$$\frac{d}{dt}\begin{bmatrix} T_1 \\ T_2 \\ T_3 \end{bmatrix} = \begin{bmatrix} \frac{-1}{C_1}\left(\frac{1}{R_{A1}} + \frac{1}{R_{12}}\right) & \frac{1}{C_1 R_{12}} & 0 \\ \frac{1}{C_2 R_{12}} & \frac{-1}{C_2}\left(\frac{1}{R_{12}} + \frac{1}{R_{23}}\right) & \frac{1}{C_2 R_{23}} \\ 0 & \frac{1}{C_3 R_{23}} & \frac{-1}{C_3}\left(\frac{1}{R_{23}}\right) \end{bmatrix} \begin{bmatrix} T_1 \\ T_2 \\ T_3 \end{bmatrix}$$

$$+ \begin{bmatrix} \frac{1}{C_1}\left(\frac{T_A(t)}{R_{A1}} + \dot{Q}_{1,D}\right) \\ \frac{1}{C_2}\left(\dot{Q}_{2,D}\right) \\ \frac{1}{C_3}\left(\dot{Q}_{3,D}\right) \end{bmatrix}$$

The analytical solutions for T_1, T_2, and T_3 in this system of ODEs are meticulous. The homogeneous portion of the general solution is the sum of multiple exponential terms, $e^{-\lambda_n t}$, where λ_n is one of three eigenvalues. The solution is complicated by the linear time-dependence of $T_A(t)$ and potential time-dependence of $\dot{Q}_{n,D}(t)$ in the nonhomogeneous portions of the ODEs. Therefore, the analytical solution is skipped in lieu of the numerical solution, which is the focus of this manuscript.

Numerical model

The ODEs for each node in the three-resistor network can be solved numerically by constructing a finite difference scheme that can be executed in an Excel (Microsoft) spreadsheet or MATLAB (MathWorks) script. The finite difference algorithm temporally discretizes the temperature derivative with respect to time (dT/dt) via a Taylor series expansion and solves the governing equation numerically by marching in time at discrete and uniform time steps (Δt) based on known parameters from the previous time step. For iteration i, the first-order ordinary derivative is discretized using a forward Euler explicit finite difference scheme as:

$$\left.\frac{dT}{dt}\right|_i \approx \frac{\Delta T}{\Delta t} = \frac{T^{i+1} - T^i}{\Delta t}$$

Here, a forward Euler explicit finite difference scheme was implemented because of its simplicity. Many other more sophisticated finite difference schemes exist that are reviewed in a numerical analysis course [3].

This numerical discretization is applied to each nodal governing equation (energy balance and Ohm's law) in the three-resistor, four-node network.

For nodes N_1, N_2, and N_3, the energy balances yield:

$$T_1^{i+1} = T_1^i + \frac{\Delta t}{C_1}\left(\dot{Q}_{12}^i - \dot{Q}_{A1}^i + \dot{Q}_{1,D}^i\right)$$

$$T_2^{i+1} = T_2^i + \frac{\Delta t}{C_2}\left(\dot{Q}_{23}^i - \dot{Q}_{12}^i + \dot{Q}_{2,D}^i\right)$$

$$T_3^{i+1} = T_3^i + \frac{\Delta t}{C_3}\left(-\dot{Q}_{23}^i + \dot{Q}_{3,D}^i\right)$$

Note that at iteration $(i + 1)$, all quantities at the prior iteration i have been calculated and are therefore known. Therefore, the above equations are used to solve numerically for nodal temperatures at iteration $(i + 1)$ based on solved parameters from the previous iteration i.

Next, for resistances R_{A1}, R_{12}, and R_{23}, applying Ohm's law yields:

$$\dot{Q}_{A1}^{i+1} = \frac{(T_1^{i+1} - T_A^{i+1})}{R_{A1}}$$

$$\dot{Q}_{12}^{i+1} = \frac{(T_2^{i+1} - T_1^{i+1})}{R_{12}}$$

$$\dot{Q}_{23}^{i+1} = \frac{(T_3^{i+1} - T_2^{i+1})}{R_{23}}$$

These six equations collectively comprise the numerical model of the three-resistor, four-node network that are calculated at each iteration and can be entered directly into a spreadsheet or analysis software. Here, the ambient temperature profile over time ($T_A(t)$) is represented as a column vector (T_A^i).

Prior to the first iteration ($i = 0$, $t = 0$), the system is assumed to be in thermal equilibrium, that is, all nodal temperatures are assumed to be equal to the initial condition (i.e., initial soak temperature of $T_A^{(0)} = T_\infty$) and heat flow quantities across resistances are consequently assumed to be zero. To initiate the numerical solution, the initial perturbation is felt with $T_A^{(1)}$ at the first iteration ($i = 1$, $t = 1\Delta t$). Hence, small time steps are recommended to accurately resolve the diffusion of heat across the network, which improves numerical stability and accuracy (discussed in proceeding paragraphs). The numerical solution is represented in tabular format in Table 1, which illustrates the propagation of heat flow based on nodal temperature differences starting with the initial temperature perturbation at $i = 1$.

For the solution presented herein, no heat dissipation is assumed for any node for simplicity, that is, $\dot{Q}_{1,D} = \dot{Q}_{2,D} = \dot{Q}_{3,D} = 0$. This pertains to a "dead mass" case for the hardware

tested in which all electronic systems are off and therefore have no heat dissipation.

Finally, numerical stability and accuracy are important considerations in any numerical model. Numerical instability is generally characterized by unbounded growth or numerical oscillations and is typically caused by implementing a time step that is too large. The stability of a numerical scheme is based on the resulting ODE's eigenvalue(s), λ. Here, a forward Euler explicit finite difference scheme was selected for time integration, whose time step is inherently limited to the following [3]:

$$\Delta t \le \frac{2}{|\lambda|}$$

The three coupled, first-order ODEs result in three eigenvalues (λ_1, λ_2, λ_3). Eigenvalues were not determined analytically due to the cubic nature of the resulting polynomial but instead were calculated using predefined functions in MATLAB (MathWorks) version R2019b, a commercial numerical software, after RC values were defined (presented later).

Moreover, the forward Euler explicit finite difference method is first-order accurate. Numerical accuracy is a measure of the finite difference scheme's approximation error resulting from truncating higher order terms in the Taylor series expansion approximating the derivative (dT/dt). The truncation error, and therefore accuracy, is directly proportional to the magnitude of the time step. For sufficiently small time steps, higher order terms in the Taylor series expansion become negligible. Here, a first-order scheme was sufficiently accurate as shown by the model validation exercise (presented later).

Iteration	Time	T_A^i	T_1^i	T_2^i	T_3^i	$\dot{Q}_{A1}^i$	$\dot{Q}_{12}^i$	$\dot{Q}_{23}^i$
Known	*Known*	*Known*	*Calc.*	*Calc.*	*Calc.*	*Calc.*	*Calc.*	*Calc.*
$i = 0$	$t = 0$	T_∞	T_∞	T_∞	T_∞	0	0	0
$i = 1$	$t = 1\Delta t$	$T_A^{(1)}$	T_∞	T_∞	T_∞	$\dot{Q}_{A1}^{(1)}$	0	0
$i = 2$	$t = 2\Delta t$	$T_A^{(2)}$	$T_1^{(2)}$	T_∞	T_∞	$\dot{Q}_{A1}^{(2)}$	$\dot{Q}_{12}^{(2)}$	0
$i = 3$	$t = 3\Delta t$	$T_A^{(3)}$	$T_1^{(3)}$	$T_2^{(3)}$	T_∞	$\dot{Q}_{A1}^{(3)}$	$\dot{Q}_{12}^{(3)}$	$\dot{Q}_{23}^{(3)}$
$i = 4$	$t = 4\Delta t$	$T_A^{(4)}$	$T_1^{(4)}$	$T_2^{(4)}$	$T_3^{(4)}$	$\dot{Q}_{A1}^{(4)}$	$\dot{Q}_{12}^{(4)}$	$\dot{Q}_{23}^{(4)}$
...	...	...	...	...	...	...	...	...
$i - 1$	$t = (i-1)\Delta t$	T_A^{i-1}	T_1^{i-1}	T_2^{i-1}	T_3^{i-1}	$\dot{Q}_{A1}^{i-1}$	$\dot{Q}_{12}^{i-1}$	$\dot{Q}_{23}^{i-1}$
i	$t = i\Delta t$	T_A^i	T_1^i	T_2^i	T_3^i	$\dot{Q}_{A1}^i$	$\dot{Q}_{12}^i$	$\dot{Q}_{23}^i$

Table 1: Tabular representation of the finite difference numerical algorithm illustrating the propagation of heat resulting from an initial ambient temperature perturbation. Parameter values were either known or calculated ("Calc.") at each time step.

Experimental data collection and RC-network calibration

In order to perform thermal characterization testing, strategic locations within the hardware were selected that fully encompassed the hardware's transient thermal response from

quickest (N₁) to slowest (N₃) as outlined earlier. Each nodal location was measured by placing a temperature sensor at the selected location and measuring the temperature over time. Experimental temperature data were recorded as the environmental temperature was raised or lowered to the relevant extrema based on the hardware's thermal requirements; the hardware was given sufficient time to reach a steady state temperature as determined by the data, representing a fully soaked condition. It is recommended to maximize the temperature extrema selected in order to minimize data instrumentation error and uncertainty.

Accordingly, various thermocouples were attached across the multiple assembly levels within the hardware system in order to fully capture its transient thermal response. Having temperature data from various locations enables the thermal engineer to determine the most appropriate locations for the exterior, intermediate, and interior nodes during post-processing. Here, all three nodes were assigned to different assembly levels within the hardware and were therefore sufficiently insulated from each other. Ambient temperature data served as the boundary condition (BC) for the numerical model. Note: Experimental data were collected but are not published herein due to U.S. export control restrictions. However, experimental data for a simplified case are qualitatively compared to the numerical model and sample RC input values are henceforth presented as an illustrative example.

Next, after collecting temperature data over time for each node, RC input values for each resistor-node pair of the numerical model (R_{A1}, R_{12}, R_{23}, C_1, C_2, C_3) that aligned well to the experimental data were determined iteratively. Trial-and-error iteration of RC values that macroscopically matched the data was sufficient for this analysis, as presented in the proceeding paragraph. These RC values are outlined in Table 2. Hence, the numerical model was calibrated to experimental data corresponding to the hardware. The experimental temperature data served as the ground truth for which to match nodal temperature responses from the numerical model. Experimental data are necessary to perform a thermal characterization. Note that the units of the thermal capacitance values shown reduce to [J/K] but are displayed as [W-s/K] to emphasize that numerical time steps were taken on the order of [s]. These RC values were applied to the input variables to the numerical model in Excel (Microsoft) and to the computational thermal model in Icepak (ANSYS).

Thermal resistance [C/W]	Thermal capacitance [W-s/K]	Time constant RC [s]
$R_{A1} = 0.015$	$C_1 = 30,000$	$R_{A1}C_1 = 450$
$R_{12} = 0.4$	$C_2 = 8,000$	$R_{12}C_2 = 3,200$
$R_{23} = 2.0$	$C_3 = 3,000$	$R_{23}C_3 = 6,000$

Table 2: Thermal resistance and capacitance values were calibrated to experimental temperature data. Sample (obfuscated) values are presented herein as an illustrative example due to U.S. export control restrictions.

Experimental data are presented herein for a simplified, two-resistor, three-node model (i.e., N_A, N_1, and N_2 only) corresponding to a different, unpublished set of RC values due to U.S. export control restrictions. Although the data are not presented quantitatively, Figure 2 provides a qualitative comparison of the finite difference numerical model to the data.

The numerical model represented the data reasonably well. This comparison served to experimentally validate the methodology used to develop the finite difference model.

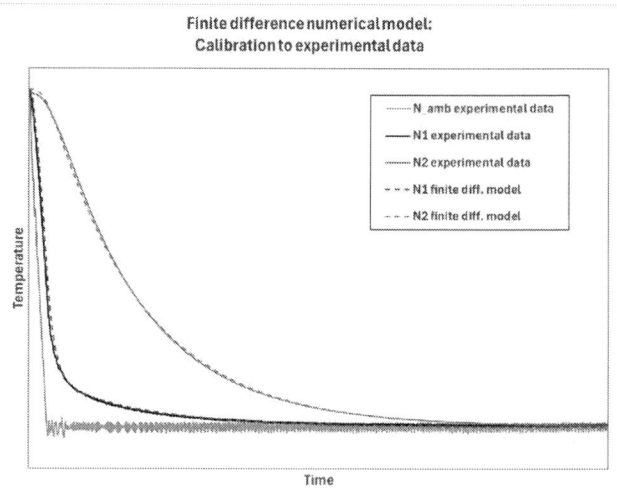

Figure 2: Qualitative comparison of finite difference numerical model results to experimental temperature data for a simplified two-resistor, three-node model. The numerical model was calibrated to experimental data.

Finally, with defined RC values in Table 2, eigenvalues were calculated (via MATLAB) to be:
$$\lambda_1 \approx \text{-2.3190E-03 s}^{-1}$$
$$\lambda_2 \approx \text{-4.0500E-04 s}^{-1}$$
$$\lambda_3 \approx \text{-1.2323E-04 s}^{-1}$$

This resulted in a maximum allowable time step of approximately 860 seconds in the finite difference model for numerical stability. However, these three eigenvalues were approximations due to the following sources of uncertainty: first, measurement error in the test instrumentation used to collect experimental data; second, approximation error in matching the numerical model's RC values to experimental data; third, the numerical method used to calculate eigenvalues from the system of ODEs. Therefore, care was taken to ensure that the time step selected was well below this numerical stability limitation, which would also improve the accuracy of the finite difference scheme.

Computational thermal model

A simple computational thermal model of the lumped capacitance, three-resistor, four-node network was constructed in Icepak (ANSYS) version 2023R1 to validate the results of the finite difference numerical solution.

The Icepak thermal model consisted of three objects as outlined in Figure 3: (1) a network object (purple-colored object) that included the three-resistor network, including all resistance and thermal capacitance values outlined in Table 2; (2) a wall object (cyan-colored object) that incorporated the ambient temperature profile over time (T_A^i), which represented the boundary condition at each time step; and (3) an arbitrarily thin block object (gray-colored object) with an arbitrarily high thermal diffusivity (i.e., $k = 10,000$ W/m/K, $\rho c_P = 1$ J/m^3/K) to diffuse heat efficiently from the boundary condition to the resistor network with a negligible temperature gradient. This thin conducting block object was necessary due to Icepak's inherent modeling structure.

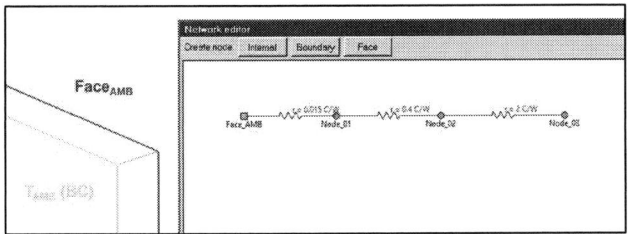

Figure 3: A computational thermal model in Icepak (ANSYS) of the three-resistor, four-node network was constructed to validate the finite difference numerical model.

Results

The numerical model's solution for nodal temperatures (T_1, T_2, T_3) was developed on a spreadsheet as illustrated in Table 1. The computational thermal model's solution was also simulated. Results of both models were nearly identical as plotted in Figure 4. A maximum absolute error of <0.8°C was observed between models at any given time step, with an average error of <0.02°C across all time steps. Hence, the finite difference model (solid line) was validated by the computational thermal model (dashed line).

A uniform time step of 20 seconds ($\Delta t = 20$ s) was implemented in both models, which was more than an order of magnitude lower than the limit for numerical stability in the finite difference model. (Interestingly, as the time step got closer to the numerical limit of approximately 860 seconds, nodal temperature responses in fact exhibited oscillatory behavior as outlined earlier.) This time step was also sufficiently small for reasonable numerical accuracy based on the error observed in this validation exercise.

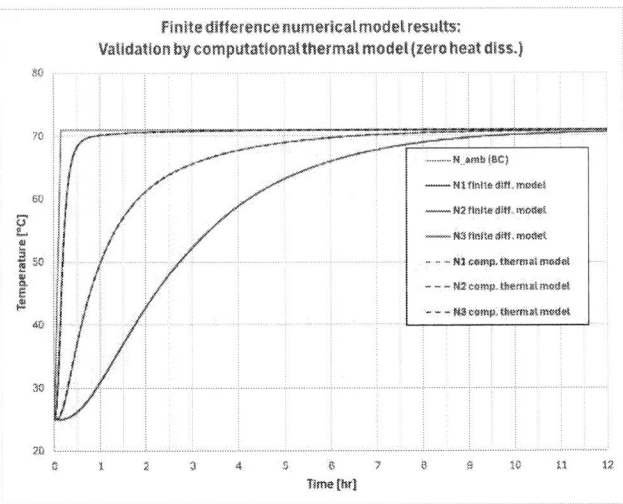

Figure 4: Comparison of the finite difference numerical model and Icepak thermal model for the zero heat dissipation case. The Icepak thermal model validated the accuracy of the finite difference numerical model.

Finally, while results assumed zero heat dissipation for all hardware nodes (N_1, N_2, N_3), the framework outlined herein

pertains to the general cases with time-independent constants or time-dependent constants. The numerical model was similarly validated by the computational thermal model for the heat dissipation case with non-zero, time-independent constants. These results assumed the same RC values outlined in Table 2 along with the following arbitrary nodal heat dissipations: $\dot{Q}_{1,D} = 10$ W, $\dot{Q}_{2,D} = 20$ W, $\dot{Q}_{3,D} = 50$ W. This validation is presented in Figure 5. In these non-zero heat dissipation cases, the thermal engineer should verify that the temperature gradient in each lumped capacitance node is still negligible.

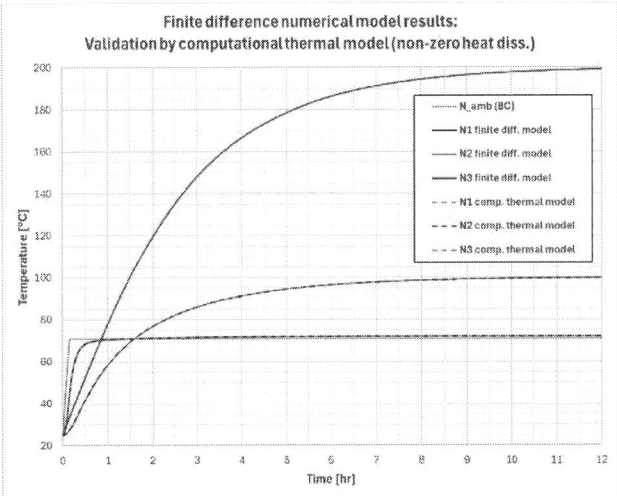

Figure 5: Comparison of the finite difference numerical model and Icepak thermal model for the constant heat dissipation case. The Icepak thermal model validated the accuracy of the finite difference numerical model.

Discussion

One primary motivation for performing a thermal characterization (or thermal survey) of a hardware system is that the thermal engineer can quickly and easily predict the expected transient thermal response of the hardware under any environmental condition or stimulus. This characterization, in turn, enables the thermal engineer to determine time-efficient temperature profiles (i.e., temperatures and dwell times) during temperature cycling and burn-in operations for deliverable hardware by incorporating temperature overshoots. When properly designed, small temperature overshoots can substantially reduce test time without changing the temperature extrema of the equipment under test. Temperature cycling and burn-in are common operations during environmental stress screening (ESS) that are intended to identify flaws, defects, and latent infant mortality of electronic components in the hardware. During temperature cycling, the hardware undergoes multiple cycles alternating between the upper (hot) and lower (cold) temperature extrema based on the hardware's thermal requirements. During burn-in, the hardware is elevated to the upper temperature limit for an extended duration of time to ensure the hardware's performance and functionality are failure-free. It is not uncommon for both operations to be combined during ESS.

As an illustrative example of the potential time savings, two sample temperature cycling profiles are considered for the characterized hardware outlined in Table 2. Each of the temperature profiles was constructed with the same finite difference equations presented herein. Thermal requirements for ground, airborne, and space hardware in the defense industry are typically based on, or derived from, MIL-STD-810, MIL-HDBK-781, or SMC-S-016, each of which provides a standard range of environmental temperature extrema: -54°C for a cold soak temperature and +71°C or +85°C for a hot soak temperature (depending on reference and type of equipment) [4][5][6]. Here, the -54°C to +71°C temperature range was selected to demonstrate the value of performing a thermal characterization of a hardware system.

The first temperature cycling profile is presented in Figure 6 without temperature overshoots. The ambient temperature is illustrated with a blue-colored line and represents the temperature setpoint of the environmental chamber. The nodal locations (N_1, N_2, N_3) are represented by black-, green-, and red-colored lines, respectively. A chamber ramp rate of 5°C per minute was assumed for all temperature transitions, which is a typical limit for environmental temperature chambers.

In Figure 6, the chamber ambient temperature was set to the hot soak temperature of +71°C for 10.75 hours. This profile allowed the hardware's location of greatest thermal inertia, represented as interior nodal location N_3, to be fully soaked at the hot soak temperature of +71°C for approximately 1 hour. Then, after completing the hot cycle, the chamber ambient temperature was set to the cold soak temperature of -54°C for 13 hours (23.75 elapsed hours) in order to allow the hardware to be fully soaked at the cold soak temperature for approximately 1 hour again. Finally, after completing one temperature cycle, the chamber was set to room temperature of +25°C for 8.25 hours (36 elapsed hours) for retrieval of the hardware from the chamber. The total run time was 36 hours without any temperature overshoots.

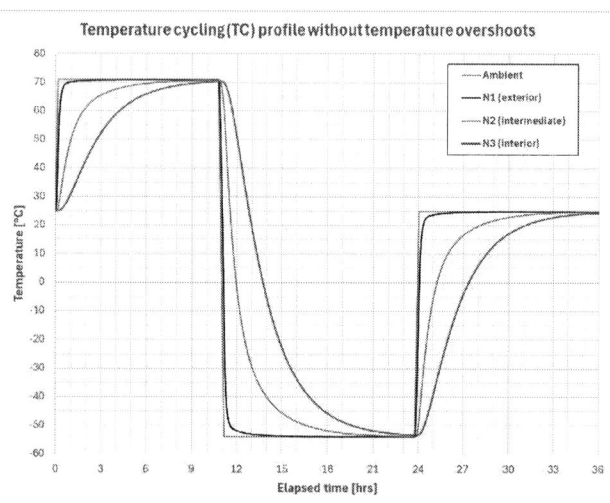

Figure 6: Temperature cycling profile without temperature overshoots for the characterized hardware, resulting in a total run time of 36 hours.

By contrast, Figure 7 presents a temperature cycling profile incorporating temperature overshoots during the hot and cold cycles. A hot temperature overshoot of +75°C with a dwell time of 6 hours was selected to expedite a hot soak temperature of +71°C for the hardware. (Note that a greater hot overshoot temperature may be considered based on the specific chamber's capacity, but the temperature response of the intermediate node N_2 should be closely tracked.) The chamber ambient temperature was then reduced to +71°C with a dwell time of 2 hours, allowing the hardware to be fully soaked at the hot soak temperature of +71°C for approximately 1 hour, corresponding to 8 elapsed hours of run time. Therefore, the hot temperature overshoot allowed for a time savings of 2.75 hours during the hot cycle. Then, the chamber ambient temperature was set to the cold temperature overshoot (-60°C) and subsequently the cold soak temperature (-54°C) with a 1 hour cold soak in similar fashion. The cold temperature overshoot enabled a time savings of 3.5 hours during the cold cycle. The hardware was finally brought back to room temperature (+25°C), expedited by a final hot overshoot to save 5.75 hours. The total run time was 24 hours with temperature overshoots.

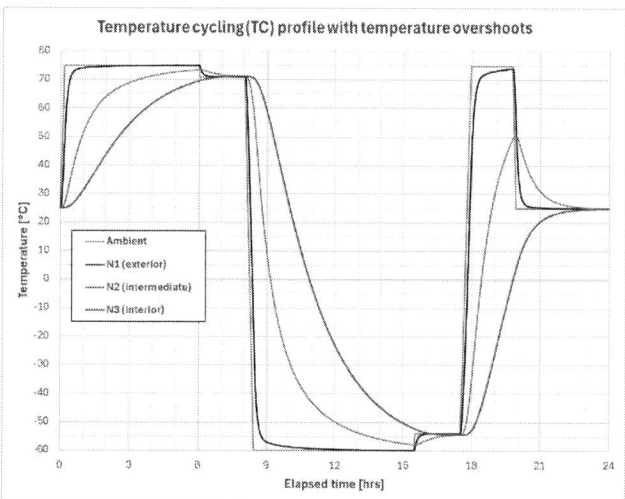

Figure 7: Temperature cycling profile with temperature overshoots for the characterized hardware, resulting in a total run time of 24 hours (12 hours of time savings due to temperature overshoots).

Comparing Figure 6 with Figure 7 shows that a time savings of 12 hours was realized by incorporating temperature overshoots on hardware that had been thermally characterized. The time savings would be compounded by the number of cycles and quantity of hardware systems screened. This simple example demonstrates the improved time-efficiency by having characterized a hardware system's transient thermal response.

Finally, expanding the one-dimensional series resistor network described herein to include additional resistor-node pairs is relatively straightforward; the intervariable relationships clearly emerge from the analytical equations in matrix form. However, a limitation to this resistor-node formulation is that it only accounts for internal heat conduction between hardware nodes; internal convection and radiation exchange within the hardware are neglected in this model. Therefore, cases in which nodal temperatures are significantly impacted by internal convective or radiative heat transfer within the hardware system would not be appropriately characterized by this finite difference model. In these cases, although the thermal engineer may attempt to calibrate the model to experimental data—which would inherently include internal convective and radiative effects—to obtain "effective" RC values, the model may not accurately resolve the transient thermal response of each node in these cases.

Conclusions

The three-resistor, four-node finite difference numerical model presented herein offers a simple approach to performing a transient thermal characterization of a hardware system. The formulas outlined can be readily applied and propagated on a commonplace spreadsheet software application like Excel (Microsoft). Implementing a numerical model that is calibrated to experimental data can significantly reduce hardware test time and can facilitate a transient thermal analysis of a hardware system.

Acknowledgments

The author acknowledges Dr. Philip S. Schmidt, Professor Emeritus of Mechanical Engineering at The University of Texas at Austin, for his academic mentorship and for his constant emphasis to perform an energy balance of the control volume as proper analysis methodology.

References

[1] Incropera, F. P. et al., Fundamentals of Heat and Mass Transfer, 6th ed., John Wiley & Sons, 2007.

[2] Schmidt, P. S. et al., Thermodynamics: An Integrated Learning System, 1st ed., John Wiley & Sons, 2006.

[3] Moin, P., Fundamentals of Engineering Numerical Analysis, 1st ed., Cambridge University Press, 2001

[4] MIL-STD-810H with Change 1, "Environmental Engineering Considerations and Laboratory Tests," U.S. Department of Defense, 18 May 2022

[5] MIL-HDBK-781A, "Handbook for Reliability Test Methods, Plans, and Environments for Engineering, Development, Qualification and Production," U.S. Department of Defense, 1 April 1996

[6] SMC-S-016, "Test Requirements for Launch, Upper-Stage and Space Vehicles," U.S. Air Force Space Command, 5 September 2014

Calibration of the Performance Measurement for Thin Vapor Chambers by Photonics Technologies

Kuang-Yu Hsu, Yi-Jing Chu, Andhi Indira Kusuma, Krishn Patel, Rajveer G. V., and Ming-Hsien Hsiao
T-Global Technology Co.
No. 33, Ln. 50, Daren Rd., Taoyuan Dist., Taoyuan City, Taiwan
kyhsu@tglobalcorp.com

Abstract

Characterization of thin vapor chambers is performed using laser heating and non-contact infra-red pyrometers for temperature measurement. This is found to be a fast and accurate measurement for mass production applications. The temperature measurement at the heater position on the evaporator side is very critical but it is seriously affected by the heat laser light. For accuracy considerations the measurement requires a pulsed heating mode to allow the measurement to be made while the laser is off. The heater operates in an on-off heating mode rather than a continuous heating mode. It is necessary to analyze the deviation from the continuous heating mode. Two methods, including matching the heat and matching the temperature, are proposed and experimentally measured. The temperature accuracy of the on-off laser heating is less than ±1 °C. The measurement data also show the transient response of the vapor chamber to a heat surge is temperature stabilizing. In this study the effective heat capacity is proposed and used for the assessment of temperature stabilization performance. The experimental data also show the effective thermal capacity of the vapor chamber is 2.5 times larger than that of the copper plate.

Keywords

Thin vapor chamber, vapor chamber characterization, vapor chamber transient analysis

1. Introduction

Traditionally, graphite sheets with a high thermal conductivity in the transverse direction have been used as the heat spreaders. The vapor chamber (VC), utilizing the two-phase mechanism rather than material thermal conduction, is an even more effective heat spreader. The thin VC, with the thickness less than 1.0 mm, has the potential for relieving the thermal issues in portable device applications with limited available space. The working range of the heating power of the thin VCs is usually less than 10 W, limited by the amount of the working fluid. The performance characterization of the thin VCs, using the contact heater and the thermocouples, suffers from the large thermal capacity as well as the large heat loss of the metal conductor to deliver the heat to the device under test [1]. The measurement time usually takes several minutes and the heat loss is yet to be improved. A laser heater needs no metal conductor to deliver the heat current [2], but the temperature measurement, using the infra-red pyrometer at the heater position on the evaporator side, is seriously affected by the scattered and reflected heat laser light. To take the temperature with accuracy, the output beam of the heating laser is shut off for a short period of time [3]. The laser heating is much more effective with much smaller heat loss than the contact heating so a measurement time of only 60 seconds is good enough. However, the off time in the on-off heating

scheme introduces a deviation in the system from the continuous heating. In this study, two calibration methods of the on-off laser heating scheme by matching the heat and by matching the temperature are proposed and experimentally demonstrated. A good temperature stabilizing performance of the vapor chamber is also found from the transient response data during the 50-ms laser off time. The effective thermal capacity parameter is proposed and used for the assessment of temperature stabilization performance.

2. Measurement System

The setup of the measurement system is shown in Fig. 1. An 808-nm infra-red (IR) laser with a collimated output from the multi-mode fiber is used as the heating source. The non-contact temperature measurement is done using the IR thermometers or pyrometers (CTlaser 4ML, optris). The thermal radiation from an object is related to its emissivity and its temperature. The emissivity is dependent on the material and also wavelength dependent. A pyrometer receives the thermal radiation in the infra-red range from the object at a distance through a suitable optical lens. By careful calibration with the emissivity of the object, the received signal is converted to the temperature. Conventionally, a thermopile detector has been used in the pyrometer, but the response speed is quite slow. Recently, much faster photodiodes have been developed and adopted in the pyrometers, facilitating millisecond level temperature measurement. The heating and temperature measurement in this system are both non-contact.

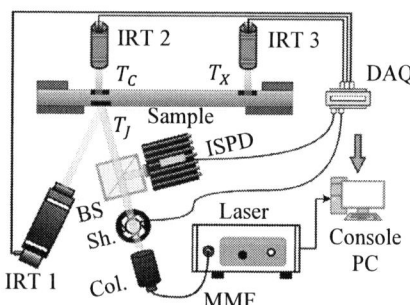

Fig. 1. Schematic system set-up. IRT: infra-red thermometer. BS: beam splitter. ISPD: integrated sphere with a photodiode detector. Sh.: shutter. Col.: collimator. MMF: multi-mode fiber.

The temperatures, T_J on the evaporator side and T_C and T_X on the condenser side of the sample, are measured. The temperatures T_C and T_X, at positions with a separation of 50 mm, are for the temperature uniformity. The position of T_J is in contact with the heat source, i.e., the chip. The non-contact measurement of T_J using a pyrometer is seriously affected by the scattered 808-nm laser light. To reduce the interference, the laser output beam is periodically shut off for a short period of

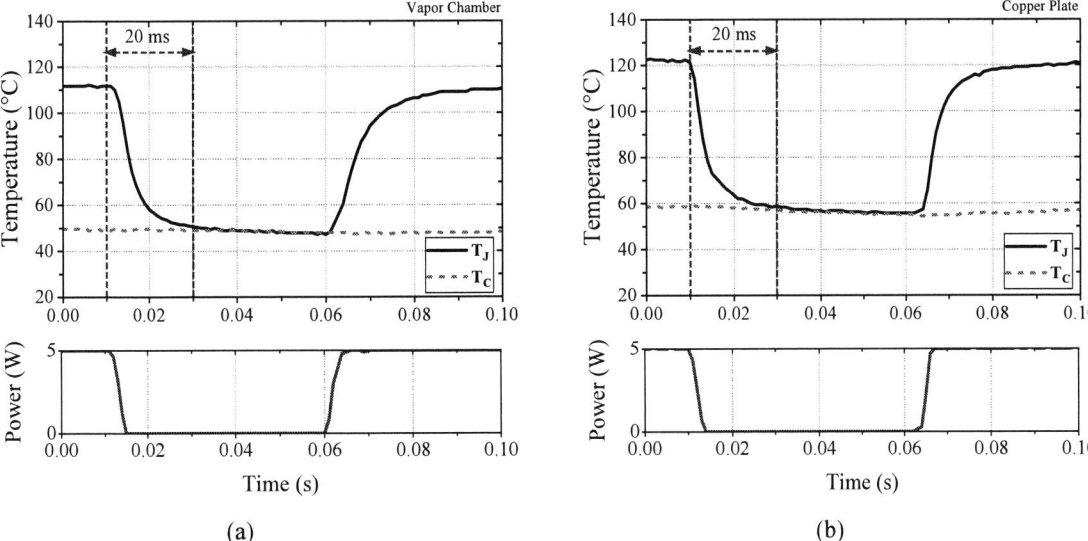

Fig. 2. The original T_J, T_C, and the heat current curves of (a) the VC and (b) the copper plate.

time using a mechanical shutter (SH05R, Thorlabs) with a high optical isolation. Meanwhile, T_J is taken with accuracy. Details of the system is described in [3]. A copper VC and a copper plate with the same dimensions of 90 mm × 60 mm × 0.4 mm were used as the testing samples.

A higher duty cycle of laser heating is preferred for minimized perturbation from the off time. For a cycle time Λ of 1 s, the laser off time is set at 50 ms. The corresponding duty cycle χ of the laser heater is 95%. The original T_J curve, T_C curve, and the heat current (power) curve of the vapor chamber and the copper plate samples in the final 60th cycle are shown in Fig. 2. The initial fake T_J temperature with interference is higher than 100 °C, but it decreases very fast after the heating (the laser beam) is shut off because of the fast response of the IR pyrometer. The temperatures should be taken at a delay time after the heat current is shut off and the fake signal has discharged sufficiently. The copper plate sample, with predictable thermal properties, is used as the calibration sample. In this way the delay time is set at 20 ms.

3. Theoretical Grounds of Calibration

The temperature, T_J, at the heating position on the evaporator of the thin vapor chamber in contact with the heater (the semiconductor chip), is an important performance parameter. The infra-red pyrometer detects the thermal radiation signal from the object for temperature measurement. However, the scattered heat laser light is much stronger than the real thermal radiation signal. The non-contact temperature measurement of T_J is severely affected. To suppress the interference, the output beam of the heating laser is switched off for a short period of time. The temperature, T_J, is taken with accuracy while there is no interference laser light. The heating is on-and-off with a cycle time Λ and a duty cycle χ. The T_J is taken once in each cycle. The heat current (power) functions of the continuous heating (CH) scheme, $H_{CH}(h^0, t)$, and the on-off heating (OOH) scheme, $H_{OOH}(h, t)$, are described as:

$$H_{CH}(h^0, t) = h^0, \text{ for } t = 0 \text{ to } \Lambda \qquad (1)$$

$$H_{OOH}(h, t) = \begin{cases} h \\ 0 \end{cases}, \text{ for } t = \begin{cases} 0 \text{ to } \chi \cdot \Lambda \\ \chi \cdot \Lambda \text{ to } \Lambda \end{cases} \qquad (2)$$

where h^0 and h are constant numbers.

Calibration of the heat current of on-off heating is necessary for reducing the difference with the continuous heating scheme.

3.1. Calibration Method by Matching the Heat

The heat matching method matches the heat of the on-off heating scheme of $h = h^{(1)}$ to the continuous heating scheme in a cycle:

$$\int_0^\Lambda H_{OOH}(h^{(1)}, t)dt = \int_0^\Lambda H_{CH}(h^0, t)dt \qquad (3)$$

$$h^{(1)} = \frac{h^0}{\chi} \qquad (4)$$

$h^{(1)}$ should be larger than h^0 to compensate for the off time in the on-off heating scheme. For duty cycle of 90% and 95%, $h^{(1)}$ is $1.11 \cdot h^0$ and $1.05 \cdot h^0$, respectively. For a higher duty cycle χ, a smaller perturbation by the off time is expected.

3.2. Calibration Method by Matching the Temperature

The measured temperatures, using the on-off heating scheme, are also related to the device characteristics, the cycle time and the duty cycle, and the delay time for temperature acquisition after the heat power is OFF. The corresponding heat power of the on-off heating scheme may not be as simple as just equating the heat. The temperature matching method matches the device temperature of the on-off heating scheme to the continuous heating scheme. The device under test is considered as a system with the heat current as the input and the temperatures T_J, T_C, and T_X as the output. We assume when one temperature output is matched, the other two temperature outputs are also matched. The temperature matching relation is schematically shown in Fig. 3. T_J is seriously affected by the fake signal, and T_X is relatively insensitive to the heat current input. Both are not suitable for the matching parameter.

Therefore T_C is chosen as the parameter for calibration. The heat power of the on-off heating scheme h is used as the tuning parameter. When $h = h^{(2)}$ for H_{OOH}, T_C is matched:

$$T_C\big(H_{OOH}(h^{(2)}, t)\big) = T_C(H_{CH}(h^0, t)) \qquad (5)$$

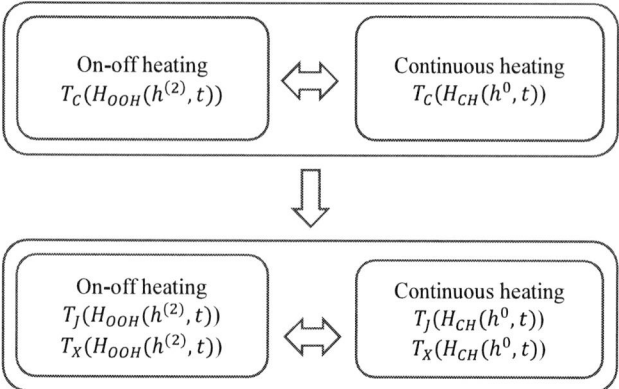

Fig. 3. Calibration procedures of the heat power of on-off heating scheme by matching T_C temperature.

The correction factor γ is the ratio of $h^{(2)}$ of the temperature matching to $h^{(1)}$ of the heat matching:

$$\gamma = \frac{h^{(2)}}{h^{(1)}} = \frac{\chi \cdot h^{(2)}}{h^0} \qquad (6)$$

3.3. Effective Heat Capacity of the Vapor Chamber

The vapor chamber is an effective heat spreader so the heater (the semiconductor chip) can be maintained at a lower temperature. The transient response data show a smaller temperature change and a faster heat propagation along the transverse plane to the heat current surge input for the VC compared to the copper plate. The higher surge power, the better temperature stabilization. The vapor chamber showed a peak temperature 9 °C lower than that of the copper plate, to a 1-s 10-W rectangular heat pulse input [3]. The electrical capacitance of an electronic or optoelectronic device is related to its response speed. A high-speed device should have a small capacitance. But a voltage stabilizer circuit requires a large capacitor. It is the same for thermal engineering. The heat source output may not always be a constant. Heat surge can also take place. Here we follow the same concept and use the effective heat capacity C_{eff} for the assessment of temperature stabilization. The effective heat capacity at the T_C position of the device is the variation (functional derivative) of the heat to the temperature:

$$C_{eff} = \frac{\delta Q}{\delta T_C} = \frac{h \cdot dt}{\delta T_C} \qquad (7)$$

where δQ is the heat difference, h is the heat current, dt is the time span of the heat current pulse, and δT_C is the temperature difference during dt. The effective heat capacity C_{eff} is more like a system function concept. The heat dissipation of the device is not excluded, whilst for the typical heat capacity measurement the sample is insulated and in a thermal equilibrium state. The heat capacity is inversely proportional to the temperature difference. Using the laser heating source, it is possible to generate the short heat power pulse with a small pulse width dt. The vapor chamber structure is composed of the top and bottom metal plates, the metal wick structure, and the working fluid and the vapor inside the chamber. It is complicated to calculate its heat capacity directly. However, we can estimate the effective thermal capacity of the VC with respect to the copper plate's at the T_C position using Equation (7):

$$\frac{C_{eff}^{VC}}{C_{eff}^{CP}} = \frac{\delta T_C^{CP}}{\delta T_C^{VC}} \qquad (8)$$

where δT_C^{VC} and δT_C^{CP} are the temperature differences of the VC and the copper plate, respectively.

4. Experimental Results and Discussion

4.1. Measured T_C Curves

The T_C curves of the VC and the copper plate samples using the continuous heating scheme at the heat current of h^0 and the on-off heating scheme at the heat current of $h^{(1)}$ (the heat match calibration) and $h^{(2)}$ (the temperature match calibration) are shown for comparison in Fig. 4. The duty cycle is 90% in Figs. 4(a) and 4(b), and is 95% in Figs. 4(c) and 4(d). The temperature differences of the VC are smaller than those of the copper plate. The VC is less sensitive than the copper plate for the on-off heating for duty cycle of both 90% and 95%.

4.2 Discussion of the Calibration Methods

The quantitative data of the temperature deviation, ΔT_C between the continuous heating and the on-off heating at $h^{(1)}$ and $h^{(2)}$, for the VC and the copper plate samples, at heat current levels of 3 and 7 W, and duty cycles of 90% and 95% are summarized in Table I. The temperature deviation ΔT_C at $h^{(1)}$ is relatively small 0.5 °C for the VC, and 1.2 °C for the copper plate, at corresponding h^0 of 7 W and 90% duty cycle. At 95% duty cycle, ΔT_C at $h^{(1)}$ is 0.7 °C for the VC, and 1.7 °C for the copper plate. The dependence of ΔT_C on duty cycle is not obvious for the VC. But for the copper plate, 95% duty cycle is better than 90%. ΔT_C of the vapor chamber is less than 1 °C at both $h^{(1)}$ and $h^{(2)}$, and for h^0 from 3 to 7 W. The heat matching calibration method is useful for most applications. If for high accuracy, the temperature matching calibration method can be useful. The dependence of the correction factor γ on the heat current level is also not obvious for the VC.

4.3. Effective Heat Capacity

Measurement of the transient response of a vapor chamber and a copper plate to a 1-s single rectangular heat current pulse was reported in [3]. A pulse heat current with an even shorter width is preferred for achieving a better accuracy using the variation (the functional derivative) method as Equation (7). The heat current function of the on-off heating scheme $H_{OOH}(h, t)$ can be considered as the superposition of a constant value term of 7 W and a periodic ($\Lambda = 1000$ ms) rectangular function $\Pi(t)$ with an amplitude of -7 W (the heat current with a negative amplitude is equivalent to cooling) and a pulse width dt of 50 ms for $t = 0$ to 1000 ms as:

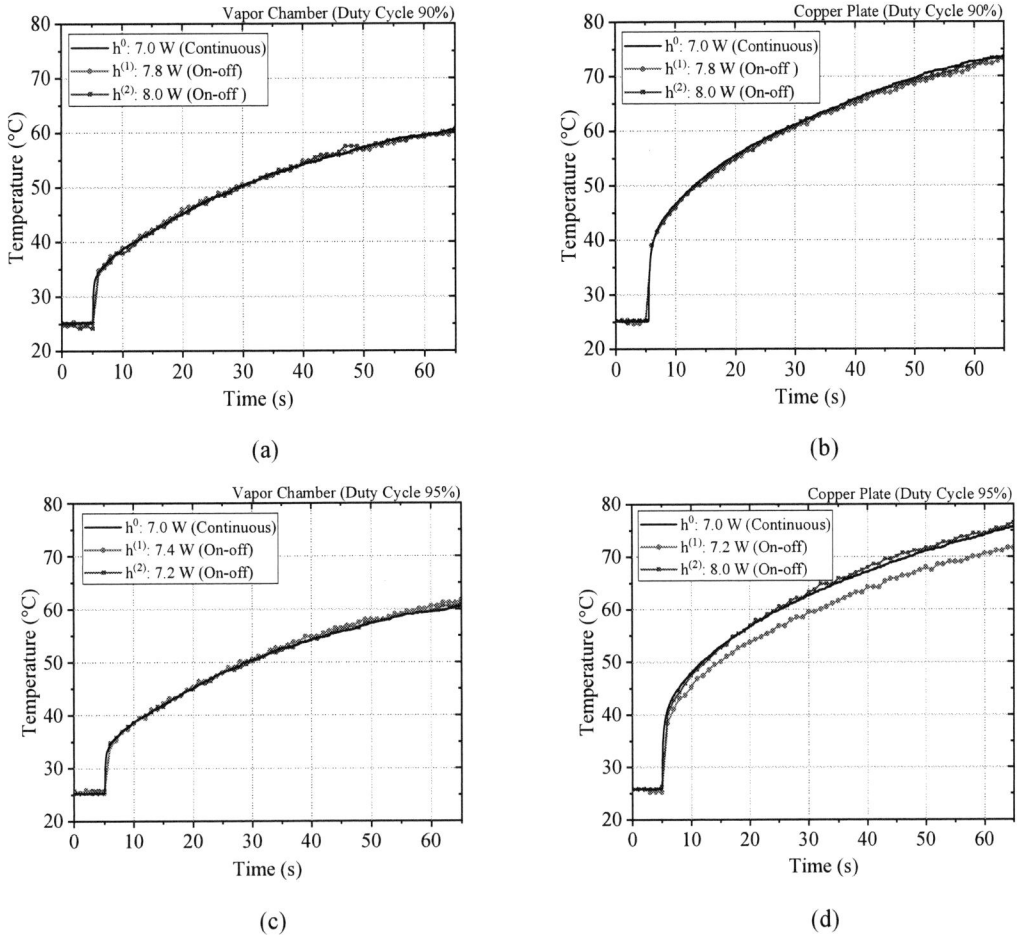

Fig. 4. T_C curves of (a) the VC and (b) the copper plate at 90% duty cycle, and (c) the VC and (d) the copper plate at 95% duty cycle.

TABLE I
RESULTS OF THE TWO CALIBRATION METHODS. $\Delta T_C = T_C(H_{CH}) - T_C(H_{OOH})$.

Parameters	Vapor Chamber				Copper Plate			
χ (duty cycle)	90%		95%		90%		95%	
h^0 (continuous heating)	3.0 W	7.0 W	3.0 W	7.0 W	3.0 W	7.0 W	3.0 W	7.0 W
$h^{(1)}$ (on-off heating)	3.4 W	7.8 W	3.2 W	7.4 W	3.4 W	7.8 W	3.2 W	7.2 W
ΔT_C at $h^{(1)}$	0.6 °C	0.5 °C	0.5 °C	0.7 °C	0.3 °C	1.2 °C	0.2 °C	1.7 °C
$h^{(2)}$ (on-off heating)	3.2 W	8.0 W	3.2 W	7.2 W	3.4 W	8.0 W	3.2 W	8.0 W
ΔT_C at $h^{(2)}$	0.4 °C	0.5 °C	0.5 °C	0.3 °C	0.3 °C	0.7 °C	0.2 °C	0.5 °C
γ (correction factor)	0.94	1.02	1.00	0.97	1.00	1.02	1.00	1.08

$$H_{OOH}(7\text{ W}, t) = 7 - 7 \cdot \Pi\left\{\frac{t-975}{50}\right\} \text{ (W)} \quad (9)$$

The T_C curve and the heat current curve in the 60th cycle are shown in Figs. 5(a) and 5(b) for the VC and the copper plate, respectively. During the 50-ms heating off time dt, the temperature drop δT_C is -2.1 °C for the VC and -5.3 °C for the copper plate. The temperature drop of the VC is obviously smaller than the copper plate's. This is also consistent with the

ΔT_C results of Fig 4. The experiment of applying positive heat pulses of the same conditions to a VC and a copper plate shows similar temperature changes as those in the negative pulse cases. The effective heat capacity of the vapor chamber is 2.5 times larger than that of the copper plate, using Equation (8). The VC utilizing the two-phase mechanism has a larger effective heat spreading area and therefore a larger thermal capacity. For the metal plate utilizing simply the material thermal conductance

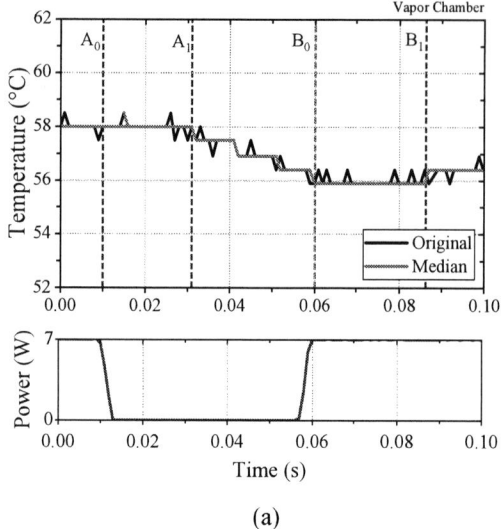 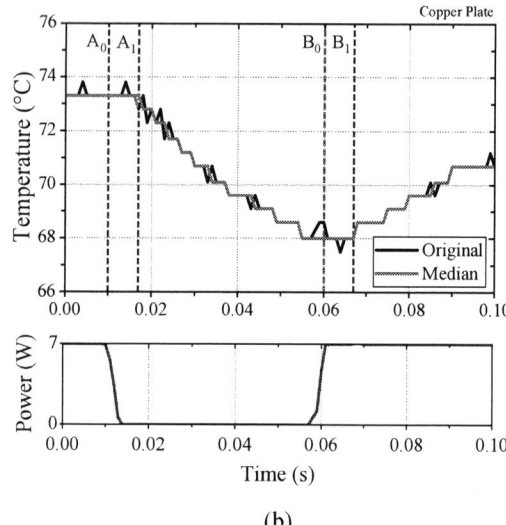

Fig. 5. T_C curve and the heat current curve of (a) the VC, and (b) the copper plate. The heat current is 7 W. The median is used for noise reduction. Thicknesses of the VC and the copper plate are both 0.4 mm.

the temperature gradience in the lateral directions is large so the effective heat spreading area and the capacity are smaller. Therefore, the VC serves as a much better heat spreader and also a temperature stabilizer than the copper plate.

Comparing the turn point of the heat current curve and the turn point of T_C curve as shown in Fig. 5, the delay time of the change of T_C temperature (from A_0 to A_1, and from B_0 to B_1) to respond to the heating on the hot side is estimated. It is 21 to 26 ms for the VC, and but only 7 ms for the copper plate. It is much slower for the T_C temperature to respond to the heating for the VC than the copper plate. The slower response is consistent with the previous conclusion that the vapor chamber has a larger effective heat capacity by comparing the temperature difference as Equation (8). However, the heat spreading in the transverse direction is faster for the VC than the copper plate, as reported in [3]. The thermal convection is faster than the thermal diffusion.

5. Conclusions

Using laser heating and infra-red pyrometer temperature measurement for characterization of thin vapor chambers is accurate and fast for mass production applications, compared with the contact measurement method. Two calibration methods of the performance characterization of the thin vapor chambers by photonics technologies were proposed and the experimental results were discussed. With the calibration, the temperature accuracy was within ± 1 °C. The transient response of vapor chamber showed temperature stabilizing. The temperature deviation using pulsed heating from constant heating was not quite significant for the vapor chamber. The effective heat capacity was proposed and used for the assessment of temperature stabilizing transient performance. By analyzing the experimental data during the 50-ms laser off time, the effective heat capacity of the vapor chamber is 2.5 times larger than that of the copper plate.

Acknowledgments

This project was partially supported by Bureau of Energy, Ministry of Economic Affairs, Taiwan (ROC) under grant 111-D0706.

References

1. G. Patankar, *et al.*, "A method for thermal performance characterization of ultrathin vapor chambers cooled by natural convection," *Journal of Electronic Packaging*, vol. **138**, 010903, 2016.
2. X. Jiang, *et al.*, "A non-contact thermal testing system for ultra-thin vapor chamber," *Review of Scientific Instruments*, vol. **92**, 124902, 2021.
3. K. Y. Hsu, *et al.*, "A Thermal Performance Characterization Method for Thin Vapor Chambers by Photonics Technologies," *2024 40th Semiconductor Thermal Measurement, Modeling & Management Symposium (SEMI-THERM)*, San Jose, CA, USA, pp. 76-82. https://ieeexplore.ieee.org/document/10535014

Thermal Impedance Measurement of SiC MOSFETs Under Strong Negative Gate Bias

Szilárd Zsigmond SZŐKE, Márk LÁZÁR, Henrik SEBŐK
Robert Bosch Kft., Engineering Center Budapest, 104. Gyömrői út, Budapest 1103, Hungary
Phone: +36 70 683-0820, e-mail: szilard.szoke@hu.bosch.com

Extended Abstract

SUMMARY

Due to ever increasing demand for better efficiency, more and more power electronics solutions adopt SiC MOSFETs as their main switching element. Thermal performance is of great importance, so reliable experimental characterization methods (i.e. thermal impedance, or Zth measurement) of SiC power semiconductors are required. In comparison to classic Si MOSFETs, SiC devices present a number of novel challenges to the users, one of which is the instability and hysteresis of the gate threshold voltage (Vgsth) [1], which also limits the ability to extract meaningful temperature estimates based on electrically measurable parameters. Numerous circuit configurations for Zth measurement were proposed in literature, but a proven "universally applicable" method, with guaranteed high accuracy, seems to be still missing.

In our paper we show the applicability of the already reported body-diode (source-drain) voltage recording method under negative gate bias voltage can be extended, by using gate bias voltages beyond datasheet limits, with special care. Our investigations suggest some of the modern SiC MOSFETs need extremely low bias voltage (-10 V … - 15 V) to get good enough Zth measurement results, which is outside of absolute maximum ratings specified in their datasheet (usually - 5 … -10 V is the lower limit for Ugs). Such severe gate bias conditions may be applied to SiC MOSFETs only for limited time (few hours), during measurements, without device destruction. We propose a simple gate leakage current test to assess how far negative one can safely push the gate bias voltage for the sole purpose of Zth measurement.

The resultant Zth curves are stable and reproducible enough in the early time range (< 1 ms) to be used for non-destructive monitoring of ageing of the thermal stack in reliability investigations spanning several months. An estimation of the expected measurement uncertainties is possible.

We plan to continue our experimental study on a multitude of commercially available samples from several vendors in the next 2 months and publish the results in this proposed article. We hope to generate a detailed overview of recommended measurement conditions, and compare behavior of different SiC technologies in Zth measurements. At the moment we only have the results from 2 different SiC MOSFET types.

1. INTRODUCTION

In case of SiC MOSFETs the application of the exact same Zth measurement methods as for Si (recording body diode voltage when Ugs = 0 V, or Gate-Source shorted) lead to erroneous results, due to the undesired influence of the trapped charges at SiC – SiO2 interface on the recorded body diode voltage, used to calculate chip temperature [2]. To work around this problem several different electrical operating modes were already presented in literature for heating and cooling during a transient thermal measurement of SiC FETs (e.g. using Rdson or Vgsth as alternative TSEP), with various degrees of success and difficulty [3]. One suggested workaround is to keep using body diode voltage as TSEP but apply negative Ugs of moderate value (e.g. -5 V), to suppress the consequences of the Vgsth hysteresis effect [2] [3]. (We found this method has serious limitations on latest generation SiC MOSFETs, because the suggested voltage levels were not enough to stabilize the entire time-range of the Zth curves in our measurements.)

The time range when Zth response distortions occur due to Vgsth hysteresis is the first 1..10 ms, depending on the measurement parameters [3], which will affect the outcome of any TDIM evaluation, e.g. reported as sensitivity to gate bias voltage by [4].

Also any significant distortion of this early time range of Zth makes reliability investigations of chip solder or silver sinter joints impossible via non-destructive Zth monitoring, because the unaged and aged measurements of the same sample happen several weeks or months apart in time, and any distortion will be misinterpreted as change due to ageing. At best, random distortions may show up as measurement noise, but systematic errors due to Vgsth instability cannot be excluded, if the effect is not fully suppressed. The rigorous definition of evaluation criteria for a good or unacceptable measurement condition is demonstrated in [5], and can be summarized as: the Vgs bias is sufficiently negative, if the body diode conductance (g_diode) and the temperature sensitivity of Vsd do not change with small changes of Vgs. However, no measurement is perfectly accurate, and [5] does not give an estimate of Zth uncertainty if the above-mentioned quantities are measured with known, >0 uncertainty.

The best advice in [5] is to use a "large negative gate turn-off voltage", but it also mentions that due to large differences in chip design and manufacturing processes across vendors, the value cannot be universally fixed. A flowchart is given to aid in selecting the proper measurement parameters, including Vgsoff and Isense, but no method is given how to handle the situation if the Vgsoff is to be pushed beyond the datasheet absolute maximum ratings.

The most negative reported gate biasing we found in literature was -10 V, within absolute maximum limits of the datasheet of their device under test [6].

2. EXPERIMENTAL/NUMERICAL METHODS

Calibration of TSEP was carefully carried out at different negative Vgs bias voltages (-7, -10, -15 V) in a temperature range from 20 to 120 °C, and reverse diode (I_d) currents of -1 mA to -120 mA. Sample results can be found in Figure 1.

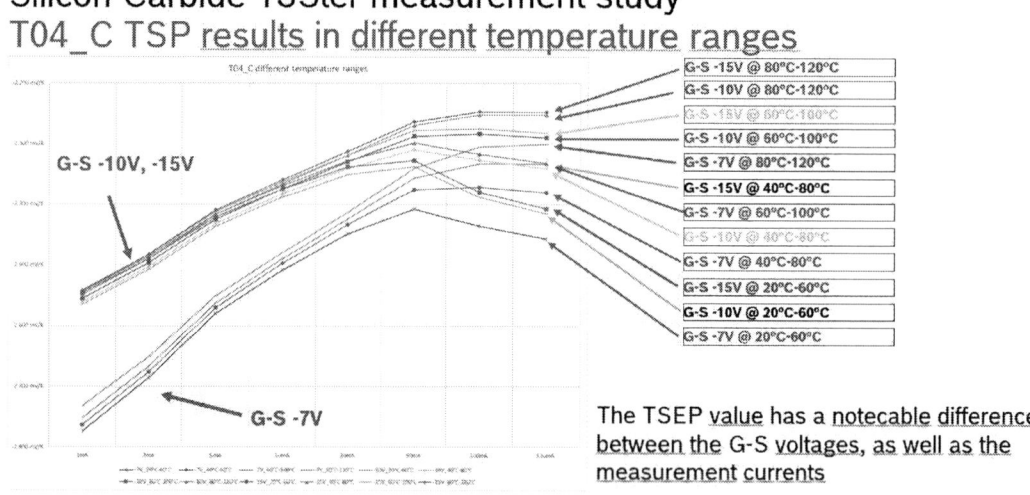

Figure 1.: TSEP values of V_sd voltage under different bias currents (horizontal axis), Vgs bias (groups indicated by arrows), and temperature range used for linearization.

Measurements of gate leakage current versus negative Vgs bias was done using an SMU, and evluated to set safe limits of applicable bias voltage range. Sample results are shown in Figure 2.

We have run many Zth measurements on 2 different SiC MOSFET types, using T3Ster equipment, with additional external voltage supply for different levels of Vgs negative bias, and different levels of I_d measurement bias currents. We examined the resultant raw responses in search for transient anomalies in the early (< 1 ms) time region (Figure 3.). Any change in the responses' shape due to bias current variation or due to progressive lowering of Gate-Source bias voltage was interpreted as undesired electrical interference with the pure thermal response.

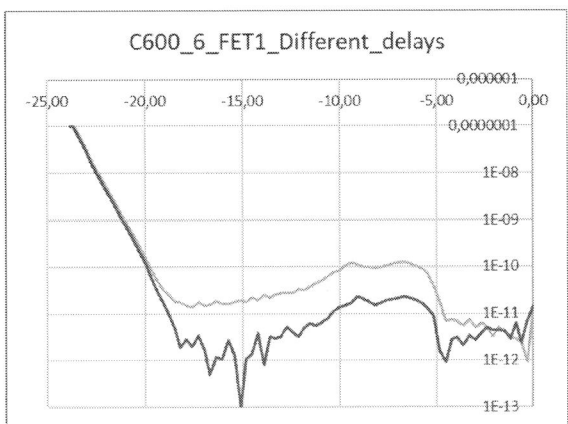

Figure 2.: Igss = f(Vgs) . Measurement shows negligible gate leakage current for Vgs = -19 ... 0 V.

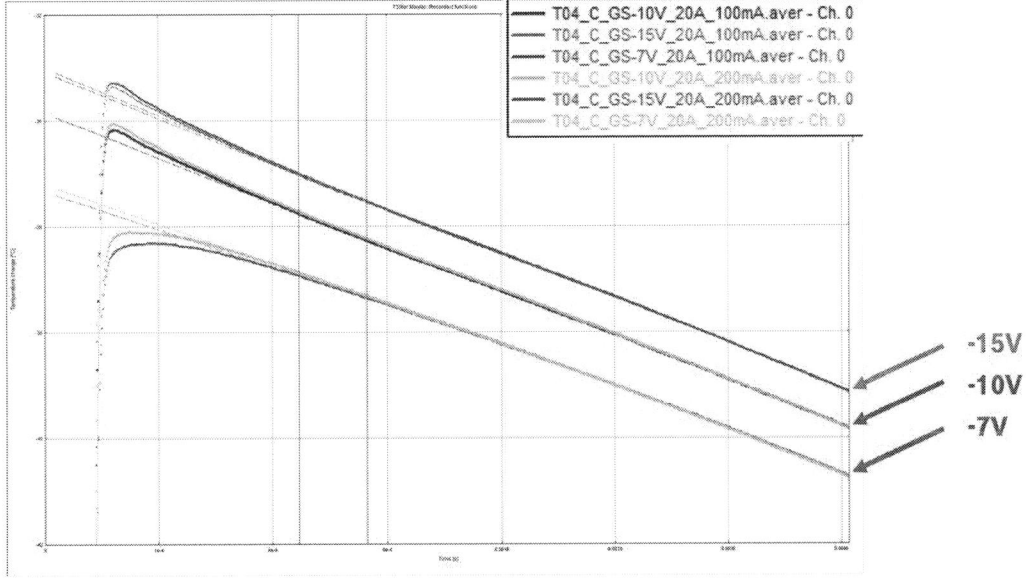

Figure 3.: Raw transient response after heating current switches off. This SiC MOSFET can be measured with Vgs = -10 V, but should not be measured with Vgs = -7 V.

3. RESULTS & CONCLUSIONS

The required negative gate voltage to get meaningful Zth measurements is very type-dependent in case of SiC MOSFETs. In at least one occasion the required Vgs value was beyond the absolute maximum ratings defined in the datasheet of the respective device. In the final version we will cover a multitude of commercially available device families.

Our experience confirms the conclusions of [5] and point to use large negative gate bias to avoid Vgsth hysteresis' distortion effects, and large measurement currents to speed up the initial transient.

To quantify the necessary negative Vgs bias we describe a slightly different method than [5], starting from a desired maximum error limit in the Zth, and using the Vsd = f(Vgsth) sensitivity measured by high-resolution oscilloscope and pulsed SMU.

We also propose a set of validation tests (by varying the 3 crucial Zth-test parameters Vgs, Imeas, Iheat and comparing results) to be performed on each measured device type, on the one hand to establish the optimum negative gate bias voltage, on the other hand to quantify the amplitude of the Vgsth hysteresis' effect on the Zth measurement. Acceptance criteria can be defined for the evaluated results, and potential errors may be detected.

4. REFERENCES

[1] D. Peters, T. Aichinger, T. Basler, G. Rescher, K. Puschkarsky and H. Reisinger, "Investigation of threshold voltage stability of SiC MOSFETs," 2018 IEEE 30th International Symposium on Power Semiconductor Devices and ICs (ISPSD), Chicago, IL, USA, 2018, pp. 40-43, doi: 10.1109/ISPSD.2018.8393597.

[2] Quang Chuc Nguyen, Patrick Tounsi, Jean-Pierre Fradin, Jean-Michel Reynes. "Investigation on the use of the MOSFET SiC body diode for junction temperature measurement." IEEE THERMINIC 2020, Fraunhofer Berlin, Sep 2020, Berlin, Germany. hal-03655865

[3] T. Funaki and S. Fukunaga, "Difficulties in characterizing transient thermal resistance of SiC MOSFETs," 2016 22nd International Workshop on Thermal Investigations of ICs and Systems (THERMINIC), Budapest, Hungary, 2016, pp. 141-146, doi: 10.1109/THERMINIC.2016.7749042.

[4] J. Knoll, C. DiMarino and C. Buttay, "A Guide for Accurate and Repeatable Measurement of the RTH, JC of SiC Packages," 2022 IEEE Energy Conversion Congress and Exposition (ECCE), Detroit, MI, USA, 2022, pp. 1-7, doi: 10.1109/ECCE50734.2022.9947814.

[5] Y. Zhang et al., "A Guideline for Silicon Carbide MOSFET Thermal Characterization based on Source-Drain Voltage," 2023 IEEE Applied Power Electronics Conference and Exposition (APEC), Orlando, FL, USA, 2023, pp. 378-385, doi: 10.1109/APEC43580.2023.10131449.

[6] C. Herold, J. Sun, P. Seidel, L. Tinschert and J. Lutz, "Power cycling methods for SiC MOSFETs," 2017 29th International Symposium on Power Semiconductor Devices and IC's (ISPSD), Sapporo, Japan, 2017, pp. 367-370, doi: 10.23919/ISPSD.2017.7988994.

Thin Film Thermal Conductivity Measurements

V. Linseis, S. Feulner, and H. Wang
Linseis Inc., 109 North Gold Drive, Robbinsville, NJ 08691, USA
(609)-223-2070, h.wang@linseis.com

Abstract

Thermal conductivity measurement is critical in the semiconductor industry for heat dissipation, device performance, reliability, miniaturization, material selection, and efficiency in power devices. Accurate thermal conductivity data is essential for both material development and the thermal design of semiconductor systems. Various methods have been developed to measure the thermal diffusivity and thermal conductivity of thin films, each offering unique advantages based on the thickness, material properties, and temperature range of the films being analyzed. Some of the most widely used techniques include Time-Domain Thermoreflectance (TDTR), the 3-Omega Method, and Frequency-Domain Thermoreflectance (FDTR).

We have previously developed a Time-Domain Thermoreflectance (TDTR) instrument that can be applied to thin layers on various substrates (e.g., non-transparent substrates or very thin films). In this method, a heating pulse is applied to the front side of the sample, and the temperature rise at this spot is measured using a detection laser from the same side. The thermal diffusivity of the sample layer is calculated using the falling edge of the normalized temperature rise, in combination with a multilayer model developed in collaboration with David G. Cahill [1].

In recent years, we have advanced the FDTR technique [2,3,4] which operates in the frequency domain by modulating the heat input and analyzing the thermal response in the frequency spectrum. As shown in Figure 1, a continuous wave laser (Probe laser) with 532 nm wavelength is used, while heating with a harmonically modulated pump laser at a different wavelength (405 nm). Local heating induces changes in the reflectivity and the phase lag between the thermal excitation and the detection is measured using a lock-in amplifier. Modelling the response in the frequency domain with a diffusive heat transport model allows us to determine the thermal conductivity, volumetric heat capacity, thermal diffusivity, thermal effusivity and thermal interface conductance. A thin metallic transducer layer (60 -70 nm in thickness) is deposited on top of the surface of the samples to enhance the temperature coefficient of reflectance, dR/dT, and at the same time to reduce the optical penetration depth in the material.

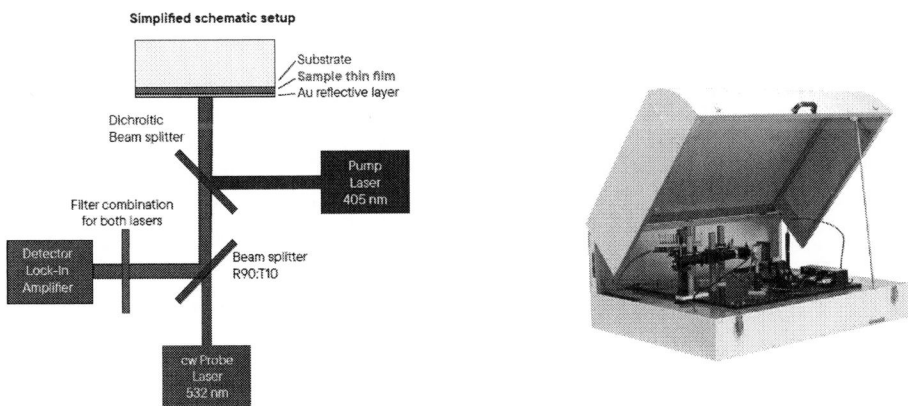

Fig. 1: FDTR system: left: schematic; right- instrument

In contrast to TDTR, where the probe laser must be manually adjusted due to slight changes in reflection when the sample is altered, our FDTR system includes automatic focusing. This feature continuously adjusts the probe laser's focus to accommodate any changes in the sample, ensuring optimal measurement conditions without manual intervention. Additionally, with aligned lasers in the FDTR system, there is no need to adjust the probe laser beam, simplifying the sample setup and resulting in more stable measurements. The measurement results for thin films, such as SiO2 with a thickness of 504 nm, Cr2AlC MAX Phase with a thickness of 1500 nm, are presented.

For in-plane thermal conductivity measurements of thin films, we developed the TFA (Thin Film Analyzer), system based on the 3-omega method [5], In this method a metal line serves as both the heater and temperature sensor. An alternating current generates heat at 2ω, causing temperature oscillations. These oscillations affect the metal line's resistance, producing a third-harmonic (3ω) voltage signal. By analyzing this signal, the thermal conductivity of the material is determined.

To minimize uncertainties from sample preparation, prefabricated chips are provided for thin film deposition, with prebuilt hot-stripe heaters and temperature measurement connections, simplifying sample preparation [6,7]. Figure 2 shows the TFA chamber with the thin film chip. The instrument and chip structure are detailed, and thermal conductivity measurement of Bi87Sb13 thin films, prepared via thermal evaporation in vacuum, are presented for the temperature range 110 K to 400 K.

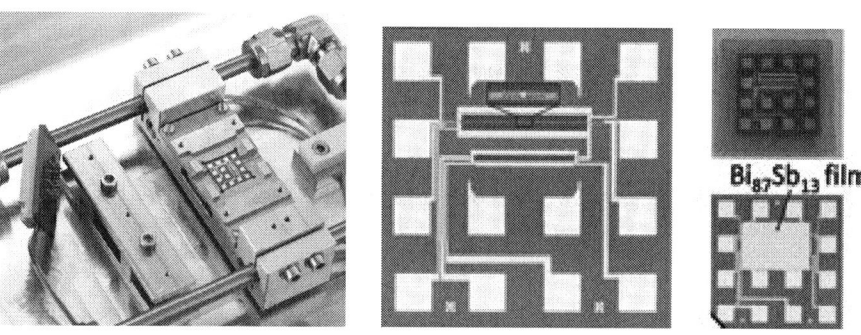

Fig. 2: Instrument setup and thin film chip for the Thin Film Analyzer

REFERENCES

1. David G. Cahill, *Rev. Sci. Instrum, 75, Number (2004)*
2. Miguel Goni, Maciej Patelka, Sho Ikeda, Toshiyuki Sato, and Aaron J. Schmidt, *Rev. Sci. Instrum,* 89, 074901 (2018)
3. Aaron J. Schmidt, Ramez Cheaito and Matteo Chiesa, *Rev. Sci. Instrum, 80, 094901 (2009)*
4. Luis A. Pérez, Kai Xu; Markus R. Wagner, Bernhard Dörling, Aleksandr Perevedentsev, Alejandro R. Goñi, Mariano Campoy-Quiles, M. Isabel Alonso, Juan Sebastián Reparaz, Rev. *Sci. Instrum. 93, 034902, (2022)*
5. David G. Cahill, M. Katiyar, and J. R. Abelson, *Phys. Rev.* B 50, 6077 (1994)
6. V. Linseis, F. Völklein, H. Reith, K. Nielsch, and P. Woias, *Rev. Sci. Instrum, 89, 015110 (2018)*
7. V Linseis, F Völklein, H Reith, R Hühne , L Schnatmann, K Nielsch1,3 and P Woias, *Semicond. Sci. Technol.* 33, 085014 (2018)

Thermo-Optical Plane Source (TOPS) for Thermal Property Characterization and Package-Level Failure Analysis

Jeffrey L. Braun, Bryan N. Baines, and John T. Gaskins, Patrick E. Hopkins
Laser Thermal
937 2nd St. SE
Charlottesville, VA 22902
jeff@laserthermal.com

SUMMARY

We demonstrate a Thermo-Optical Plane Source (TOPS) technique to measure the thermal conductivity of materials. This high-throughput, simple, and efficient method measures thermal conductivity of materials with minimal sample preparation and limited restrictions on sample shape and geometry. Moreover, the technique is applied to solids, liquids, gels, and pastes with no change in implementation. Using laser heating and thermography, TOPS leverages Fourier's law to directly measure the thermal conductivity of a material using steady-state conditions.

1. INTRODUCTION

Optical thermometry has enabled non-contact measurement of the thermal properties of materials when combined with laser heating. The laser flash analysis (LFA) technique pioneered this approach by introducing a way to measure thermal diffusivity.[1] However, LFA requires a standardized sample geometry and optical access to both front and back of a sample for heating and temperature sensing, respectively. More recently, modulated photothermal radiometry has been used to measure thermal diffusivity of materials using a single side heating and sensing methodology.[2] However, this technique requires a lock-in infrared camera and is limited in the types of materials it can measure. Moreover, both of these techniques are transient in nature, measuring thermal diffusivity rather than thermal conductivity. To rectify these issues, we introduce the TOPS technique.

2. EXPERIMENTAL/NUMERICAL METHODS

The TOPS technique uses localized surface heating of a sample via a continuous wave laser to induce a steady-state temperature rise in a material; infrared thermography is used to measure the corresponding temperature rise at the sample surface. The TOPS technique leverages Fourier's law to directly measure thermal conductivity, rather than thermal diffusivity or effusivity, eliminating the need for prior knowledge of density and specific heat. The thermal model captures air and radiation effects, allowing for accurate measurement of thermal conductivities on par with that of air (<0.03 W/m-K) without the need for environmental control of the experiment.

3. RESULTS

We demonstrate the ability to measure thermal conductivities of highly insulating materials such as porous foams, woods, and plastics. We measure the thermal conductivity of foams in open air with a measurement time on the order of 1 minute and a sample size of only a few millimeters in each dimension needed; by comparison, the standard guarded hot plate technique requires hours for a single test and a sample size on the order of 100s of mm in each dimension. On the other end of the spectrum, we measure thermally conductive materials including metals, sapphire, and germanium.

We show the robustness of the TOPS technique by demonstrating it is insensitive to sample thickness down to ~2x the laser beam diameter, allowing for direct measurement of materials without the need to measure sample dimensions in almost all cases. We experimentally demonstrate this by measuring samples of varying thicknesses from 10s of cm down to 100s of microns, both suspended and placed on a heat sink, to quantify this effect. Additionally, we show that the technique is completely insensitive to measurement location on a sample, including the edge of a sample, enabling the ability to measure multiple locations on the same sample for statistical sampling without the need for multiple samples. Furthermore, in the case of inhomogeneous materials, the localized nature of the technique enables an understanding of the variation in thermal conductivity across a sample. This allows for thermal conductivity "mapping" or "imaging" with a pixel size on the order of the laser spot diameter. Finally, we demonstrate that TOPS is highly resilient to sample roughness by systematically varying the roughness of an acrylic sample to capture when the thermal conductivity measured begins to deviate from that of a control sample.

Beyond thermal conductivity measurement, TOPS can be applied at the package level for the assessment of buried interfaces to understand the spatial variation of thermal contact conductance across an area. This enables us to visualize the effects of how a thermal interface material spreads or settles when placed as a link between a device and heat sink. We demonstrate the feasibility of TOPS as a failure analysis tool for identifying where heat dissipation is impeded below a surface.

4. CONCLUSIONS

Altogether, TOPS offers an approach for simple, accurate, and high-throughput thermal conductivity measurements with minimal sample preparation and limited restrictions on sample shape and geometry.

5. REFERENCES

[1] ASTM E1461-13: Standard Test Method for Thermal Diffusivity by the Flash Method.

[2] F. Cernuschi, P. G. Bison, A. Figari, S. Marinetti, and E. Grinzato. Ther-786 mal Diffusivity Measurements by Photothermal and Thermographic Tech-787 niques. *International Journal of Thermophysics*, 25(2):439–457, March788 2004.

Heterogeneous Liquid Metal – Silver – Polymer Composites for Thermal Interface Materials

Aastha Uppal, Wilson Kong, Ashish Rana, Jae Sang Lee, Matthew Green, Konrad Rykaczewski, and Robert Y. Wang

Arizona State University, School for Engineering of Matter, Transport and Energy
501 East Tyler Mall, ECG 301, Tempe, AZ, 85287
rywang@asu.edu

SUMMARY

This presentation reports on the thermal conductivity of heterogeneous liquid metal-silver-polymer composites for thermal interface materials. More specifically, we report on how the polymer viscosity during composite formation impacts the ability to form thermally conductive percolation pathways of the liquid metal and silver fillers. We achieve a thermal conductivity of ~15 W m^{-1} K^{-1} for samples made with a 100 cSt viscosity polymer in a volumetric ratio of 37:13:50 (liquid metal:silver:polymer) and an applied curing pressure of 2 MPa.

1. INTRODUCTION

Liquid metal microdroplets offer unique thermal-mechanical characteristics that make them attractive as a filler material for polymeric thermal interface material (TIM) composites. Gallium-based liquid metal is simultaneously thermally conductive due to its metallic nature and mechanically soft due to its liquid nature. In addition, even higher thermal conductivity fillers (e.g., silver microparticles) can be incorporated into the composite to further boost overall thermal transport. In this work, we systematically explore the impact of the composition (filler type, volume fraction, and matrix viscosity) and applied pressure on the composite's thermal conductivity and multiscale structure.

2. EXPERIMENTAL/NUMERICAL METHODS

We use a five-component vinyl-terminated polydimethylsiloxane (V-PDMS) elastomer system to enable fine control of the pre-cure viscosity of the polymer system. These five components are the adjustable base polymer, crosslinker, inhibitor, catalyst, and chain extender, and they were subsequently mixed in-house with the desired ratios. We mixed gallium microdroplets (prepared via sonication) with the base polymer, catalyst and inhibitor. We subsequently added the silver particles into the mixture, stirred, and lastly added the crosslinker and chain extender. Samples were cured under pressure to promote thermally conductive percolation pathways.

Samples were characterized using optical microscopy, scanning electron microscopy, and rheology. The thermal conductivity of the samples was measured using an in-house built stepped bar apparatus that is based on a modified ASTM D5470 standard. A variety samples were prepared using different filler types (liquid metal fillers only; liquid metal and silver co-fillers), different curing pressures spanning 0 – 4 MPa, and different pre-cure polymer viscosities (100 cSt and 2300 cSt).

3. RESULTS

We demonstrate that it is important to overcome two physical barriers to achieve large amounts of thermally conductive percolation pathways in the polymer matrix. These two physical barriers are displacement of the polymer between fillers as well as rupture of the native oxide shell on the liquid metal droplets. We overcome these barriers through a synergistic combination of solid silver and liquid metal fillers, viscosity-tuning of the polymer matrix precursor, and sample compression. We select silver as the solid additive because it rapidly alloys with gallium to form microscale needles that could act as additional paths that aid in connecting the liquid metal droplets.

Our experimental results demonstrate that the viscosity of the polymer matrix precursor plays a major role in the resulting thermal conductivity of these composites. More specifically, samples made with low viscosity ~100 cSt solutions achieve a high thermal conductivity of ~15 W m^{-1} K^{-1} at a curing pressure of 2 MPa. This thermal conductivity is double that achieved with high viscosity ~2,300 cSt solutions.

4. CONCLUSIONS

This study demonstrates that the viscosity of the polymer prior to sample curing is a critical parameter in the formation of solid pad-like thermal interface materials. Since many polymer systems employed in industry and research have a high precursor viscosity, this insight has important implications for next-generation TIMs.

Design & Evaluation of Liquid Metal-based TIM for Laptop Application

Hing Jii Mea, Benjamin F. Dorau, and Navid Kazem,
Arieca Inc.

Abstract— **The prevalence of high-power computing in consumer products has increased the need for high performance thermal interface materials (TIM) that are compatible with high-volume manufacturing methods. Liquid metal-based materials have limited adoption due to inherent difficulty in material application and containment. In this work, we sought to demonstrate the use of a hybrid liquid metal embedded elastomer (LMEE) material that has low thermal resistance and is compatible with stencil printing. In-package thermal characterization results of the hybrid LMEE show demonstrable improvement. Finally, we showcase a straightforward dam design to provide effective containment of the CPU die.**

Keywords—Liquid Metals, Thermal Interface materials, (key words)

I. INTRODUCTION

For compact consumer electronics, thermal management usually involves removal of excess heat from the highest power density components, typically the CPU, directly via a heatsink coupled to a cooling solution (e.g. forced convection). Within this relatively compact setup, the thermal interface material (TIM), typically referred to as a TIM1.5, is crucial for establishing an effective thermal coupling of the CPU die directly to the heatsink. Desirable properties for TIM1.5 include low thermal resistance, ease of application and thermomechanical reliability over the course of the expected product lifetime that is subject to varying degrees of thermal cycling and mechanical shock. With the rising prevalence of high-power computing for consumer products such as gaming consoles, materials based on liquid metals (LM) have gained increased consideration as a potential TIM1.5 solution for such demanding applications, primarily due to their superior thermal performance. However, LM materials for semiconductors are typically marked by difficulty in handling and poor in-package reliability. In particular, the displacement or migration of LM poses a severe electrical risk to the device due to its conductivity. Consequently, any LM-based TIM1.5 solution would require significant effort to mitigate and minimize risks traditionally associated with LM.

A. Methodology

In this presentation, we report the use of liquid metal embedded elastomer (LMEE) as a high performance TIM1.5 for use on a consumer gaming laptop. We provide insight into material thermal properties, compatibility with stencil printing as well as system level evaluation and architectural adjustment (i.e. using a dam) to ensure no liquid metal migration during severe drop test (5x).

Figure 1. Microscopy image of hybrid LMEE as TIM1.5

II. LMEE AS A TIM1.5 SOLUTION

Recent progress in embedding microdroplets of liquid metals inside variety of polymer systems have shown high promise to provide a thermal interface material that is highly versatile and adoptable in different applications from TIM1 inside packaging to TIM2/1.5 interfacing between external heatsink and the package IHS or silicon die [1]. Here we designed a non-silicone polymer system that shows excellent conformity with the gallium-based liquid metal and allows embedding extremely high volume loading of fillers (volume loading (φ) of more than 80%). In order to further enhance thermal conductivity of the LMEEs and provide additional stability at low bond-line thicknesses (BLT < 50μm) we also embed copper particles. See Fig. 1 for the microscopy of the hybrid-LMEE material.

III. THERMAL PERFORMANCE

We characterize the thermal performance of this formulation using ASTM D-5470 test setup (TIMA5 Nanotest) with copper test heads (area of 132.75 mm^2). The sample is loaded in an X or cross pattern between the test heads, which are brought together to an initial bondline thickness of 300 microns. The gap is then reduced at a rate of 2 micron/s until the pressure on the test heads is 30 psi. Once this pressure is achieved, the temperature of the heater is raised to 100°C. The lower test head is connected to a chilled water circulator and kept at 15°C. The temperatures across the test heads are recorded and used to determine heat flow and thermal resistance.

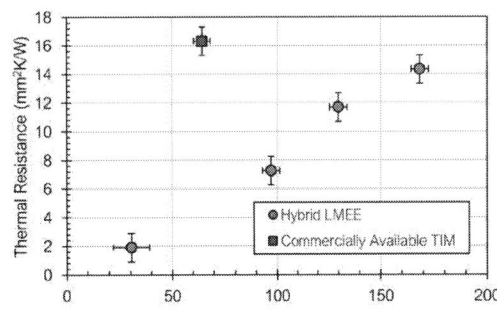

Figure 2. Thermal resistance of Hybrid-LMEE as a function of BLT. Red data point is using a commercially available thermal interface material as a calibration material.

Hing Jii Mea is with Arieca Inc., 201 N Braddock Ave STE 334, Pittsburgh, PA 15208 (412-409-9019;; e-mail: tmea@arieca.com).

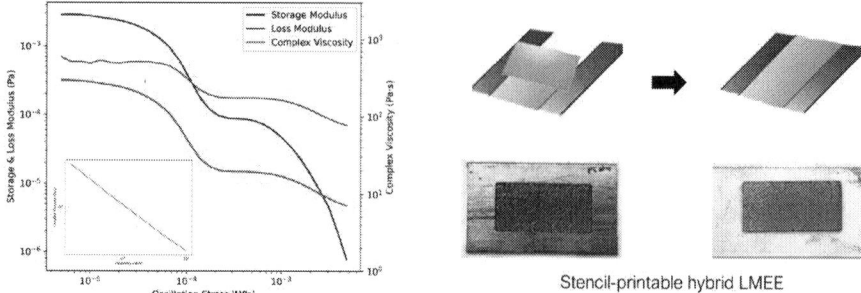

Figure 3. Evaluation of stencil printing compatibility. (Left) Oscillatory rheology curves of the hybrid-LMEE material stress sweep. Inset shows the frequency sweep at 5% strain. (Right) Schematic of stencil printing process and resulting traces on different substrates.

We used a commercially available thermal grease as a calibration material for our measurements. Given the consistency between our measurement and the technical data sheet which gives an R_{th} of 13 to 17 mm²K/W at a BLT of between 65 – 85 µm, we report a minimum thermal resistance of ~ 2 mm²K/W for the hybrid LMEE material (Figure 2).

IV. RHEOLOGY AND STENCIL PRINTABILITY PERFORMANCE

Stencil printing is a fabrication method involving application of material through spaces cut out of a mask onto a substrate beneath. Based on the physical characteristics of the material being applied, the thickness of the mask, and the surface features of the substrate, there is a range of rheological properties in which stencil printing becomes viable and consistent. We performed an oscillatory stress sweep to determine the stress value at which storage modulus is equal to the loss modulus. The cross-over stress measured for the material is 110 Pa, meaning the material behaves more as a flowable liquid when subject to an applied stress of equal magnitude or greater. This value is similar to those reported in the literature for solder pastes commonly used in a stencil printing application [2]. Traces produced using stencil printing show minimal pinholes and consistent appearance (Figure 3).

V. SYSTEM LEVEL (LAPTOP) TESTING

To demonstrate the comparative performance of the hybrid LMEE in package, we selected a commercially available gaming laptop with an Intel Core i7, 13th generation CPU to conduct benchmarking tests. We applied the hybrid LMEE over the heatsink via stencil printing, re-assembled the laptop package, and performed 50 cycles of Cinebench R23 testing, with each cycle having a duration of 20 minutes and a duty cycle of 33%. Comparison of the CPU or package temperatures of the POR material with the hybrid LMEE reveals the ability of the hybrid LMEE material to dramatically delay the onset of CPU thermal throttling (30 seconds vs 9 seconds in the POR) (Figure 4).

VI. DROP TEST EVALUATION AND OPTIMIZATION

Having verified the improved thermal management capability of the hybrid LMEE material in package, we sought to design a means to manage the risk of LM migration during mechanical shock or impact. Our approach was to engineer a foam dam structure that would surround the CPU die using widely available materials such as polyurethane foam and acrylic adhesive tape. To evaluate the ability of the foam dam structure to contain LM migration, we performed drop testing of a mechanical test vehicle from a height of 5 ft. With a foam dam structure in place, LM migration from the CPU area even after consecutive drops was effectively eliminated (Figure 5).

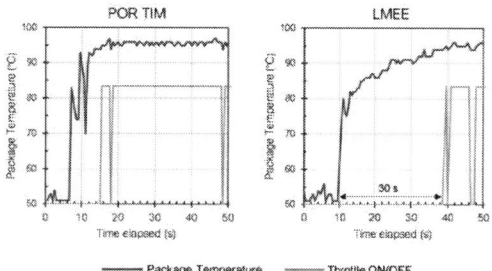

Figure 4. System level testing System-level CPU benchmarking data showing the comparative CPU temperature and onset of thermal throttling of the laptop POR TIM versus hybrid LMEE.

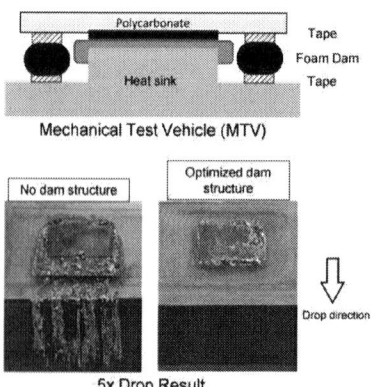

Figure 5. Drop test evaluation using mechanical test vehicle. (Top) Schematic of mechanical test vehicle. (Bottom) Comparative results of drop test with vs without optimized dam structure.

70

VII. CONCLUSION

LM-based materials are well positioned to be a high-performance TIM1.5 solution. This work demonstrates how embedding LM within a polymer system can result in a hybrid material is easy to apply and exhibits low thermal resistance. To eliminate the most notable drawback of LM (i.e. migration) during severe drop test (5x), the design and optimization of a foam dam structure proved to be a sufficient solution. Next-generation hybrid LM-based materials would benefit from further innovations in polymer formulation to more comprehensively address risks associated with LM and enable wider adoption for a TIM1.5.

REFERENCES

[1] M.D. Bartlett, N. Kazem, M.J. Powell-Palm, X. Huang, W. Sun, J.A. Malen and C. Majidi, "High thermal conductivity in soft elastomers with elongated liquid metal inclusions", *Proceedings of the National Academy of Sciences*, vol. 114, 2017, pp. 2143-2148

[2] R. Durairaj, S. Ramesh, S. Mallik, A. Seman and N. Ekere, "Rheological characterization and printing performance of Sn/Ag/Cu solder pastes", *Materials and Design*, vol. 30, 2009, pp.3812-3818

Anisotropic and heterogeneous thermal conductivity in programmed liquid metal composites through direct ink writing

Ohnyoung Hur[1†], Eric J. Markvicka[2,3,4*], and Michael D. Barlett[1,5*]

[1] Mechanical Engineering, Soft Materials and Structures Lab, Virginia Tech, Blacksburg, VA 24061, USA.
[2] Mechanical & Materials Engineering, Smart Materials Robotics Lab, University of Nebraska-Lincoln, NE 68588
[3] Electrical & Computer Engineering, University of Nebraska-Lincoln, Lincoln, NE 68588
[4] School of Computing, University of Nebraska-Lincoln, Lincoln, NE 68588
[5] Macromolecules Innovation Institute, Virginia Tech, Blacksburg, VA 24061, USA.
[†] Presenter, * correspondence
Email: [†]ohnyoung@vt.edu, *eric.markvicka@unl.edu, *mbartlett@vt.edu

Effective thermal management in electric vehicles, electronics, and robotics requires the ability to dissipate and direct heat flow systematically. Thermally conductive soft composites show promise for thermal management due to their high thermal conductivity and mechanical flexibility. However, conventional composites often exhibit a uniform microstructure throughout a film, limiting the directional and spatial control necessary for managing distributed heat loads in emerging systems. In this study, we demonstrate how to program directional and spatially tunable thermal properties into liquid metal (LM) soft composites using a direct ink writing (DIW) process. By locally controlling the aspect ratio and orientation of LM droplets, we achieve a thermal conductivity of 9.9 W/m·K in the direction of LM elongation, which is approximately 40 times greater than that of an unfilled elastomer (0.24 W/m·K). This approach introduces anisotropic and heterogeneous thermal conductivity in compliant films, enabling precise control over both the direction and magnitude of heat transfer. This methodology and the resulting materials offer engineered thermal management solutions for both rigid and soft devices.

Keywords: direct ink writing, liquid metal, heterogeneous, thermal conductivity

I. INTRODUCTION

The demand for effective thermal management in electric vehicles, robotics, and soft electronics is becoming increasingly critical [1]. To meet this need, ensuring both spatial and directional heat dissipation has become paramount [2,3]. Thermal interface materials (TIMs), generally consisting of a flexible polymer matrix embedded with thermally conductive rigid fillers, are essential for enhancing heat transfer [4]. However, when these fillers are randomly dispersed, the resulting composites often exhibit isotropic thermal properties [5]. To achieve directional heat management, controlled microstructures with aligned inclusions offer a promising solution [6]. Techniques such as mechanical [7], magnetic [8], and chemical [9] alignment of solid particles have been explored to enhance anisotropic thermal conductivity. Despite their success in improving directional heat flow, the incorporation of solid fillers can reduce the material's flexibility, which is crucial for effective thermal management in soft systems [10]. Furthermore, these strategies are typically limited to unidirectional heat control, constraining their ability to manage heat distribution across multiple directions.

Elongated microstructures of liquid metal (LM) droplets can be formed to enhance thermal conductivity using methods like mechanical deformation [11] and magnetic alignment [12]. However, these techniques often rely on external forces or require pre- or post-treatments to stabilize the microstructures, which typically leads to uniform, homogeneous structures throughout the material. This uniformity can hinder the ability to control heat dissipation in multiple directions. As an alternative, material extrusion additive manufacturing offers a more flexible approach to programming material properties during the fabrication process. For instance, direct ink writing (DIW) has been employed to fabricate LM microstructures with varying inclusions, from spherical to elongated, without the need for external forces or additional treatments [13]. By adjusting both material and printing parameters, this method enables precise control over thermal conductivity, allowing for advanced thermal management solutions that can be tailored during the manufacturing process.

II. EXPERIMENTAL/NUMERICAL METHODS

A. Fabrication of LMSC by DIW process

A material extrusion system (DIW, Hyrel SR 3D printer) was used to print the material. Printed sample size is $25 \times 25 \times 2$ mm³. The extrusion velocity was set at 4.1 mm/s for both V^* (print head speed / extrusion rate) = 2 and 12. The print heights were 210 μm and 50 μm for $V^* = 2$ and 12, respectively. After completing the multilayer printing, the samples were fully cured overnight in a convection oven set to 80 °C.

B. Measurement of thermal conductivity

The anisotropic thermal conductivity (k_{axial} and $k_{transverse}$) was determined using the transient hot wire (THW) method. From these values, k_y and $k_{x=z}$ were calculated as described in previous studies [11]. The sample size for this measurement remained at $25 \times 25 \times 2$ mm³. This measurement process was repeated five times with 1-minute intervals to allow the samples to dissipate any residual heat from prior tests.

III. RESULTS 1

Liquid metal soft composite (LMSC) films with precise control over LM microstructures, such as droplet aspect ratio, orientation, and direction, are fabricated using DIW process. This method produces a range of LM microstructures, including both spherical and elongated droplets throughout the film (Figure 1a). By incorporating elongated droplets, the LMSCs achieve a high anisotropic thermal conductivity of 9.9 W/m·K, approximately 40 times higher than that of unfilled elastomers (Figure 1b) while preserving mechanical flexibility (Figure 1c). Adjustments to the DIW printing path allow for anisotropic and heterogeneous LM microstructures to be strategically positioned within the film, enabling tunable thermal conductivity based on location (Figure 1d,e). This control is demonstrated via infrared (IR) imaging of a high-power LED, where heat dissipation occurs directionally, channeling heat toward a designated area on the film (Figure 1f). This technique, which allows for customizable and selective heat management by manipulating LM microstructures, provides a versatile solution for applications in thermal diodes, soft electronics, and soft robotics.

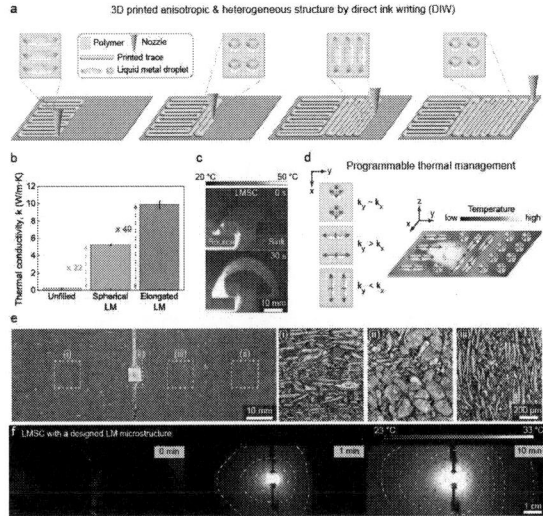

Figure 1. Programmable thermal conductivity achieved by tailoring LM microstructure through DIW printing. a) Schematic of the DIW printing process for creating anisotropic and heterogeneous LM microstructures. b) Comparison of thermal conductivity across three samples. c) IR images demonstrating the flexibility and heat dissipation of LMSCs with elongated LM inclusions. d) Schematic illustrating isotropic and anisotropic thermal conductivity in LM microstructures, with expected heat dissipation along a designed path (yellow square indicates an LED). e) Optical microscopy image of a DIW-printed LMSC showing three regions: (i) horizontally elongated, (ii) spherical, and (iii) vertically elongated LM microstructures. f) Time-sequence IR imaging of the LMSC with designed microstructures from e, highlighting the boundary of directed heat dissipation.

IV. CONCLUSION

This study showcases the development of LMSC films with anisotropic and heterogeneous LM microstructures, enabling spatial and localized heat control for thermal management applications. This method allows for the local design of LM microstructural characteristics such as aspect ratio and orientation, overcoming the limitations of prior techniques that produced uniform or spherical microstructures. This innovation paves the way for customized thermal management systems by programming varying thermal properties within a single material and process, offering new possibilities for controlling heat dissipation in next-generation device architectures.

ACKNOWLEDGMENT

OH and MB acknowledge support through NSF (No. CMMI-2054409). EM acknowledges support through the NSF (No. CMMI-2054411).

REFERENCES

[1] Eren Sevinchan, Ibrahim Dincer, and Haoxiang Lang. A review on thermal management methods for robots. Applied Thermal Engineering, 140:799–813, 2018.

[2] Amol R. Dhumal, Atul P. Kulkarni, and Nitin H. Ambhore. A comprehensive review on thermal management of electronic devices. Journal of Engineering and Applied Science, 70(1):140, 2023.

[3] Zhihao Zhang, Xuehui Wang, and Yuying Yan. A review of the state-of-the-art in electronic cooling. e-Prime-Advances in Electrical Engineering, Electronics and Energy, 1:100009, 2021.

[4] Junaid Khan, Syed Abdul Momin, and M. Mariatti. A review on advanced carbon-based thermal interface materials for electronic devices. Carbon, 168:65–112, 2020.

[5] Mengmeng Qin, Yuxiao Xu, Rong Cao, Wei Feng, and Li Chen. Efficiently controlling the 3D thermal conductivity of a polymer nanocomposite via a hyperelastic double-continuous network of graphene and sponge. Advanced Functional Materials, 28(45):1805053, 2018.

[6] Weifei Wu, Tianli Ren, Xueqing Liu, Ryan Davis, Kai Huai, Xin Cui, Huaixiao Wei, Jinjin Hu, Yuming Xia, Shuohan Huang, et al. Creating thermal conductive pathways in polymer matrix by directional assembly of synergistic fillers assisted by electric fields. Composites Communications, 35:101309, 2022.

[7] Cuiping Yu, Wenbin Gong, Wei Tian, Qichong Zhang, Yancui Xu, Ziyin Lin, Ming Hu, Xiaodong Fan, and Yagang Yao. Hot-pressing induced alignment of boron nitride in polyurethane for composite films with thermal conductivity over 50 W/mK. Composites Science and Technology, 160:199–207, 2018.

[8] Chao Yuan, Bin Duan, Lan Li, Bin Xie, Mengyu Huang, and Xiaobing Luo. Thermal conductivity of polymer-based composites with magnetic aligned hexagonal boron nitride platelets. ACS Applied Materials & Interfaces, 7(23):13000–13006, 2015.

[9] Minh Canh Vu, Won-Kook Choi, Sung Goo Lee, Pyeong Jun Park, Dae Hoon Kim, Md Akhtarul Islam, and Sung-Ryong Kim. High thermal conductivity enhancement of polymer composites with vertically aligned silicon carbide sheet scaffolds. ACS Applied Materials & Interfaces, 12(20):23388–23398, 2020.

[10] Ravi Tutika, Shihuai H. Zhou, Ralph E. Napolitano, and Michael D. Bartlett. Mechanical and functional tradeoffs in multiphase liquid metal, solid particle soft composites. Advanced Functional Materials, 28(45):1804336, 2018.

[11] Michael D. Bartlett, Navid Kazem, Matthew J. Powell-Palm, Xiaonan Huang, Wenhuan Sun, Jonathan A. Malen, and Carmel Majidi. High thermal conductivity in soft elastomers with elongated liquid metal inclusions. Proceedings of the National Academy of Sciences, 114(9):2143–2148, 2017.

[12] Seoyeon Kim, Sihyun Kim, Kyeongmin Hong, Michael D. Dickey, and Sungjune Park. Liquid-metal-coated magnetic particles toward writable, nonwettable, stretchable circuit boards, and directly assembled liquid metal-elastomer conductors. ACS Applied Materials & Interfaces, 14(32):37110–37119, 2022.

[13] Ohnyoung Hur, Ravi Tutika, Neal Klemba, Eric J. Markvicka, and Michael D. Bartlett. Designing liquid metal microstructures through directed material extrusion additive manufacturing. Additive Manufacturing, 103925, 2023.

Ultra-low Resistance Bonded Flexible Metal (FlexiMetal™) Thermal Interface Materials for Fast Transient Response in Mobile Computing Devices

Himanshu Pokharna
Deep Materials Inc.
14654 Placida St
Saratoga, CA 95070
pokharna@deep-materials.com

Abstract

Thermal solution to cool the CPU/GPU in a system is created to tackle the steady state at the worst case system power scenario, which means often times for various work loads, there is an available thermal margin on the CPU or the GPU. This paper describes a way to run the CPU and /or the GPU in a transient fashion such that it can operate at much higher power for a sustained time without thermal throttling. This is enabled by very high performance TIM between the heat source and the heat sink.

Keywords

Metallic Thermal Interface Materials, Transient Performance, High performance TIMs

1. Introduction

With the continuous rise in integration levels and assembly density, electronic components produce significant heat that is getting harder and harder to dissipate effectively and efficiently. This leads to elevated device temperatures in these components, which negatively impacts their lifespan and stability. Therefore, effective thermal management is essential to maintain a safe and stable operating temperature. In addition, modern computing systems demand bursty workloads. This is especially true for AI based applications where the AI application may suddenly demand a huge workload and in response CPU may go in turbo mode, creating a large spike in power.

Minimizing thermal contact resistance between the interfaces, such as between CPUs and heat sinks, is critical for effective thermal management. When two solid surfaces make contact, air with low thermal conductivity fills the gaps, creating high thermal contact resistance that obstructs heat transfer. Utilizing thermal interface materials (TIMs) effectively reduces this resistance, enabling more efficient heat dissipation. Myriad different kinds of TIMs are used for different applications [1]. For high power components such as GPUs, CPUs, Networking processors etc, a dedicated attach with typically four screws is used to create uniform and high pressure on the TIM to create a thin bond-line and low thermal resistance.

In mobile devices such as laptops and smartphones, the problem is compounded as the hot components are often very close to the casing of the device. To make matters even worse, these mobile devices are in intimate contact with its user, thus the hot skin temperature can lead to user discomfort or safety related issues. Most of the modern Laptops these days are skin temperature-limited as much as CPU temperature-limited due to the need to prioritize user comfort and safety.

In mobile devices such as laptops, skin temperature at the hot spot is very closely correlated with the CPU temperature as well as the temperature of the heat sink fins. In order to reduce the skin temperature, the focus generally is on better thermal spreaders such as graphite, or on achieving higher airflow through the system. In this paper, we present another way to ameliorate the CPU cooling challenge in mobile systems, while providing high compute performance from the device.

To accomplish higher transient workloads, there have been many research attempts, mostly utilizing phase change materials including organic and ingoranic materials such as paraffins, mico-encapsulated materials in various morphologies and salt hydrates [2].

We describe a newly developed all metallic thermal interface material with significantly lower resistance than the existing TIMs such as greases and PCMs. This material has thermal conductivity of 35 W/mK and can achieve very low resistance, which helps in reducing the CPU temperature, which directly translates into lower skin temperature in a system.

The metallic TIM is a non-eutectic mixture of metals which has a solidus temperature of ~20 °C and a liquidus temperature of ~300 °C, i.e. it is a solid below 20 °C, liquid above 300 °C and a partially melted mixture between 20 °C and 300 °C. In fact, between the temperatures of ~40-90 °C, which is typically the range within which CPU/GPU TIM temperature is, this material behaves like a highly viscous liquid, similar to a thermal grease. This is in contrast with the typical liquid metal TIMs used in systems such as the Sony PS5 or gaming laptops such as the Asus ROG Mothership which are low viscosity liquids in operating conditions (Typical liquid metal TIM such as Galinstan or liquid Gallium has a viscosity of ~ 2 centi-Poise [3]. For comparison, water has a viscosity of 1 centi-Poise and typical thermal grease viscosity is greater than 100 K centi-Poise. Another feature of the technology is unique diffusion bonding process between the metallic TIM and heat sink which results in two key advantages – (a) Interfacial resistance between TIM and the heat sink is almost eliminated completely and (b) Bonding keeps the TIM attached the heat sink, resulting in low pump out and eliminating the need for a dam that is typical of all currently available liquid metal TIMs (to ensure that the low viscosity liquid metal does not spill in other sections of the system, shorting it). Since this material is bonded to a heat sink, and is flexible we have named it **FlexiMetal™**. Table below shows some of the key performance parameters of this material. At 35 W/mK it is at least 5x better than any other organic material such as a PCM or a grease, while having similar viscosity.

Table 1: Typical properties of FlexiMetal

Property & Units	Values
Thermal conductivity (W/mK)	35
Density (g/ml)	7.0
Appearance	Silvery
Viscosity at Shear rate of 10 s-1 (Pa-s)	~100
Young's Modulus (25 C−material in solid state) Mpa	0.25
Liquidus point for non-eutectic alloy (°C)	~20
Solidus point for non-eutectic alloy (°C)	~300
Boiling point (°C)	>2200
Thermal Impedance, BLT~80 micron; °C-cm2/W	~0.025

2. Experimental Method

To understand the impact of this metallic TIM, we have designed a thermal test vehicle with a copper stub heated with cartridge heaters with thermocouple to measure the "CPU" junction temperature and ambient temperature. Same heat sink and fan combination was used to compare various different TIMs. The copper heat source was 1.5x1.5 cm and the pressure applied by the heat sink was about 30 psi (based upon pressure paper reading). Figure 1 shows the picture and the schematic of the setup.

This comparative test was also repeated in an Acer laptop using an Intel 13th Generation Core I5 13500H CPU. Figure 2 shows the laptop and the thermal module used in the laptop. Various different TIMs were compared in the actual application, including some of the market leading grease and PCM. In addition, Deep Materials competing grease, TCG-Grease 6000 was also tested against the FlexiMetal TIM. To stress the CPU, Intel Power and Thermal Analysis Tool (PTAT) was used with power set at 55 Watts.

For the transient tests, we used the same PTAT program to change the power in a single step from 20W to 100W. CPU, power and Temperature was logged to study the temperature evolution of the CPU as well as any power throttling of the CPU.

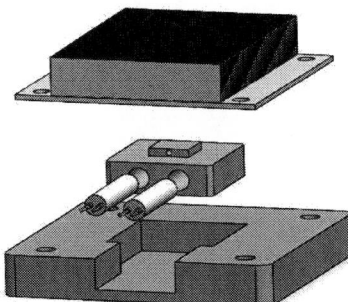

Figure 1: Test setup for testing TIM (Fan not shown). Heat source is 1.5 cm x 1.5 copper stub with a total area of

2.25 cm². Two cartridge heaters provide the heat and a thermocouples is inserted in the center of the heat source via a small blind hole to measure the heat source temperature

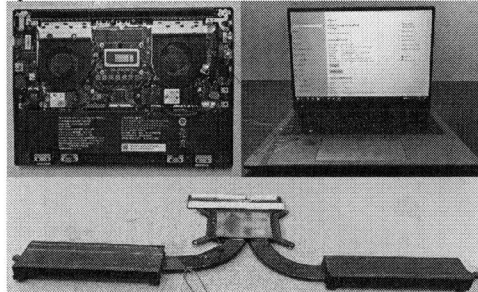

Figure 2: Actual laptop for testing of TIMs & its thermal module with two heat pipes forming a wing shape

3. Results

In this section first we present the steady state test results, followed by transient analytical results and lastly we will show the transient experimental results.

3.1 Steady State Test Results

Figure 3 shows the thermal resistance data from the uniformly heated copper heat source and figure 4 shows the relative performance of the various TIMs in the actual laptop. Properties of various leading conventional TIM products (based upon their published data sheets) and Deep Materials products used are shown in the table below:

Table 2 Properties of various TIMs compared

	Leading Competing TIMs		Deep Materials TIMs	
	PCM	Grease	TCG-Grease 6000	FlexiMetal™
Thermal Conductivity (W/mK)	8	6	6	35
Bond Line Thickness (@ 20- 30 psi) μ M	40	38	35	80
Thermal Resistance (°C-cm2/W)	0.058	0.073	0.055	0.025

As can be seen, the performance difference in thermal resistance is significantly accentuated due to highly non-uniform nature of the heat from the bare die processor.

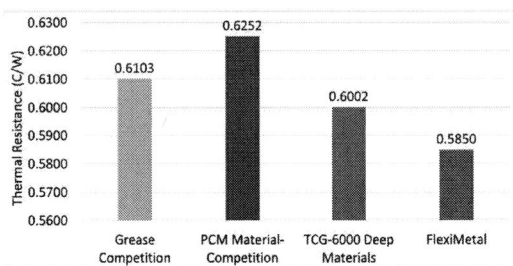

Figure 3: Performance data of various TIMs in the TIM test setup of Figure 1. Data shown is thermal resistance in °C /W for a 1.5 cmx 1.5 cm copper heat source

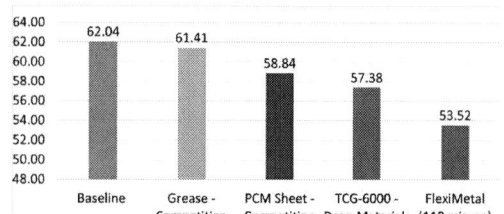

Figure 4: Temperature rise of CPU with different TIM materials at steady state. CPU Power = 55 W, Ta = 25 °C. Baseline represents the performance of the laptop as received in a brand new condition

In addition to benefits in steady state performance, we show that low TIM resistance also helps in quickly diffusing the heat from CPU to the thermal solution, thereby allowing for a large spike in power in the CPU, without the CPU overheating.

3.2 Transient Analytical Model and Results

Figure 5 shows a simple resistive and resistive-capacitive electrical equivalent of the thermal system. The first is a steady state resistive network for the system that every thermal engineer is familiar with. In this the case-to-air resistance is broken into two parts – resistance through the TIM and resistance through the heat exchanger. The bottom part of the figure 5 shows a transient network of the same thermal system, taking into account heat capacity of both of the components. Using the elementary electrical circuit analysis methodology, we can analyze thermal transient processes by extending thermal elements from thermal resistances to thermal resistances and heat capacitors in parallel. The nodes are T_{case} (the case temperature), T_{HP} (Heat sink temperature) and T_{air} is the ambient temperature.

Steady-State Model :

Transient Model :

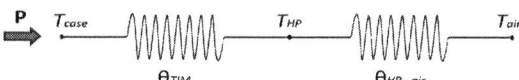

Figure 5: Electrical equivalent of the resistive network (for steady state) and resistive and capacitance network (for trasient analysis)

$$P = P_1 + P_2 \quad (1)$$
$$P = P_3 + P_4 \quad (2)$$

Here P is the total heat from the CPU. P_1 and P_2 is the distribution of that power into the resistive and the capacitive element of the TIM and the P_3 and P_4 are the distribution in the resistive and capacitive element of the heat sink.

By applying the second Kirchhoff law in the first closed loop of the thermal element of figure 5, one can arrive at

$\theta_{TIM} P_1 = \dfrac{U}{C_1}$ Where U is the energy stored in the capacitor C_1. By differentiating both sides of this equation and by using the first Kirchhoff law applied to one node of the thermal circuit (note that $P_2 = \dfrac{dU}{dt}$) we can arrive at the following simultaneous equations:

$$\frac{dP_2}{dt} = \left(\frac{P_2}{\theta_{TIM}C_1}\right) \quad (3)$$
$$P = P_1 + P_2 \quad (4)$$

And similarly for the second (heat exchanger) component, we get,

$$\frac{dP3}{dt} = \left(\frac{P3}{\theta_{HP-air}C_2}\right) \quad (5)$$
$$P = P_3 + P_4 \quad (6)$$

Substituting (4) and (6) respectively in (3) and (4) the following two uncoupled linear ordinary differential equations (ODE) are obtained:

$$\frac{dP_2}{dt} + \frac{P_2}{\theta_{TIM}C_1} - \frac{P}{\theta_{TIM}C_1} = 0 \quad (6)$$
$$\frac{dP_4}{dt} + \frac{P_4}{\theta_{HP-air}C_2} - \frac{P}{\theta_{HP-air}C_2} = 0 \quad (7)$$

Using the initial conditions of P_2 and P_4 to be zero at time zero, we can obtain the following solution to these ODEs

$$P_2 = P\left(1 - e^{-\frac{t}{\theta_{TIM}C_1}}\right) \quad (8)$$
$$P_4 = P\left(1 - e^{-\frac{t}{\theta_{HP-air}C_2}}\right) \quad (9)$$

Therefore, P1 and P3 are:

$$P_1 = P\left(e^{-\frac{t}{\theta_{TIM}C_1}}\right) \quad (10)$$
$$P_3 = P\left(e^{-\frac{t}{\theta_{HP-air}C_2}}\right) \quad (11)$$

Since $T_{case} - T_{air} = \theta_{TIM}P_2 + \theta_{HP-air}P_4$,

$$T_{case} - T_{air} = \theta_{TIM} P\left(1 - e^{-\frac{t}{\theta_{TIM}C_1}}\right)$$
$$+\theta_{HP-air} P\left(1 - e^{-\frac{t}{\theta_{HP-air}C_2}}\right) \quad (12)$$

Since the mass of the TIM is very low, its heat capacity is also very low, thus $C_1 \approx 0$.

$$T_{case} - T_{air} = \theta_{TIM} P$$
$$+\theta_{HP-air} P\left(1 - e^{-\frac{t}{\theta_{HP-air}C_2}}\right) \quad (13)$$

When $t \to \infty$; $T_{case} - T_{air} = \theta_{TIM} P + \theta_{HP-air} P$ which matches the steady state results of the thermal circuit.

Figure 6 shows a plot showing the evolution of the case temperature difference for two scenarios. One where the TIM resistance is high and one where the TIM resistance is low. The solution matches the expectation, i.e. when the TIM resistance is high, the temperature of the heat source increases more rapidly, than if the TIM resistance was lower. These charts prove that the CPU junction temperature will rise much faster for a relatively higher resistance TIM and the CPU temperature will hit its junction temperature limit much quicker. The exponentially decaying curve makes the initial part of the slope much sharper which is why the CPU throttles quickly.

This makes physical sense. As the TIM resistance increases, heat diffusion is limited and as the heat can not be rejected "fast enough", the low mass of CPU rises in temperature rapidly. With very low TIM resistance, the heat can "quickly" diffuse into the thermal solution. That has much greater mass (along with very high effective thermal conductivity within the heat pipe or the vapor chamber, though those effects are ignored in this paper) and can absorb a significant amount of heat due to its relatively large mass, thus allowing the CPU temperature to rise gradually.

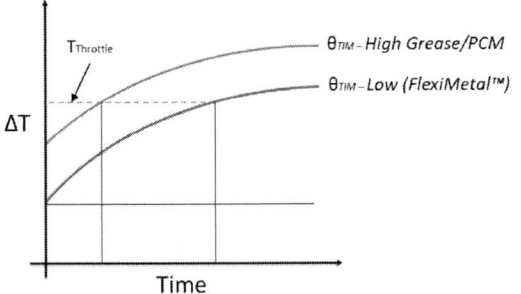

Figure 6: Solution of the transient model for two TIMs, with different thermal perforamnces

3.3. Transient Experimental Results

In addition to the analytical work, we have also tested the performance of the Acer laptop used above with two different TIMs – one is the Deep Materials' TCG Grease-6000 (which showed the best in system performance for non-metallic TIMs) and the other is Deep Materials' advanced soft metal TIM – FlexiMetal™. The results are shown in figures 7 and 8 respectively. In these tests, the CPU power was changed from an initial 20 W to 100 W (figure 7) and as can be seen, the temperature rise with TCG Grease-6000 is almost instantaneous with the CPU heating to reach its junction temperature limit in about a second (figure 8), whereas the junction temperature with FlexiMetal reached its T_j limit in 30 seconds. Lower resistance TIM allows for higher sustained workload for a significantly longer duration. Note that the temperature in both the cases reach the same eventually because the power with the TCG Grease-6000 has gone down as the laptop CPU is not able to sustain the high power due to junction temperature limitation of 100 °C.

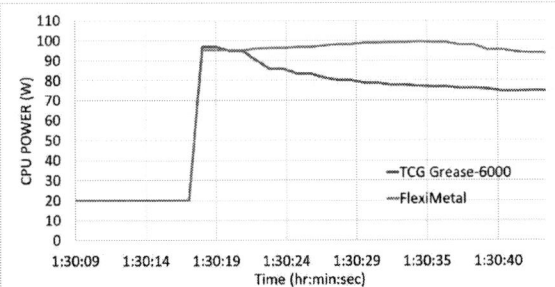

Figure 7: CPU power step change from 20 W to 100 W. With TCG Grease-6000 starts throttling almost immediately

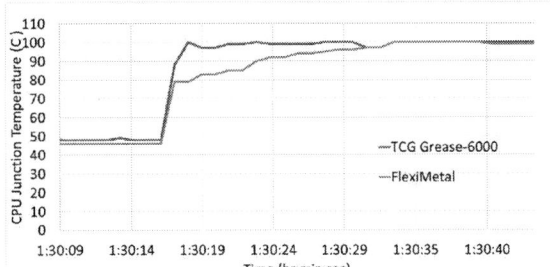

Figure 8: CPU Temperature evolution. TCG Greae-6000 reaches junction temperature limit almost immediately.

Preliminary reliability testing include:

- Power cycling on a bare silicon die in a real application for 2000 on-off cycles without any observed pumpout,
- High temperature bake at 150 °C for 1000 hours with no change in conductivity and
- High humidity testing with no detremental effect in the perforamnce.

Some of the challenges with using metallic TIMs also include their reactions with other metals such as aluminum and to a lesser extent with copper. FlexiMetal is also incompatible with aluminum, but it is compatible with nickle and copper. As a part of the bonding process, an intermetallic is created, which is similar to what happens during lead-free soldering process [4]. This intermatallic is an essential feature of the technolgy as it promotes perfect electron transport, resulting in almost zero interfacial resistance between the heat sink and the FlexiMetal TIM. As against the typical galium based liquid metal TIMs, this intermatallic formation is contained due to the semi-solid nature of the product. There has not been any observed progression of the intermatallic over six months at 120 °C.

4. Conclusion

In this paper, we present the results of a novel metallic thermal interface material (TIM) and its transformative impact on semiconductor performance. While it is widely recognized that high-performance TIMs can lower the junction temperature of semiconductor chips, our study introduces both theoretical and experimental evidence demonstrating that low thermal resistance TIMs can significantly improve the transient thermal response of semiconductor chips during sudden power surges. Traditionally, researchers have focused on incorporating high heat capacity materials, often adding substantial weight, to enhance transient response times in CPUs. However, our findings reveal a more effective strategy: utilizing the thermal mass of the existing thermal module, which often outweighs added materials by an order of magnitude, to absorb heat. This approach not only improves steady-state thermal performance but also significantly increases the capacity to sustain large heat loads over extended durations. This is particularly critical for emerging AI-based workloads, which place intense and dynamic demands on CPUs in modern PCs, redefining the requirements for effective thermal management.

5. . References

[1] Xing, W.; Xu, Y.; Song, C.; Deng, T. Recent Advances in Thermal Interface Materials for Thermal Management of High-Power Electronics. *Nanomaterials* **2022**, *12*, 3365.

[2] Mathew, J., and Krishnan, S. (August 6, 2021). "A Review on Transient Thermal Management of Electronic Devices." ASME. *J. Electron. Packag.* March 2022; 144(1): 010801.

[3] Liu, Jing. (2020). Advanced Liquid Metal Cooling for Chip, Device and System. 10.1142/12517, Shanghai: Shanghai Science & Technology Press, ISBN: 978-7-5478-4532-5

[4] Ramli, M.I.I.; Salleh, M.A.A.M.; Abdullah, M.M.A.B.; Zaimi, N.S.M.; Sandu, A.V.; Vizureanu, P.; Rylski, A.; Amli, S.F.M. Formation and Growth of Intermetallic Compounds in Lead-Free Solder Joints: A Review. *Materials* **2022**, *15*, 1451.

Acoustic Properties of Stretchable Liquid Metal-Elastomer Composites for Matching Layers in Wearable Ultrasonic Transducer Arrays

Ethan Krings, Eric Markvicka
Smart Materials and Robotics Laboratory
Department of Mechanical & Materials Engineering
University of Nebraska–Lincoln, Lincoln, NE 68588, USA
Phone: 402-472-1617; Email Address: eric.markvicka@unl.edu

Extended Abstract

Ultrasound is a safe, noninvasive diagnostic technique used to measure internal structures such as tissues, organs, and arterial and venous blood flow. Skin-mounted wearable ultrasound devices can enable long-term continuous monitoring of patients to provide solutions to critical healthcare needs. However, stretchable ultrasound devices that are composed of ultrasonic transducers embedded in an elastomer matrix are incompatible with existing rigid acoustic matching layers, leading to reduced energy transmission and reduced imaging resolution. Here, a systematic study of soft composites with liquid metal (LM) fillers dispersed in elastomers reveals key strategies to tune the acoustic impedance of soft materials. Experiments supported by theoretical models demonstrate that the increase in acoustic impedance is primarily driven by the increase in density with negligible changes to the speed of sound through the material. By controlling the volume loading and particle size of the LM fillers, we create a material that achieves a high acoustic impedance 4.8 Mrayl, (> 440% increase over the polymer matrix) with low modulus (< 1 MPa) and high stretchability (> 100% strain). When the device was mechanically strained, we observed a small decrease in acoustic impedance (< 15%) with negligible decrease in sound transmittance and impact on attenuation for all droplet sizes. The stretchable acoustic matching layer was then integrated with a wearable ultrasound device and the ability to measure motion was demonstrated using a phantom model as is performed in Doppler ultrasound.

1. INTRODUCTION

Here, we introduce the fabrication, characterization, and use of an LM elastomer composite with tunable acoustic properties for use as a matching layer in stretchable ultrasonic transducer arrays. By combining recent work on LM elastomer composites and wearable ultrasound devices, the incorporation of a liquid phase into elastomers eliminates the mechanical compliance mismatch of rigid fillers and thus preserves the mechanics of the host elastomer, offering a unique combination of low stiffness and high acoustic impedance. The integration of LM microdroplets increases the acoustic impedance to over 440%, displays a low attenuation, and maintains a soft and stretchable response. The LM-based acoustic matching layer can be integrated with an array of individually addressable ultrasonic transducers. As demonstrated, the ultrasound device can be used to detect the motion of deep tissue and organ functions such as the motion of heart valves and ventricle walls, which was demonstrated using an ultrasound phantom model.

2. EXPERIMENTAL METHODS AND RESULTS

We present a systematic study of soft composites with liquid metal (LM) fillers dispersed in elastomers and reveal a key strategy for tuning the acoustic impedance of soft materials. Soft and stretchable materials with tunable acoustic impedance are critical components for wearable, skin-mounted ultrasound devices that can enable long-term continuous monitoring of tissues, organs, and central arterial and venous blood flow, providing solutions for critical healthcare needs. Here, we explore how the properties of the LM filler influence the acoustic impedance and attenuation of soft materials. In this study, we measure the speed of sound through a gallium-based LM alloy EGaIn (eutectic alloy of 75% Ga and 25% In) for the first time. Ga-based LMs are a particularly attractive liquid filler for tuning the acoustic impedance of soft materials due to the relatively high density $\rho =\$ 6.25 g cm^{-3}) and speed of sound (c = 2.753 mm μs^{-1}) as compared to other liquids. We then show that the acoustic impedance of elastomers can be controlled by tailoring the volume loading of a LM filler dispersed in a soft silicone elastomer composite. Specifically, as the volume loading of the LM filler is increased, the acoustic impedance can be increased to over 440% (as compared to the elastomer) while exhibiting low attenuation and maintaining a soft and stretchable response. The increase in acoustic impedance is in good agreement with the Wood's model, which predicts the speed of sound through composite materials based on the average density and compressibility of the continuous and discontinuous phases. Signal attenuation is another important factor that is considered when designing acoustic matching layers. As expected, LM particles reflect the sounds waves and increase the attenuation of the ultrasound signal, especially for the smaller LM droplets. However, the attenuation generally decreased as the LM droplet size increased.

A matching layer integrated into a stretchable ultrasound device will undergo elastic deformation perpendicular to the direction of sound wave propagation resulting in changes to the shape of the LM droplets. To examine the influence of the particle shape, the acoustic properties of the LM-elastomer composite was studied as a function of strain. When mechanically strained, the acoustic impedance decreased (< 10%) for all particle sizes considered. The small decrease in acoustic impedance resulted in a negligible decrease in sound transmittance (< 1%) as compared to the unstrained samples. The mechanical deformation also has a negligible impact on the attenuation for all droplet sizes tested. These new insights into the relationship between LM filler size and loading will provide a quantitative design guide for incorporating soft and stretchable acoustic matching layers with wearable ultrasound devices.

The LM-elastomer composite acoustic matching layer was then integrated with a stretchable ultrasound device. We observe a short A-line pulse duration (about 4 cycles at 2 MHz yields), which is desired for good axial resolution in diagnostic medical ultrasound imaging. A phantom model was then created to mimic heart valve motion (i.e., tissue Doppler). The ultrasound device with acoustic matching layer was able to detect the motion of the object as indicated by the set of consecutive A-lines, which can be used to calculate the velocity of motion or flow towards or away from a transducer as a function of time.

3. CONCLUSIONS

Acoustic matching layers are typically composed of rigid particles embedded in a polymer matrix, which limits their use in stretchable ultrasound devices. In this work, we describe and demonstrate a soft and stretchable acoustic matching layer that was created by embedding LM inclusions within a soft elastomer matrix. The relatively high density (6.25 g cm^{-3}) and fluidity of the room-temperature LM allows for the fabrication of soft composites with acoustic impedance as high as 4.8 Mrayl, which represents more than 440% increase in acoustic impedance as compared to the unfilled elastomer. However, the LM inclusions do increase the attenuation of the ultrasound signal, especially for the O(1) LM droplets. Under strains typically observed in wearable devices, we observed negligible changes to transmitted sound or attenuation. The performance of the acoustic matching layer could be further improved by functionalizing the surface of the LM droplets to improve the size distribution and interface between the polymer matrix or considering LM mixtures with high impedance particles.

A Multi-Method Approach for Optimizing and Accelerating Thermal Conductivity Characterization of Thermal Interface Materials

A. Hakimian

C-Therm Technologies Ltd., 40 Crowther Ln Suite 200, Fredericton, NB E3C 0J1

(506) 457-1515 arya.hakimian@ctherm.com

Abstract

Thermal conductivity plays a pivotal role in the efficiency and reliability of electronic devices, where Thermal Interface Materials (TIM) serve as essential components for managing heat dissipation. The diversity of TIM formats presents significant challenges in standardizing test methods, often complicated by filler dispersion issues that lead to heterogeneity and non-uniform heat transfer. This, in turn, affects the overall thermal management of the device which can significantly impact performance as well as considerations around safety. It is, therefore, imperative to have reliable and sensitive test methods that can detect and quantify these variations to ensure the consistent performance of TIMs.

1. INTRODUCTION

The C-Therm Modified Transient Plane Source (MTPS) method is a powerful tool in the research and development of TIMs. It offers quick feedback on the thermal conductivity of materials, including solids, liquids, powders and pastes, allowing for rapid iteration and improvement. The MTPS method is particularly adept at detecting issues related to filler dispersion, making it an indispensable asset in the development and quality control processes of TIM manufacturing. In addition to the MTPS method, the Transient Plane Source (TPS) method is also widely recognized for its benefits. The TPS method conforms to the ISO 22007-2[1] standard, providing a robust framework for TIM characterization, and is often used for product performance certification on technical data sheets. Testing under representative conditions is also important for properly understanding material performance related to the specific, real-world operating conditions of interest.

2. EXPERIMENTAL/NUMERICAL METHODS

The Modified Transient Plane Source (MTPS) Method employs a one-sided interfacial heater/sensor surrounded by an integrated heated guard ring. The heater/sensor applies a transient heat pulse to the sample. The heated guard ring provides a thermal barrier surrounding the heater/sensor and negates lateral heat transfer at the sensor/sample interface. The rate of increase in the sensor voltage is used to determine the thermal properties of the sample. The voltage is factory-calibrated to temperature. The thermal conductivity is inversely proportional to the rate of increase in the temperature at the point of contact between the sensor and the sample[2].

The Transient Plane Source Method (TPS) ISO 22007-2 uses a heated disc composed of a bifilar nickel spiral (contains two closely spaced spiral windings) encapsulated in a dielectric material such as polyimide. During a measurement, the sensor is positioned between two identical samples of the material being measured; a pulse of constant electrical power is applied to the spiral heater and the increase in temperature in the spiral is measured. The spiral in the sensor serves as both the source of heat and a dynamic temperature sensor during a measurement[2].

3. RESULTS

To illustrate the impact of filler inconsistency, an epoxy with a conductive filler was tested using the MTPS at multiple locations across the material surface. All testing was performed at RT. Results indicated large differences (ranging from 0.3-0.9 W/mK) based on the localized differences related to the filler. TPS testing was performed on a 2k adhesive TIM provided by an external collaborator. Testing was performed at RT, and thermal conductivity was measured to be 2.1 W/mK, which aligns well with the expected TDS value of 2 W/mK.

4. CONCLUSIONS

By leveraging both the MTPS and TPS methods, researchers and manufacturers can gain a comprehensive understanding of the thermal properties of TIMs. This knowledge is crucial for optimizing the design and application of TIMs, ultimately leading to more efficient and reliable electronic devices. As the demand for high-performance electronics continues to grow, the importance of advanced TIM testing methods cannot be overstated. These methods not only contribute to the development of superior TIMs but also support the broader goal of enhancing thermal management solutions in the electronics sector

5. REFERENCES

1- ISO 22007-2. 2022. Plastics-Determination of Thermal Conductivity and Thermal Diffusivity – Part 2: Transient Plane Heat Source (Hot Disc) Method.
2- C-Therm Technologies Ltd. 2023. Trident Thermal Conductivity Instrument User Manual.

MULTI-ZONE THERMOELECTRIC ON-CHIP COOLING UNDER DIFFERENT HEAT LOAD SCENARIOS ON A THERMAL TEST VEHICLE

Dan R. Wargulski, Torsten Nowak, and Mohamad Abo Ras

Berliner Nanotest und Design GmbH

Volmerstraße 9B, 12489 Berlin, Germany

+49 30 6392 3611, wargulski@nanotest.eu

Extended Abstract

INTRODUCTION

CPUs and the increasingly important AI chip assemblies exhibit heat loads with highly variable local distributions at different power densities and at different times depending on the current computing task. Such localized heat loads, known as hot-spots, can lead to high temperature peaks and non-uniform temperature distribution across the die and chipset, reducing the performance and reliability of the assembly. Thermoelectric coolers (TECs) can be used to provide on-demand cooling controlled by the actual temperatures occurring on the chip. By applying an array of TECs, on-demand cooling is not limited to temporal but also to local on-demand cooling [1]. The application of a thermal test vehicle (TTV) [2] allows the simulation of different thermal load scenarios by individually heating its heater zones or smaller hot spots. Recording the measured values of the temperature sensors distributed across the entire TTV allows a detailed analysis and comparison between the application of e.g. one large TEC, a 2x2 array of TECs or no TEC regarding energy efficiency, inhibition of temperature peaks and uniformity of temperature distribution across the chip.

METHODOLOGY

The TTV is comprised of four independent heater zones, each of which makes up a quarter of the TTV, as well as heatable smaller hot spots and 16 resistance temperature detectors (RTDs) (Figure 1a).

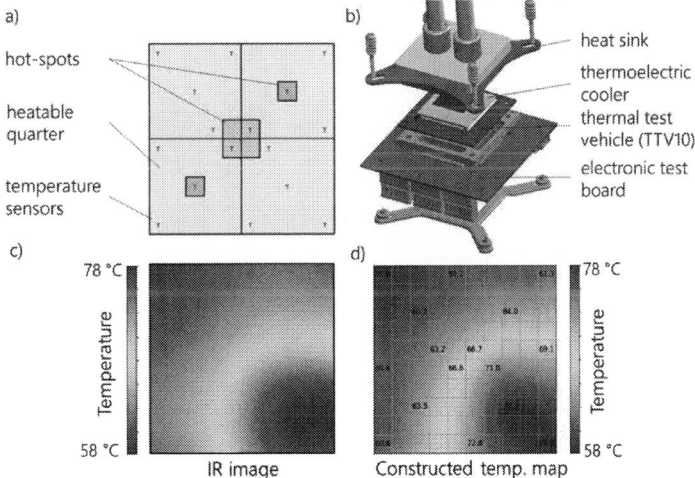

Figure 1: a) schematic of the 24.9 x 24.9 mm² TTV with 4 individual heater zones, smaller hot-spots and 16 temperature sensors marked with small T's. b) shows the complete assembly of the TTV mounted on its electronic test board, the single TEC as on-chip cooling and a heat sink with water cooling. c) and d) show a comparison of an IR image acquired on the blackened TTV without heat sink with the lower right heater zone active and a temperature map constructed by the interpolation of the sensor temperatures, exhibiting well matching results.

The cooling was realized in three versions:

1. One 30 x 30 mm² TEC with 40 W cooling power + heat sink (Figure 1b)
2. Four 15 x 15 mm² TECs with 40 W cooling power (10 W each) + heat sink
3. No TEC, heat sink only

The efficiency of the three distinct cooling concepts was evaluated through the examination of a multitude of heat load scenarios and control implementations. The single TEC cooling was controlled in two ways: first, corresponding to the mean temperature of all sensors; second, corresponding to the maximum measured temperature. In the 2x2 TEC array cooling, only the TECs above active heater zones were in operation and controlled corresponding to the maximum measured temperature. The parameters for the heat sink were kept steady for all three cooling implementations.

RESULTS

First investigations exhibited advantages of using several TECS instead of one TEC regarding the efficiency of the cooler and the temperature distribution in the chip. The temperature mapping resulting from the RTD readings on the TTV allow a direct evaluation of the temperature uniformity across the chip. Figure 2 shows the comparison of a TTV with only one quarter of the area heated and cooled by one 30 x 30 mm² TEC (Fig. 2a) and cooled by one 15 x 15 mm² TEC of a 2x2 array of TECs (Fig. 2b). The larger and more powerful TEC is capable to cool the chip down even to temperatures below 0 °C, but it cannot inhibit the temperature spread ΔT across the chip which amounts to $\Delta T = 24$ K for one TEC vs. $\Delta T = 14$ K in case of the 2x2 TEC-array. With increasing heat loads we assume an increasing trend of this ΔT with increasing efficiency of the 2x2 TEC-array cooling concept.

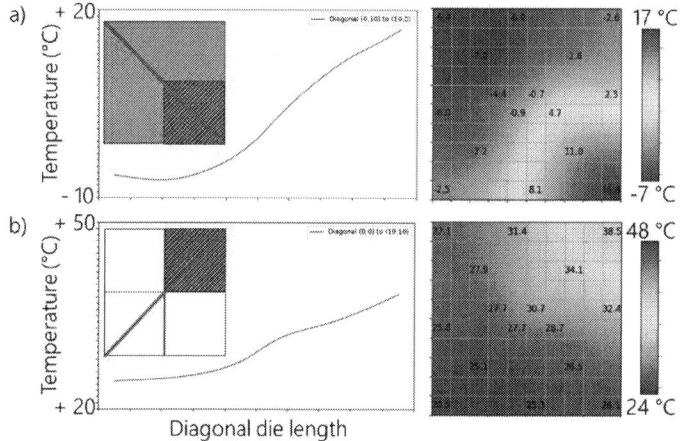

Figure 2: a) shows measurement data of an operating (at I_{max}) full-die-sized TEC cooling a TTV with one heated quadrant (lower right) and b) an array of 2x2 TECs with only the upper right one in operation (at I_{max}) and cooling the upper right heated quadrant of the TTV. The graphs show the temperature profiles across the diagonal of the test vehicle and the inlets a schematic of the cooled (blue) and heated (red shaded) zones. The temperature maps are constructed by the sensor data exhibiting the different temperature gradients.

[1]   O. Sullivan et al., "Array of Thermoelectric Coolers for On-Chip Thermal Management", Journal of Electronic Packaging, Vol. 134, pp. 021005-1, 2012, doi: 10.1115/1.4006141

[2]   M. Sternberg et al., "Thermal Test Vehicle for HPC–System Level Approach for Investigation of the Thermal Heat Path Signature with the Property of Spatial Resolution", 29th Int. Workshop on Therm. Invest. of ICs & Syst. (THERMINIC), 2023, doi: 10.1109/THERMINIC60375.2023.10325887.

Impedance Matching and Flowrate Requirements for Heterogeneous Liquid Cooling Servers and Racks

Javier Avalos, Nvidia Corporation
Travis Gaskill, Nvidia Corporation
Umut Z. Uras, Nvidia Corporation

Abstract

Liquid cooling is considered to provide a far better cooling capability compared to air cooling in electronics cooling applications. This document investigates the design methodology and compares thermal analysis of liquid cooling networks for pressure matching liquid cooling systems and to meet flowrate requirements. Basic governing equations are presented, and application examples are provided to demonstrate the choices a thermal engineer will have to evaluate.

The objective of this paper is to describe the relationship between the fluid inlet temperature, the required liquid flowrate, and the liquid cooling loop design to successfully cool down a system of electronic devices with respect to their temperature limits. As well, this study highlights the considerations needed to properly design for a common pressure operating point for the liquid cooling network within the larger system.

Keywords

Datacenter, Liquid cooling, flowrate requirements, pressure matching.

Nomenclature

ΔP: Pressure drop (Pa)

Q: Flowrate (l/min)

a, b: flowrate resistance coefficients.

A, B, G: flowrate requirement coefficients.

ρ: liquid density (kg/m^3)

Cp: liquid heat capacity (J/kg·°C)

CDU: Cooling Distribution Unit

QD: Fluid Quick Disconnect

UQD: Universal Quick Disconnect

Pwr: Component power being cooled (W)

T_{inlet}: Temperature entering the component (°C)

1. Discussion

The era of GPU centric AI clusters has arrived, enabled via high TDP chips with parallelized workloads across tens of thousands of GPUs installed in densely configured racks pushing over 100 kW, all cooled via liquid cooling. The high-powered chips are often cooled via skived microchannel cold plates that effectively remove the heat [1,2,3,4,5]. The fluid is supplied to the servers via blind-mate quick disconnects and is pumped by a CDU [6].

Today, these large GPU clusters are supported by a plethora of support infrastructure like storage / management servers, PDUs, power supplies, and batteries. As liquid cooling becomes adopted en masse, the industry will be looking for liquid cooling solutions for this supporting infrastructure to increase cooling efficiency and reduce energy consumption due to cooling. This support infrastructure will look to be installed in heterogeneous racks, where multiple systems from different suppliers are installed in a single rack. The fluid is provided in parallel via rack manifolds and the liquid cooling loops inside them must be designed to play nicely with the others, so as not to disrupt the liquid cooling capabilities of the other systems within the rack and to maintain an efficient use of the cooling fluid.

1.1. Discussion on Heterogenous Racks

Setting a common pressure of heterogenous servers in parallel requires balancing the multiple hydraulic impedances within liquid cooling network.

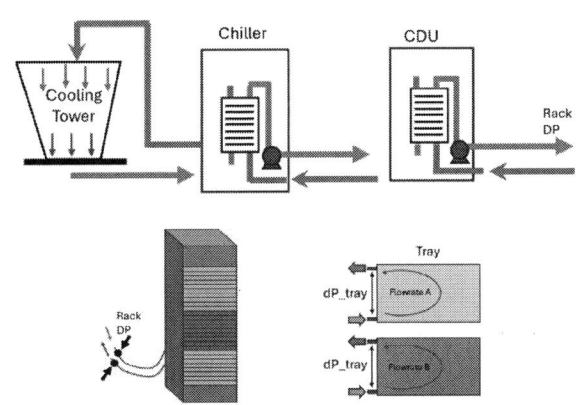

Figure 1: Conceptual liquid cooling rack with different liquid cooling server trays connected in parallel to a rack manifold fed by single supply / return hose.

As presented in Fig. 1, a full liquid cooled rack consisting of two or more different types of liquid cooling server trays will have different flowrates requirements. Ideally, they should be designed to operate at a common pressure to avoid unnecessary and liquid flowrate overflow beyond the needs of the necessary to cool components. Likewise, such a circumstance would bring unnecessary burden to the liquid cooling infrastructure in a datacenter.

Heat load within a rack defines flowrate allocation. A generally accepted value in the industry is 1.5 L/min per kW. Pressure and flowrate are limited resources constrained by CDU specifications and facility infrastructure such as piping and control valve pressure drops. Fig. 2 and Fig. 3 shows typical components within a cooling network at a rack level that generate pressure losses.

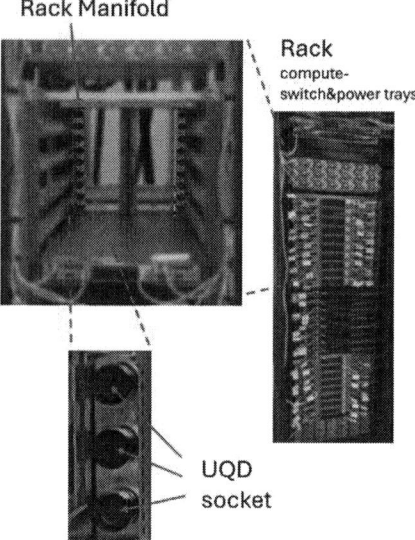

Figure 2: A Heterogeneous liquid cooled rack, front view

Figure 3: A Heterogeneous liquid cooled rack, back view

2. Component Level and Rack Level Liquid Cooling Applications

Liquid cooling loops often cool multiple components either in series or parallel. How the cooling loop is designed plays an important role in how the cooling loop interacts with other loops at a rack level.

A liquid cooled rack's fluid is supplied / returned via rack manifolds which are designed to support servers that operate with an expected flow impedance. This impedance expectation is often termed the pressure budget or rack pressure drop. Systems within the rack must be designed such that they operate normally when their flow impedance matches the rack pressure budget. All the systems in the rack receive fluid in parallel and the inlet / outlet pressures are equal.

The impedance of a cooling loop is a function of its design, especially when it contains multiple cold plates and the designer has a choice of series or parallel flow. Like the principle of impedance matching at the rack level, impedance matching of multiple cold plates in parallel is necessary to optimize the performance and efficiency of the liquid cooling loop. By balancing the impedance of different components in

the cooling loop, the designer can ensure that the cooling fluid flows with the intended ratio to effectively remove heat from all critical areas.

To exemplify the basic governing equations, two cold plates in series and in parallel are evaluated including hose, QD, and manifold pressure losses.

2.1. Series vs. Parallel in a Server

To exemplify the basic governing equations, as illustrated in Fig. 4, two cold plates in series and in parallel with be evaluated including hose, Quick Disconnect (QD), and manifold pressure losses.

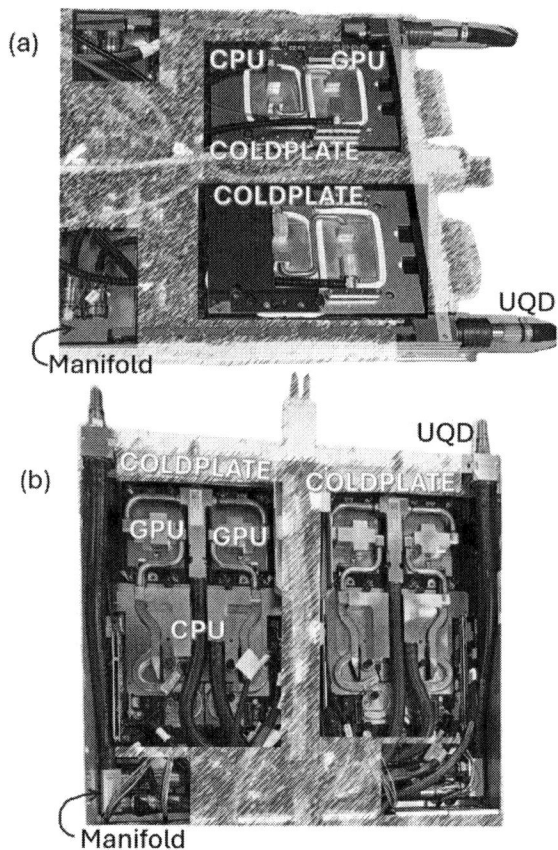

Figure 4: Tray internal liquid cooling components. Fluid is supplied / returned via UQDs and manifolds. a) A GPU and CPU in series. b) Two GPUs in parallel and a CPU in series with the GPUs

2.1.1. Physics on the Series vs Parallel Flowrate Impedance

Pressure loss in a liquid cooling system is described using both linear and quadratic terms due to the presence of both laminar and turbulent flow within the system. In laminar flow, the pressure drop is linearly proportional to velocity and is described by the Hagen-Poiseulle equation. For turbulent flow, the relationship is described by the Darcy-Weisbach equation. This relationship between flowrate, Q, and the pressure drop, ΔP, can be generalized using a linear term and a quadratic term with constant coefficients a and b, respectively, as shown in Eq. 1.

$$\Delta P = a \cdot Q + b \cdot Q^2 \tag{1}$$

Fig. 5 shows an analogous resistance network to Fig. 1, where a_1 and b_1 represent the coefficients for cold plate 1 and a_2 and b_2 represent the coefficients for cold plate 2.

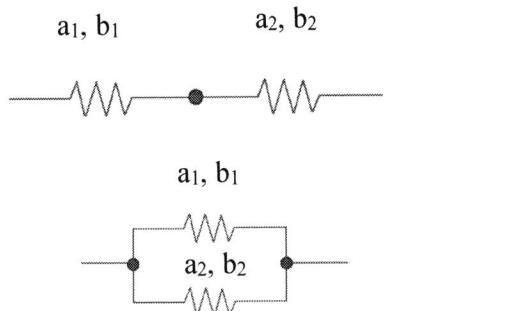

Figure 5: Analogous cold plate resistance networks with coefficients

To derive the equivalent resistance for either series or parallel, the equations for pressure balance as well as mass conversation are utilized at two flowrate conditions. Two liquid cooling resistances in series is straight forward, the solution being the summation of both coefficients in the corresponding linear and quadratic terms.

Equivalent flowrate resistance in Series:

$$a = a_1 + a_2 \tag{2}$$

$$b = b_1 + b_2 \tag{3}$$

For parallel equivalent resistance, a closed form solution is derived using the same basic principles of mass conservation and pressure balance. The equivalent single resistance coefficients for parallel flow are shown in Eqs. (4-5)

Equivalent flowrate resistance in Parallel:

$$a = \left[\begin{array}{c} (a_1 + b_1) \cdot \left(2 + \dfrac{\sqrt{a_2^2 + 4b_2(2a_1 + 4b_1)} - a_2}{2b_2} \right)^2 - \cdots \\[2em] (2a_1 + 4b_1) \cdot \left(1 + \dfrac{\sqrt{a_2^2 + 4b_2(a_1 + b_1)} - a_2}{2b_2} \right)^2 \end{array} \right] \frac{1}{M} \tag{4a}$$

$$b = \left[\begin{array}{c} (2a_1 + 4b_1) \cdot \left(1 + \dfrac{\sqrt{a_2^2 + 4b_2(a_1 + b_1)} - a_2}{2b_2} \right) - \cdots \\[2em] (a_1 + b_1) \cdot \left(2 + \dfrac{\sqrt{a_2^2 + 4b_2(2a_1 + 4b_1)} - a_2}{2b_2} \right) \end{array} \right] \frac{1}{M} \tag{4b}$$

$$M = \left[\begin{array}{c} \left(1 + \dfrac{\sqrt{a_2^2 + 4b_2(a_1 + b_1)} - a_2}{2b_2} \right) \times \cdots \\[2em] \left(2 + \dfrac{\sqrt{a_2^2 + 4b_2(2a_1 + 4b_1)} - a_2}{2b_2} \right) \times \cdots \\[2em] \left\{ \dfrac{\sqrt{a_2^2 + 4b_2(2a_1 + 4b_1)} - a_2}{2b_2} - \cdots \\[1.5em] \dfrac{\sqrt{a_2^2 + 4b_2(a_1 + b_1)} - a_2}{2b_2} + 1 \right\} \end{array} \right] \tag{5}$$

In the special case where two parallel resistances are equal the equivalent parallel resistance simplifies to the following equivalent coefficients.

$$a = \frac{a_1}{2} \tag{6}$$

$$b = \frac{b_1}{4} \tag{7}$$

As shown by Eqs (2-3) and Eqs. (6-7), resistances in series are additive, while resistances in parallel result in a fractionally reduced resistance relative to its constituents.

2.1.2. Effects on Architecture

The choice of a series or parallel configuration depends on the specific requirements and constraints of the system. An obvious advantage to parallel cooling networks is that there is no thermal load preheat to be considered, in contrast to the series network that the downstream component gets all the liquid cooling preheat from the upstream components.

Fig. 6 shows two three-resistance systems, one in series and one in parallel, for which there is a constant pressure drop across them. For the purposes of simplification in this analysis, we focus only on the quadratic coefficients in the Q vs. ΔP relation. Resistance #1 and b_0 represent the hydraulic resistance in a liquid cooling network such as a QD, hoses, contractions, etc., and Resistances #2 / #3 and b_1 represent the hydraulic resistance for two identical cold plates, as in a server configuration with two GPUs.

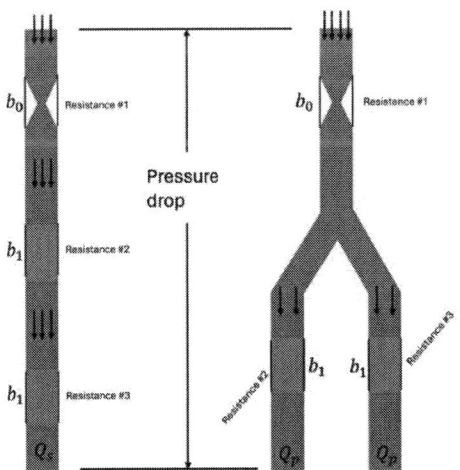

Figure 6: Cold plate assembly

Evaluating two extreme conditions of these systems helps to identify trends in Q for a given ΔP. Looking at Case (I) in Table 1, where $b_0 \ll b_1$, it has the lowest network resistance of all configurations and provides the highest flowrate ratio for an individual cold plate, Q_p/Q_s, for a given ΔP. The equations for calculating Q_s and Q_p for Case (I) are shown in Eqs. (8-9), respectfully.

$$\Delta P = 2Q^2 \rightarrow Q = Q_s = \frac{1}{\sqrt{2}}\sqrt{\Delta P} \tag{8}$$

$$\Delta P = \frac{1}{4}Q^2 \rightarrow Q = 2Q_p = 2\sqrt{\Delta P} \rightarrow Q_p = \sqrt{\Delta P} \tag{9}$$

Thus, the flowrate is 1.4x higher in each cold plate in a parallel network compared to an equivalent series network. An example of this scenario is one in which a large QD is used to supply fluid to a fine-pitched, skived microchannel cold plate.

In contrast is Case (II), where $b_1 \ll b_0$. An example scenario is one in which a small QD is used to supply fluid to an embedded tube style cold plate. This scenario results in the lowest flowrate ratio for cold plates for a given ΔP and results in 0.5x the flowrate compared to an equivalent series network.

Table 1 shows that, for a given ΔP, the effect of the lower hydraulic resistance of cold plates in parallel could result in more flowrate per cold plate with respect to series flow, even though the total flowrate splits in two paths, as seen in Case (1). In other scenarios, like Cases (II) and (III), putting the cold plates in parallel results in a lower flowrate relative to series flow. What this means is that the magnitude of a component's resistance relative to its constituents and its location in the hydraulic resistance network matters.

Table 1: Series and parallel flowrate ratios for different resistance coefficients

Scenario	b0	b1	Parallel vs. Series Flowrate ratio Q_p/Q_s	
I	0	1	$\frac{Qp}{Qs}$	1.4:1
II	1	0	$\frac{Qp}{Qs}$	0.5:1
III	1	1	$\frac{Qp}{Qs}$	0.77:1
IV	1	3	$\frac{Qp}{Qs}$	1:1

2.1.3. Inlet Temperature and Flowrate Requirement

The flowrate requirement through a cold plate depends on several factors including cold plate design, the thermal or fluid dynamics involved, heat source, and the inlet temperature of the fluid. For a given cold plate design, the relationship between inlet temperature and flowrate required, seen in Eq. (11) and shown in Fig. 8, is derived from the thermal resistance vs. flowrate curve of the cold plate, which is of the form seen in Eq. (10) and shown in Fig. 7. Coefficients α, β, γ are found by curve fitting the simulated or experimental data. T_o physically relates to thermal conductivity properties and β_o, γ from the forced convection properties of the cold plate.

$$R_{th} = \alpha + \frac{\beta}{Q^\gamma} \tag{10}$$

Let:

$$T_{rise} = Pwr \cdot R_{th}$$
$$T_o = T_{inlet} + T_{rise}$$

Then:

$$T_{inlet} = T_o - \frac{\beta_o}{Q^\gamma} \tag{11}$$

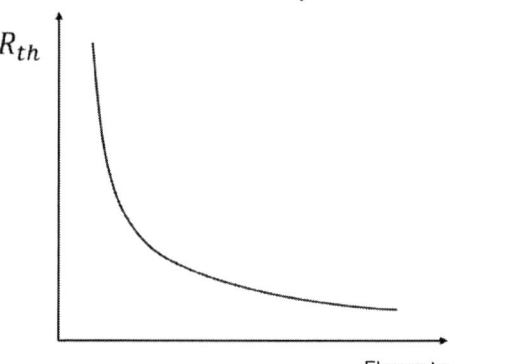

Figure 7: Generic Thermal Resistance Curve

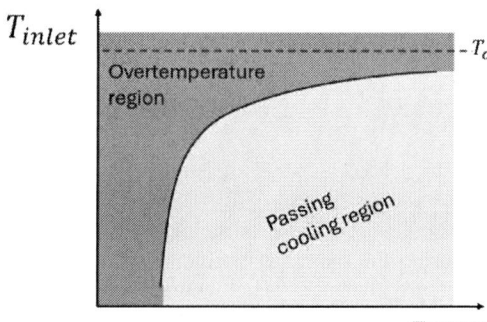

Figure 8: Inlet Temperature and Flowrate Cooling Requirement.

With no preheat from upstream components, T_{inlet} is defined as the fluid entering the system. For cold plates downstream of other cold plates, the fluid gets preheated. The preheat of the fluid from the first cold plate in series with the second cold plate for a given preheat power, $P_{preheat}$, is estimated in Eq. (9) and Eq. (10) is used to calculate the required flowrate of the downstream cold plate.

$$\Delta T_{preheat} = \frac{P_{preheat}}{Q \cdot \rho \cdot Cp} \tag{9}$$

$$Q_{required} = \left(\frac{\beta_0}{T_0 - T_{inlet} - \frac{P_{preheat}}{Q \cdot \rho \cdot Cp}} \right)^\gamma \tag{10}$$

Now that the tools for hydraulic and thermal analysis for a liquid cooling system have been established, we show an example of their application to highlight some interesting findings.

3. Examples of Application

Two separate examples are discussed. The first example highlights how cold plate performance, fluid network arrangement, and the hydraulic resistance of the components within the fluid network affect the outcome of achieving a viable solution.

In the second example, a closer to real world example of a GPU driven system is studied to reveal the nuance associated with defining the required flowrate across the span of allowable inlet temperatures for a system with multiple components. In the parallel flow scenario, it results in two separate functions driving the flowrate requirement, each applicable for only a subset of the total temperature range.

3.1. Searching the Viable Solution Space

Consider the cooling of two 1400W GPUs. While holding the pressure budget constant ($\Delta P = 69\ kpa$), two scenarios are studied. Comparing series / parallel arrangements, in the first scenario the hydraulic impedances of the QDs/hose/manifold and cold plates are equal while in the second scenario the QD/hose/manifold has twice the hydraulic impedance when compared to the cold plates. Also, for the purposes of discussion, the two scenarios have slightly different cold plate thermal performance.

1) Assume a cold plate has the flowrate requirements $T_o = 50, \beta_o = 82, \gamma = 1$, QD/hose/manifold resistance coefficient $b_0 = 1$, cold plate resistance coefficient $b_1 = 1$. Comparing series vs. parallel network arrangement, with inlet cooling fluid at 45degC and 1400W preheat.

Table 2: Series and parallel flowrate thermal results

	Parallel	Series
Flowrate per Cold Plate (l/min)	1.4	1.8
$Q_{required}$ **(l/min)**	1.35 - pass	1.85 - fail

2) Assume cold plate thermal performance with flowrate requirements $T_o = 50, \beta_o = 90, \gamma = 1$, QD/hose/manifold flowrate resistance $b_0 = 2$, cold plate resistance $b_1 = 1$, comparing series vs. parallel connection, with inlet cooling fluid at 45degC and 1400W preheat.

Table 3: Series and parallel flowrate thermal results

	Parallel	Series
Flowrate per Cold Plate(l/min)	1.05	1.58
$Q_{required}$ **(l/min)**	1.1 - fail	1.48 - pass

Both scenarios produce a per cold plate flowrate that is higher when in series which is a common scenario. However, due to the preheat of the upstream cold plate in series, Scenario 1 fails to meet the flowrate required to cool the component effectively. Conversely, due to the differences in cold plate performance and the increased resistance of the hoses and QDs, Scenario 2 fails to meet the flowrate required in the parallel network configuration.

3.2. Cold Plate Common Pressure Drop

For this study, a dual chip module consisting of a GPU and CPU is considered, as shown in Fig. 9. The GPU and CPU TDP are 2.5 kW and 0.8 kW, respectively. In this scenario, assume that the GPU cold plate is state of the art for the industry and its thermal performance is maximized within manufacturing limits. Due to this, when studying series vs. parallel arrangement, the GPU cold plate is the same for both cases and the CPU cold plate is assumed optimized for each configuration: In series, the CPU cold plate is optimized for common flowrate relative to the GPU cold plate and, in parallel, it is optimized for a common pressure drop relative to the GPU cold plate.

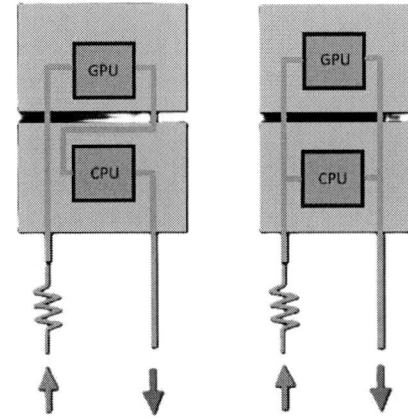

Figure 9: Optimize pressure for a) series and b) parallel

In both scenarios, assume a total module pressure budget of 14 kPa (2 psi). For the configuration in series, the GPU is cooled first. The flowrate requirement for this cold plate is shown in Fig. 10 as a solid blue line. The CPU cold plate is designed to minimize cold plate pressure drop, following as close as possible to the GPU flowrate requirement while accounting for preheat.

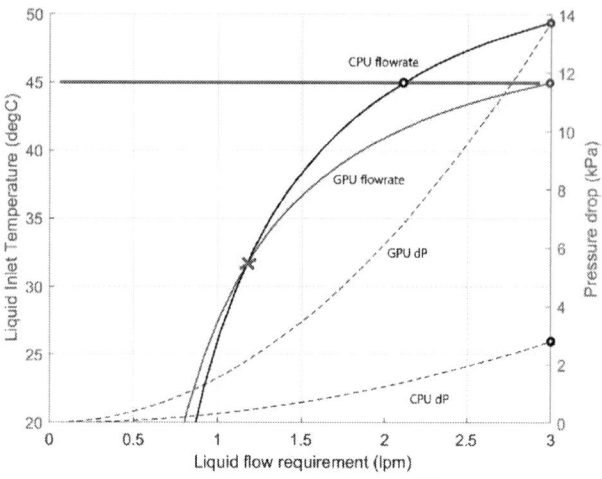

1) GPU cooling: $T_{inlet} = 51.4 - 24Q_{GPU}^{-1.2}$
2) CPU cooling: $T_{inlet} = 58 - 32Q_{GPU}^{-1.2}$

Figure 10: Series cold plate performance optimization

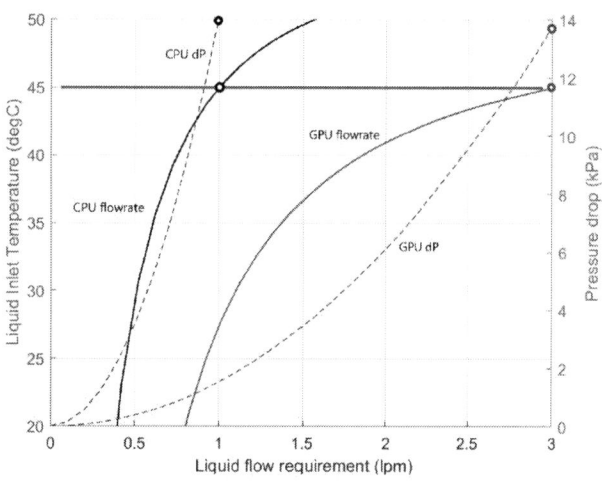

1) GPU cooling: $T_{inlet} = 51.4 - 24Q_{GPU}^{-1.2}$
2) CPU cooling: $T_{inlet} = 57 - 12Q_{GPU}^{-1.2}$

Figure 11: Parallel cold plate pressure match

In contrast is the parallel scenario using same optimized GPU cold plate design. The CPU cold plate is now optimized to pressure match for inlet water of 45°C as seen Fig. 11. The total flowrate requirement is then the summation of GPU flowrate and CPU flowrate. For example, at 45°C inlet temperature, the GPU requirement is 3 L/min and the CPU 1 L/min for a total flowrate requirement of 4 L/min.

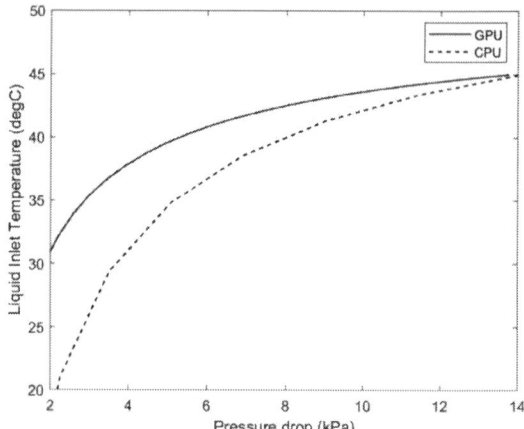

Figure 12: Parallel cold plate inlet temperature vs pressure.

For both scenarios, there is a range of inlet temperatures in which the flowrate requirement is defined by two components on this module. For the case in series, the flowrate is driven by the GPU between the 32-45°C range then there is a crossing in which CPU drives the minimum required flowrate. This also happens for the parallel flow scenario, although this one is harder to spot as the common variable is pressure as opposed to flowrate, see Fig. 12. But, for any given inlet temperature of operation, there is one component that drives the minimum required flowrate (or minimum pressure operation that meets such condition). This is relevant within components in a network as well as within rack consisting of different types of trays with different characteristics.

4. Conclusions

The choice of designing a liquid cooling network in series vs. parallel is shown to be affected by the need to operate with a common pressure budget and the ratio of the hydraulic impedances within the network delivering the cooling fluid (UQD, hoses, manifold, etc.) with respect to the hydraulic impedances of the cold plates performing the cooling.

Three coefficients define the liquid cooling flowrate requirement correlation. Two coefficients define the hydraulic impedance correlation of a liquid cooling network. For simplicity, this paper omitted the hydraulic impedance's linear term when comparing liquid network topologies which is inaccurate in real cold plate characterization. However, the conclusions highlighted remains valid when accounting for the linear term.

Liquid flowrate requirements correspond to the minimum necessary cooling flowrate to meet specification temperatures of all devices within server tray. Flowrate requirement is driven by one dominant component and a liquid cooling network flowrate impedance is designed to be balanced with other loops in parallel, but this equilibrium of flowrate impedances changes by changing the operating inlet temperature, and in the process, changes the dominant device defining the minimum flowrate requirement (or pressure operating point). These findings apply to cooling loops with multiple cold plates as well as heterogeneous racks.

References

1. Y. Fan, C. Winkel, D. Kulkarni and W. Tian, "Analytical Design Methodology for Liquid Based Cooling Solution for High TDP CPUs," *2018 17th IEEE Intersociety Conference on Thermal and Thermomechanical Phenomena in Electronic Systems (ITherm)*, San Diego, CA, USA, 2018, pp. 582-586, doi: 10.1109/ITHERM.2018.8419562.

2. F.M. Naduvilakath-Mohammed, R. Jenkins, G. Byrne, A.J. Robinson, "Closed loop liquid cooling of high-powered CPUs: A case study on cooling performance and energy optimization, Case Studies in Thermal Engineering", Volume 50, 2023, 103472, ISSN 2214-157X

3. Y. Fan, L. Ji, R. Yu and M. Fang, "The Study on cold plate liquid cooling solution for high performance server," *2024 23rd IEEE Intersociety Conference on Thermal and Thermomechanical Phenomena in Electronic Systems (ITherm)*, Aurora, CO, USA, 2024, pp. 1-8, doi: 10.1109/ITherm55375.2024.10709608.

4. M. Vuckovic and N. Depret, "Impacts of local cooling technologies on air cooled data center server performance: Test data analysis of Heatsink, Direct Liquid Cooling and passive 2-Phase Enhanced Air Cooling based on Loop Heat Pipe," *2016 32nd Thermal Measurement, Modeling & Management Symposium (SEMI-THERM)*, San Jose, CA, USA, 2016, pp. 71-80, doi: 10.1109/SEMI-THERM.2016.7458448.

5. Y. Tsai *et al.*, "An Advanced Cold Plate Liquid Cooling Rack Design for Hyperscale Data Center," *2022 21st IEEE Intersociety Conference on Thermal and Thermomechanical Phenomena in Electronic Systems (iTherm)*, San Diego, CA, USA, 2022, pp. 1-5, doi: 10.1109/iTherm54085.2022.9899555.

6. A. Heydari *et al.*, "Guidelines and Experimental Hydraulic performance Evaluation for Single-Phase CDUs Under Steady and Transient Events," *2023 22nd IEEE Intersociety Conference on Thermal and Thermomechanical Phenomena in Electronic Systems (ITherm)*, Orlando, FL, USA, 2023, pp. 1-5, doi: 10.1109/ITherm55368.2023.10177633.

7. T. Gao, H. Tang, Y. Cui and Z. Luo, "A Test Study of Technology Cooling Loop in a Liquid Cooling System," *2018 17th IEEE Intersociety Conference on Thermal and Thermomechanical Phenomena in Electronic Systems (ITherm)*, San Diego, CA, USA, 2018, pp. 740-747, doi: 10.1109/ITHERM.2018.8419519

8. A. Radmehr and S. V. Patankar, "A flow network analysis of a liquid cooling system that incorporates microchannel heat sinks," *The Ninth Intersociety Conference on Thermal and Thermomechanical Phenomena In Electronic Systems (IEEE Cat. No.04CH37543)*, Las Vegas, NV, USA, 2004, pp. 714-721 Vol.1, doi: 10.1109/ITHERM.2004.1319246.

9. L. P. Kozlova and O. A. Kozlova, "Neural Network Control of the Liquid Cooling System," *2021 XXIV International Conference on Soft Computing and Measurements (SCM)*, St. Petersburg, Russia, 2021, pp. 67-69, doi: 10.1109/SCM52931.2021.9507121

10. M. Lingling and Q. Xuefei, "The Research of Cooling Water Temperature and Flow Rate Control System Based on Heat Transfer and Neural Network," *2015 Fifth International Conference on Instrumentation and Measurement, Computer, Communication and Control (IMCCC)*, Qinhuangdao, China, 2015, pp. 655-658, doi: 10.1109/IMCCC.2015.143

11. R. Lucchese, D. Varagnolo and A. Johansson, "Controlled Direct Liquid Cooling of Data Servers," in *IEEE Transactions on Control Systems Technology*, vol. 29, no. 6, pp. 2325-2338, Nov. 2021, doi: 10.1109/TCST.2019.2942270

Development of Dielectric Oscillating Heat Pipes

Corey Wilson & Patrick Margavio

ThermAvant Technologies, LLC

2508 Paris Road, Columbia, MO 65202

573-321-1567, Corey.Wilson@ThermAvant.com

2-Page Abstract (all text from here is 10pt New Times Roman)

Abstract

The oscillating heat pipe (OHP) is an advanced passive thermal management solution that is capable of removing high heat fluxes and can be manufactured in thin form factors. OHPs operate by vapor pressure driven liquid flow through a meandering capillary channel. The structure of an OHP is inherently higher strength than heat pipes and vapor chambers enabling a wider array of applications including as a structural element.

This study focuses on integrating dielectric coatings on OHPs. The dielectric coating provides electrical insulation, which is crucial for direct integration with printed circuit boards (PCBs), particularly as a substitute for copper and aluminum PCBs enabling substantially higher performance than existing solutions. This makes them an ideal solution for high-power applications such as LED lighting, power electronics, and 5G communication devices, where compactness and efficiency are paramount. This study will present the thermal and dielectric performance of coated OHPs.

1. INTRODUCTION

The oscillating heat pipe (OHP) is an advanced passive thermal management solution that is capable of removing high heat fluxes and can be manufactured in thin form factors. OHPs operate by vapor pressure driven liquid flow through a meandering capillary channel. The structure of an OHP is inherently higher strength than heat pipes and vapor chambers enabling a wider array of applications including as a structural element. The OHP is commonly manufactured in a flat form-factor. The planar dimensions are highly adaptable for the customer's needs and the thickness is typically 2-5 mm. This thickness is similar to aluminum substrate printed circuit boards (PCBs), which are typically 2-8 mm thick [1]. This study will demonstrate the viability of integrating dielectric coatings onto OHPs to produce a high performance heat spreader integral to the PCB structure thereby eliminating the need for a secondary heat spreader.

2. EXPERIMENTAL/NUMERICAL METHODS

The dielectrically coated OHPs will be tested thermally and electrically. The experiments will validate the thermal performance of the OHPs with the coatings and compare them with an uncoated OHP. Lastly, the dielectric coatings will be measured to determine the characteristics.

3. RESULTS

The thermal and dielectric performance of the OHPs will be compared against aluminum and copper PCBs. In addition, based on other OHP designs, the expected performance for a range of heat fluxes and heat maps will be discussed to show applications where a dielectric coated OHP would be the viable solution.

4. CONCLUSIONS

The study will demonstrate that the OHP is a viable substrate for PCBs and will enable substantially higher heat fluxes than are possible with the current aluminum and copper substrates. In addition, other developments will enable these high heat fluxes in sub-2 mm thicknesses.

5. REFERENCES

[1] Technical Tips for PCBs - Copper Thickness, Controlled Impedance, and more! | PCB Universe, (n.d.). https://www.pcbuniverse.com/pcbu-tech-tips.php?a=13 (accessed October 21, 2024).

Unified Approach to Model Single-Phase Open-Loop Liquid Cooling

Albert Chan, Don Nguyen and Michael Brooks
Cisco Systems, Inc.
225 East Tasman Drive, San Jose, CA 95136
albercha@cisco.com

Abstract

Single-phase Open-loop liquid cooling (OLLC) provides an energy-efficient method for cooling high power electronics tightly packed in a chassis. OLLC is especially useful in the case where a chassis contains more than one high power ASIC, in which passive air-cooling is not feasible. The goal of this work is to put together a unified method for modeling coolant flow and heat transfer occurring in the cold plates, pump(s), heat exchanger, and interconnecting network of tubings and manifolds. The focus is on developing a set of mathematical models and solution method that can be easily implemented on desktop PC with a programming language like Python. Insights to cold plate design, CDU sizing and selection, coolant type and concentration, operating characteristics, and system optimization can be quickly determined for a specific system using the model. This is especially useful during early design stage when actual system is not available for test. The goal is to reduce the overall thermal resistance of the liquid cooling system so least energy is used for cooling.

Keywords

Single phase, Liquid cooling, open-loop, pump, cold plate, heat exchanger, operating point, thermal resistance

Nomenclature

Cv	flow coefficient	
GPM	gallons per minute	
ID	inside diameter	
N	number of chassis	
M	number of ASIC dies per chassis	
PG	propylene glycol	
Q	single ASIC power	W
R_{CP}	cold plate thermal resistance	°C/W
R_{HX}	heat exchanger thermal resistance	°C/W
R_{js}	junction to sink thermal resistance	°C/W
R_{ja}	junction to ambient thermal resistance	°C/W
R_{sa}	sink to ambient thermal resistance	°C/W
$T_{a,}$	ambient air temperature	°C
$T_{c,in}$	coolant temperature entering heat exchanger	°C
$T_{c,out}$	coolant temperature leaving heat exchanger	°C
T_s	sink (cold plate surface) temperature	°C
T_j	die junction temperature	°C
V	flow rate	m³/s
ΔP	pressure drop	Pa

1. Background

Single-phase liquid-cooling is an effective method for removing heat from high-power ASICs and CPUs. There is increasing focus from industry standards body and organizations on liquid cooling[1-6]. In systems where multiple high-power ASICs or CPUs are packed into a single chassis, the total heat dissipation can be too high for air-cooling. Even closed-loop liquid cooling (CLLC) may not be feasible, since CLLC depends on air flow through the chassis for heat removal, and there can be preheating of the air prior to reaching the heat exchanger. In such cases, OLLC provides a viable solution, since heat removed from the chassis is dissipated by an external heat exchanger, where the heat can be removed by datacenter air or by facilities water. Because of this, the chassis can be low profile, such as 1 RU, and still be able to dissipate large amounts of heat from the ASIC dies that otherwise would not be possible with conventional air-cooling or CLLC. The substantially higher thermal capacity of liquid coolants also enables cooling of ASIC dies with higher power densities.

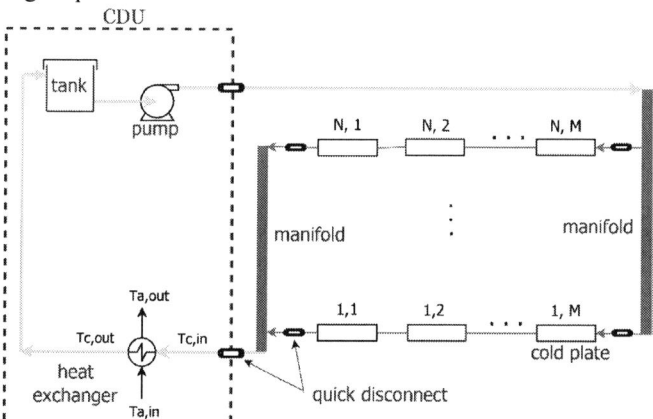

Figure 1: OLLC system with N chassis and N·M ASICs

Figure 1 shows the schematic representation of an OLLC rack system with N·M total ASICs to be cooled. There are N number of chassis in the rack, with each chassis containing M ASICs. All ASICs are of the same type and power dissipation. For the OLLC system to adequately cool the ASICs, the cold plates, pump and heat exchanger must be properly designed and sized. The heat transfer is coupled to coolant flow through the cold plates and heat exchanger, which makes the model more complex. The focus of this work is to highlight a simple, unified method to model the fluid flow and heat transfer in the OLLC system. This provides the design engineer with a set of tools for creating a viable OLLC system.

2. Model

The model follows the same approach as that described in a previous paper for single-phase CLLC[7]. The model is based on correlations developed for fluid flow and heat transfer in channels, tubes and compact heat exchangers[9-14]. Exceptions and additions to the model are outlined below.

Heat Transfer

The junction-to-ambient thermal resistance (R_{ja}) for one ASIC is given by

$$R_{ja} = R_{js} + R_{CP} + R_{HX,1} \qquad (1)$$

where R_{js} is junction to sink resistance, R_{CP} is resistance due to the cold plate, and $R_{HX,1}$ is resistance due to heat exchanger adjusted for power from one ASIC.

From figure 1, thermal resistance of the heat exchanger is given by

$$R_{HX} = \frac{(T_{c,in} - T_{a,})}{N \cdot M \cdot Q} \qquad (2)$$

where Q is the power dissipated by one ASIC. For a single ASIC die, the thermal resistance due to the heat exchanger ($R_{HX,1}$) is given by

$$R_{HX,1} = N \cdot M \cdot R_{HX} \qquad (3)$$

R_{js} can be obtained from computer simulation for a known ASIC powermap. Alternately, it can be measured if the ASIC is available for testing.

The method for calculating R_{CP} and R_{HX} is outlined in the previous paper on unified model for CLLC[7]. The main difference for OLLC is the heat exchanger dissipates the heat from all ASICs in the system as described by equation (2).

Fluid Flow

The OLLC system shown in figure 1 depicts N chassis connected to common supply and return manifolds. The manifolds serve as header to distribute and collect the coolant flowing through each chassis. The manifolds and chassis form a network flow system. The following assumptions are made to simplify the flow system.

- The manifolds are sized so the cross-section area for flow is much larger than that for each chassis.
- The pressure drop through each manifold is much smaller than that through each chassis.
- All chassis have the same design and construction.

From above assumptions and application of Kirchoff's rules, total coolant flow rate through the external loop (shown as green in figure 1) is equally divided among the N chassis (flow through chassis shown by red lines in figure 1). If V is the overall coolant flow rate, then flow rate through each chassis is V/N. This means pressure drop through each chassis is the same, therefore the total pressure drop through the system is given by

$$\Delta P = \Delta P(V) + \Delta P\left(\frac{V}{N}\right) \qquad (4)$$

where ΔP is total pressure drop, $\Delta P(V)$ is pressure drop between the manifolds on coolant distribution unit (CDU) side (as indicated by green line in figure 1), and $\Delta P\left(\frac{V}{N}\right)$ is pressure drop through one chassis.

Pressure drop through the cold plate, heat exchanger (tube side) and tubing are calculated using the method outlined in previous paper[7]. Pressure drop (psi) through one quick disconnect is calculated from

$$\Delta P = SG \cdot \left(\frac{V}{Cv}\right)^2 \qquad (5)$$

where SG is coolant specific gravity, V is total coolant flow rate (GPM) and Cv is the flow coefficient. For chassis quick disconnect, V is replaced by $\left(\frac{V}{N}\right)$. Manufacturers publish values of Cv for each quick disconnect in their portfolio. A pair of quick disconnect is used for each chassis, and a pair of larger diameter quick disconnect is used for the CDU.

Equation (4) is used to calculate the system impedance curve for the OLLC system. The operating point is determined from the intersection between the pump curve and impedance curve. The pump curve is measured by the pump vendor, typically with water at 25°C. This reference pump curve must be corrected for the glycol coolant at the operating temperature. The method of ANSI/HI 9.6.7-2010 is used to make the correction[8]. For redundant pumps, the "effective" pump curve is measured and used as the reference. Figure 2 shows the effect of temperature on the pump curve for an example pump with 25% PG coolant. Figure 3 shows the operating point for an example OLLC system.

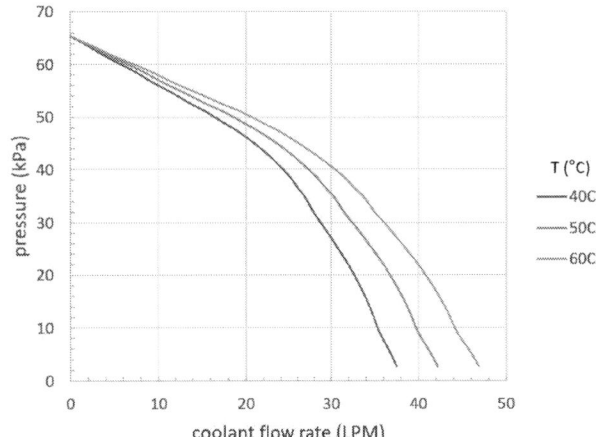

Figure 2: Example pump curve for three temperatures.

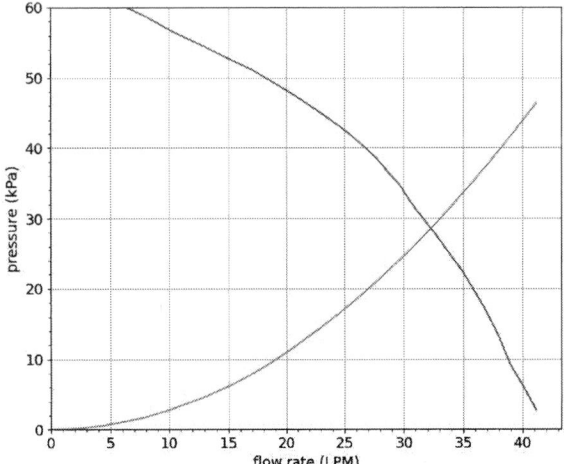

Figure 3: Operating point for example OLLC system.

Algorithm

Figure 4 shows the algorithm used to simulate the OLLC system. Known variables include the measured pump curve, heat to be removed per ASIC (Q), ambient temperature (T_a), geometry of the OLLC system, and glycol type/concentration. At the start of simulation, $T_{c,in}$ is unknown and a guess value is used. This allows coolant properties to be calculated. An iterative loop is used until the value of $T_{c,in}$ has converged. At each iteration, the new value of $T_{c,in}$ is estimated using the difference between specified heat to be removed and the calculated heat removed by the heat exchanger. Upon convergence, the new $T_{c,in}$ will differ from previous value by less than the specified tolerance. For this work, the convergence tolerance is 0.01°C.

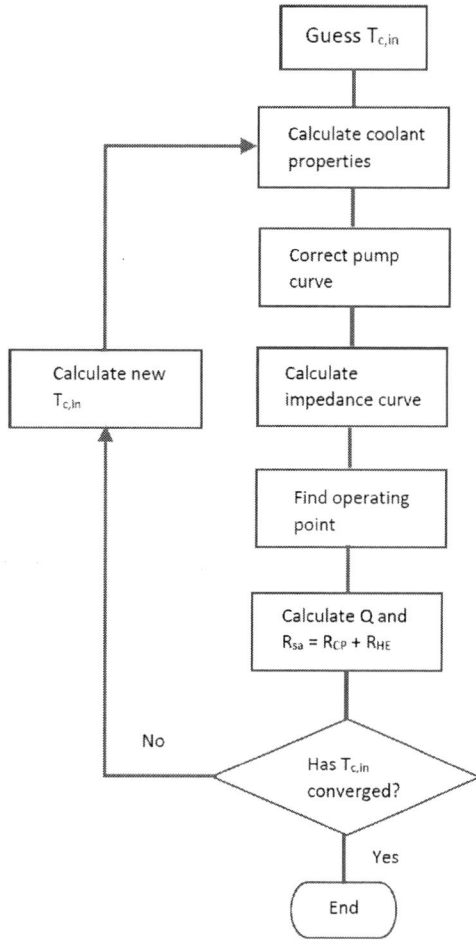

Figure 4: Algorithm for the unified model.

3. Discussion

Examples on use of the unified model to simulate OLLC cases is presented in this section. A pump with pump curve shown in figure 2, together with an air-cooled compact heat exchanger, are used in all cases. The compact heat exchanger has tube length of 400mm with 30 rows of tubes. Each row has 3 tubes, with each tube having a width of 25mm and inside gap of 0.8mm. Figure 5 shows some dimensions (mm) of the tube and fin. Fins are 0.08mm thick, 2.2mm pitch, 8mm height, and 0.9mm louver pitch.

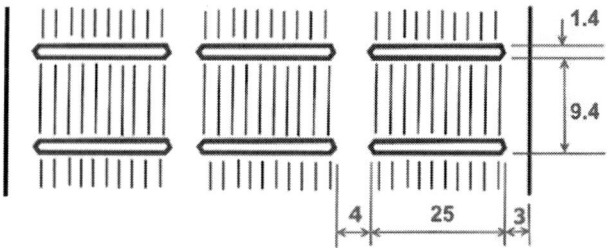

Figure 5: Heat exchanger tube and fin details.

The first case study, tabulated in table 2, compares single versus multiple ASIC dies in the chassis. The cold plate has the same fin array footprint as the die. Cold plate fin array consists of microchannel fins with 0.2mm thickness, 0.5mm pitch and 4mm height. Coolant flow is perpendicular to the wider die dimension. Air flow through compact heat exchanger is 1500CFM for both cases. Coolant is 25% PG. Tube ID is 9mm inside the chassis and is 20mm outside the chassis. Cv is 2.62 and 16.52 respectively for quick disconnects in chassis and CDU. Tubing/fitting equivalent length is 1m inside chassis and is 2.5m outside the chassis.

Table 1 Example case study

Case study	1	2
No. of chassis	10	6
No. of die per chassis	1	3
Die size (mm²)	40 x 30	30 x 25
Power (W) [per die]	1000	600
Power (W) [total]	10,000	10,800
Operating Point		
Pressure (Pa)	22640	35440
Flow-rate (LPM)	34.2	29.5
ΔP (Pa)		
cold plate(s)	3224	16140
heat exchanger	3173	2578
tubing + QD [chassis]	3718	7436
tubing + QD [CDU, manifold]	12520	9289
R_{CP} (°C/W)	0.0246	0.0356
$R_{HX,1}$ (°C/W)	0.0201	0.0369
Rsa (°C/W)	0.0447	0.0725
Rjs (°C/W)	0.0142	0.0173
Rja (°C/W)	0.0589	0.0898
Tc,in (°C)	45.1	47.1
Tc,out (°C)	40.9	41.8
Ts (°C)	69.7	68.5
Tj (°C)	83.9	78.9

Results show it is easier to cool multiple dies in a chassis with lower power than a single die with high power, even if the combined die power is higher. Value of Rsa is for a single die and is the sum of R_{CP} and $R_{HX,1}$. With multiple dies per chassis, a significant portion of the total pressure drop is through the cold plates. Figures 6 and 7 show the operating points for both cases. Higher pressure drop for case 2 results in lower coolant flow through the system.

At the same die power of 1000W, higher coolant concentration results in lower coolant flowrate and higher die temperature. A comparison of the operating points for 35% and 50% PG are shown in figures 8 and 9. As PG concentration increases, coolant viscosity increases, resulting in drastic shift in pump curve to lower flowrates.

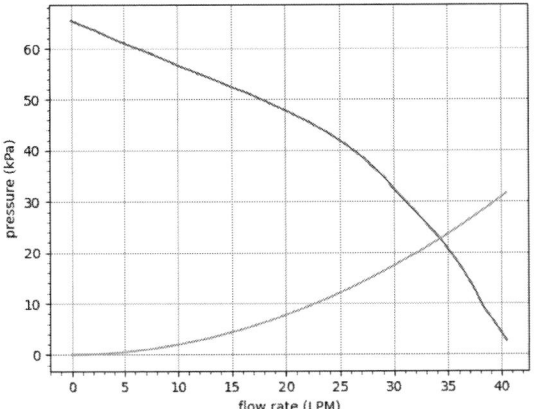

Figure 6: Case 1 operating point.

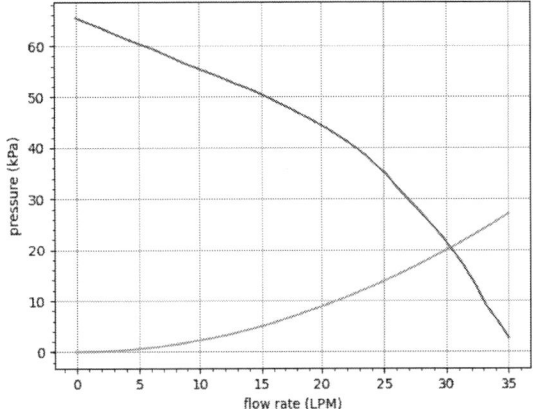

Figure 8: Operating point for 35% PG coolant.

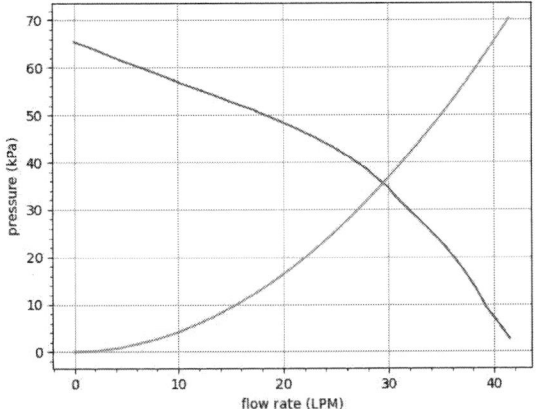

Figure 7: Case 2 operating point.

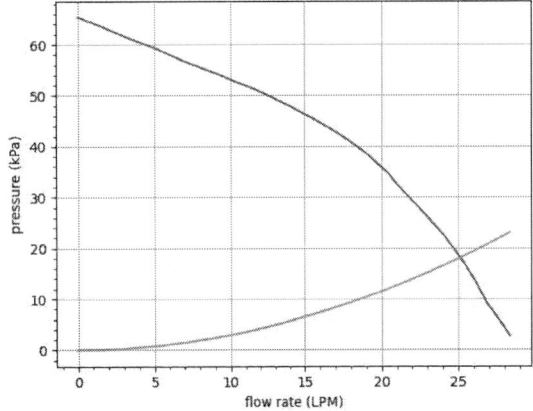

Figure 9: Operating point for 50% PG coolant.

Case study 2 explores effect of coolant concentration on OLLC performance. The baseline is case 1 from table 1. Results for 3 coolant concentrations are tabulated in table 2.

Subsequent studies will look at optimizing the cold plate fin array parameters to reduce energy consumption for fans/pumps in the CDU. Fan/pump power scales with cube of RPM. Hence, reducing thermal resistance can help reduce fan/pump energy usage.

Table 2: Effect of coolant concentration on performance.

% PG (Coolant)	25	35	50
coolant flow (LPM)	34.2	30.4	25.1
ΔP total (Pa)	22640	20360	18070
R_{CP} (°C/W)	0.0246	0.0260	0.0286
$R_{HX,1}$ (°C/W)	0.0201	0.0207	0.0219
R_{sa} (°C/W)	0.0447	0.0468	0.0505
Tc,in (°C/W)	45.1	45.7	46.9
Tc,out (°C/W)	40.9	40.9	40.8
Tj (°C)	83.9	86.0	89.7

Table 3: Effect of fin array footprint on R_{CP}.

Case	1		2	
	Model	CFD	Model	CFD
flowrate (LPM)	3.20		3.23	
fin array (mm)	40 x 30		60 x 30	
ΔP (Pa)	2988	3002	1969	2183
R_{CP} (°C/W)	0.0249	0.0242	0.0202	0.0261
fin array	0.0186	0.0184	0.0139	0.0202
base	0.0063	0.0058	0.0063	0.0058

Table 3 shows comparison of R_{CP} from model and CFD simulations. R_{CP} is further broken down into contribution from fin array, and from the base. Fin array consists of fins with 0.2mm thickness, 0.5mm pitch and 4mm height. The base includes a 1mm thick base support plus a 2mm pedestal. For lidless die packages, a metal frame is used to support the organic substrate on which the die is attached to. The frame is typically thicker than the die, so a pedestal is needed to allow the cold plate to contact the thermal interface material on top of the die.

Several observations can be made from table 3. First, when fin array footprint is close die footprint, there is good agreement in pressure drop and cold plate thermal resistance between the model and CFD. In case 2, however, the fin array is significantly wider than the die footprint, and there is sizeable difference in R_{CP} between model and CFD (difference is primarily in thermal resistance of fin array). The reason for the discrepancy can be deduced from figures 10 and 11.

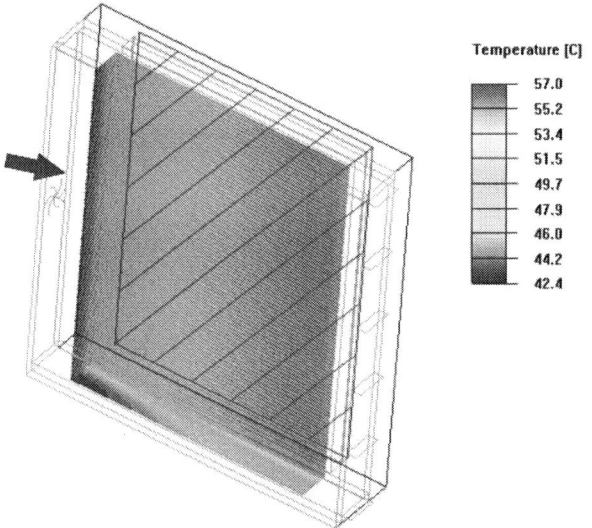

Figure 10: CFD results for fin array temperatures (case 1).

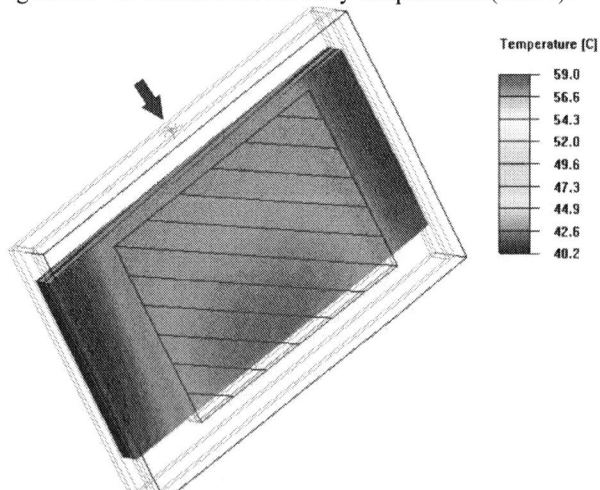

Figure 11: CFD results for fin array temperatures (case 2).

Figure 10 shows temperature field of the fin array. Along the width dimension (perpendicular to flow), temperature is fairly uniform for a given distance from the entrance. From figure 11, there is noticeable difference in fin array temperature between area within die footprint, compared to that outside of it. Coolant that by-passes the die footprint region are not well utilized. This results in a lower effectiveness for the fin array. Even though the wider fin array has much larger fin surface area, the thermal resistance is higher. Hence, having a fin array footprint that is close to die footprint provides near optimum thermal resistance for the cold plate.

The discrepancy in fin array thermal resistance between model and CFD is higher when the fin array base temperature has a large gradient, such as that in figure 11. This is because the model assumes a constant temperature at the base of the fin array. In figure 10, the temperature gradient at fin array base is much smaller than that in figure 11, leading to better agreement between model and CFD.

Figure 12 shows effect of fin height on R_{CP}. An optimum fin height exists, but the variation in R_{CP} versus fin height is small.

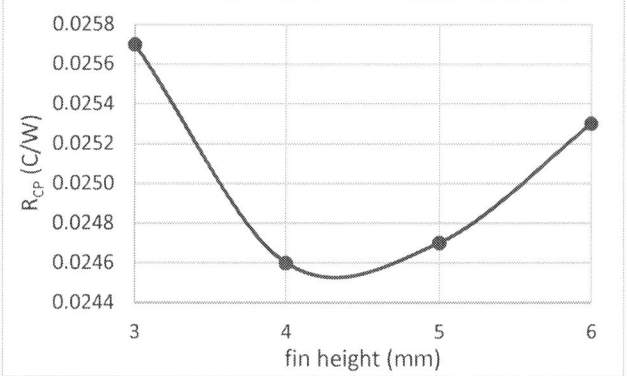

Figure 12: Effect of fin height on R_{CP}.

Optimizing the heat exchanger is another area for reducing energy usage for cooling. The ultimate in energy savings is to take the heat outside of the datacenter for recycling or discharge to the outside environment. This requires a liquid-to-liquid CDU and plumbing/equipment infrastructure to allow heat discharge outside of the datacenter. A change in heat exchanger type is needed, with a liquid-to-liquid plate heat exchanger replacing the liquid-to-air compact heat exchanger. These are subject matter for future study.

4. Conclusions

A unified model has been developed for modeling single-phase OLLC system. Unlike the case of CLLC, coolant flow in OLLC is split into a network flow system. Kirchoff's rule is used to simplify the treatment of network flow through the rack/chassis system. The method of calculation for the thermal resistances due to cold plate and heat exchanger was based on a previous model derived for CLLC. Modifications and additional equations relevant to OLLC were described and highlighted. Examples were presented to illustrate how the model is used to simulate typical OLLC systems. Examples were included to explore how various materials and operating variables, as well as geometric variables, affect cooling performance. The model is useful for preliminary design, as well as to obtain insights into OLLC performance when actual die and system are not yet available for system test.

References

1. ASHRAE, "Water-Cooled Servers Common Designs, Components, and Processes", ASHRAE Technical Committee 9.9 (2019).

2. ASHRAE, "Emergence and Expansion of Liquid Cooling in Mainstream Data Centers", ASHRAE Technical Committee 9.9 (2021).

3. ASHRAE, "Liquid Cooling: Resiliency Guidance for Cold Plate Deployments", ASHRAE Technical Committee 9.9 (2024).

4. J. Gullbrand, N. Gore, J. Matteson and E. Langer, "ACS Liquid Cooling Cold Plate Requirements Document", Open Compute Project (2019).

5. M. Gonzales, et al, "Liquid Cooling Integration and Logistics White Paper", Open Compute Project (2021).

6. C. Chen, et al, "OCP OpenAI System Liquid Cooling Guidelines", Open Compute Project (2022).

7. A. Chan, D. Nguyen and M. Brooks, "Unified Method to Model Closed-Loop Liquid Cooling", 37th Semitherm Symposium (2021).

8. ANSI/HI 9.6.7-2010, "Effects of Liquid Viscosity on Rotodynamic (Centrifugal and Vertical) Pump Performance", Hydraulic Institute (2010).

9. R.K. Shah and A.L. London, "Laminar Flow Forced Convection in Ducts", Academic Press (1978).

10. W.M. Kays and A.L. London, "Compact Heat Exchangers", Krieger Publishing Company (1984).

11. A. Achaichia and T.A. Cowell, "Heat Transfer and Pressure Drop Characteristics of Flat Tube and Louvered Plate Fin Surfaces", Experimental Thermal and Fluid Science, 147-157 (1988).

12. Davenport, C. J., "Heat Transfer and Flow Friction Characteristics of Louvered Heat Exchanger Surfaces", Heat Exchangers: Theory and Practice, pp. 397-412, Hemisphere/McGraw-Hill (1983).

13. Triboix, A., "Exact and Approximate Formulas for Cross Flow Heat Exchangers with Unmixed Fluids", Int. Comm. Heat Mass Transfer, 36(2):121–124 (2009).

14. Lienhard, J. IV and J. Lienhard V, A Heat Transfer Textbook, 5th Edition, Phlogiston Press (2020).

Bi-Phasic Nanocomposites with Liquid Metal for Thermal Management and Soft Electronics

Carmel Majidi, *Carnegie Mellon University*

Abstract— Droplets of liquid metal (LM) – i.e. metals that are liquid at room temperature – can be combined with various nanomaterials and suspended within elastomer to create soft composites with an extraordinary combination of mechanical, thermal, and electrical properties. When engineered as a film or coating, these "bi-phasic" LM-based nanocomposites can support a broad range of functions – from dielectric films and thermal interface materials to high stretchable electrical conductors. Here, I present several approaches to creating soft multifunctional films composed of LM droplets dispersed in a soft polymer or elastomer. Specifically, I use eutectic gallium indium (EGaIn) as the liquid metal alloy and show how the choice of polymer matrix material and processing conditions can influence the bulk and interfacial properties of the composite. I also present multifunctional composites that, in addition to EGaIn, also incorporate a second dispersion phase, such as Ag flakes, reduced graphene oxide, or MXenes. Such material systems have potential use in a variety of applications, including stretchable and wearable electronics, energy harvesting, bioelectronics, and thermal management for high performance computing.

Keywords—liquid metal, eutectic gallium-indium, soft electronics, stretchable electronics, thermal interface materials

I. INTRODUCTION

Emerging trends in computing and wearable electronics are increasingly reliant on electrically and thermally conductive materials that are intrinsically soft, stretchable, and mechanically robust. These needs range from printable conductive inks for smart textiles that are highly stretchable and flexible to thermal interface materials (TIMs) for high performance computing in which the TIM has both low thermal resistance and high mechanical reliability. For many of these technologies, functionality relies on soft elastic materials that combine the robust compliance and deformability of rubber with the conductivity of metal.

To date, efforts in creating these soft multifunctional materials have primarily focused on elastomer composites composed of a rubber matrix embedded with rigid conductive fillers. While successful in many applications, these particle-filled rubber composites exhibit a variety of limitations that arise from the mechanical mismatch between the compliant elastomer matrix and the rigid particle inclusions. These include increased mechanical stiffness, hysteresis, and inelasticity along with reduced elastic strain limit. Such limitations introduce a fundamental design trade-off between rubber-like mechanical properties (which favors a lower particle filler concentration) and high electrical or

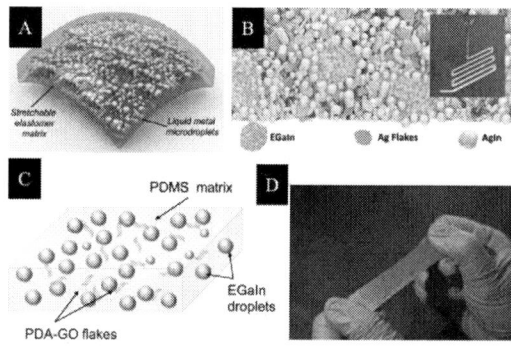

Fig. 1: (A) LMEE architecture [2]. (B) Printable biphasic ink with EGaIn, Ag flakes, and AgIn intermetallic inclusions [4,5]. (C,D) Dielectric elastomer with EGaIn and rGO [6].

thermal conductivity (which requires high filler concentrations).

One promising approach to circumvent this trade-off in particle-filled rubber composites is to replace the rigid particle inclusions with droplets of liquid metal (LM) [1]. Liquid metals are metals and metal alloys that are liquid at room temperature. For decades, room temperature liquid metals like mercury and LM alloys like gallium-indium-tin have been used in strain gauges, thermometers, tilt sensors, and other use cases that require a combination of high electrical conductivity and fluidic rheology. In recent years, eutectic gallium-indium (EGaIn) has become especially popular as a liquid metal due to its negligible vapor pressure, low toxicity, and low viscosity. As the surrounding elastomers stretches, the fluidic EGaIn inclusions can elongate and maintain electrical connectivity.

Here, I focus on one particular class of LM-elastomer material systems in which liquid metals like EGaIn or Ga-In-Sn (Galinstan) are embedded in the form of micro- or naonscale droplets (**Fig. 1A**). These so-called liquid metal embedded elastomers (LMEEs) represent a class of LM-based materials that are applicable to a variety of soft polymer matrix materials and LM droplet morphologies [2,3]. For some material properties and applications, performance can be further improved by incorporating a second dispersion phase into the LMEE architecture. Early efforts focused on suspensions of silver flakes, which combined with the LM inclusions to create highly conductive networks within a soft elastomer that could withstand extreme mechanical deformation [4,5]. More

---

C. Majidi is with the Department of Mechanical Engineering at Carnegie Mellon University, Pittsburgh, Pennsylvania, 15213 (412-268-2492 e-mail: cmajidi@andrew.cmu.edu).

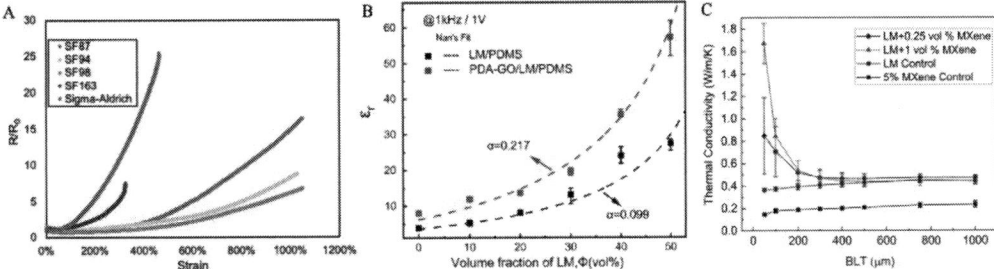

Fig. 2: (A) Electromechanical properties of EGaIn-Ag-SIS conductive ink for various Ag flakes [5]. (B) Dielectric constant of EGaIn-rGO-Silicone composite for varying LM volume fractions [6]. (C) Thermal conductivity of MXene-coated LM droplets versus BLT.

recently, there has been progress in showing enhancement in dielectric and thermal properties of LMEE composites by embedding them with reduced graphene oxide (rGO) or MXenes [6]. Henceforth, these materials are referred to as LM-based "bi-phasic" nanocomposites due to the combination of liquid phase LM inclusions and solid phase nanomaterials (**Fig. 1B**) [7,8]. Here, I will review some key features of this emerging class of material systems.

II. EXPERIMENTAL METHODS

Bi-phasic nanocomposites with liquid metal are typically synthesized using either shear mixing or probe sonication. In the case of shear mixing, the bulk liquid metal and filler particles can be dispersed within an uncured polymer using a planetary centrifugal mixer, immersion blender, or magnetic stirring bar. For probe sonication, a solvent is typically used to control viscosity and the solution is later subject to vacuum filtration or centrifuge. Details of the fabrication steps for both of these methods can be found in the following references: [4-6].

III. RESULTS

The electrical, dielectric, and thermal properties of LMEE composites with a secondary dispersion phase have been well documented in the literature and the results are not shown here. Instead, this section highlights some of the key features of LM-based biphasic nanocomposites.

A. EGaIn-Ag-SIS Conductive Inks

Dispersions of EGaIn microdroplets and Ag flakes suspended in a soft styrene-isoprene-styrene (SIS) elastomer matrix results in a printable conductive ink that can be used to create stretchable circuits. As with other LMEE composites, these biphasic conductive inks exhibit high electrical conductivity and low electromechanical coupling with strain [4,5]. **Fig. 2A** shows that for certain choices of Ag flake, the electrical conductivity remains relatively stable even for extreme strains of >500%.

B. EGaIn-rGO-Silicone Dielectric Elastomer

EGaIn-Silicone composites embedded with a dispersion of reduced graphene oxide show substantial enhancements in electrical permittivity when compared to composites without rGO filler [6]. Here, rGO is combined with polydopamine (PDA)-coated graphene oxide (GO). As shown in **Fig. 2B**, the impact of the rGO/PDA-GO is to amplify electric permittivity 2×, resulting in elastomers with a dielectric constant as high as 60 – a value that far exceeds

that of other dielectric elastomers with similar mechanical compliance and elasticity.

C. LM-MXene Composites

Coating nanoscale droplets of EGaIn with Ti_3C_2Tx MXenes results in a unique LM morphology in which MXene-LM droplets self-assemble to form semi-solid aggregate networks. This is accomplished using an ultrasonic synthesis method in which MXenes sheets are wrapped around individual LM droplets to form "sticky" particles that self-assemble into large aggregates. The size of the aggregates is controlled by MXene concentration and influence the interplay of thermal conductivity with bond line thickness (BLT; see **Fig. 2C**).

IV. CONCLUSION

Elastomers embedded with droplets of liquid metal and nanomaterial suspensions exhibit unique combinations of mechanical, electrical, and thermal properties. Applications range from printable conductive inks for stretchable electronics to TIMs for use in high performance computing. Progress depends on further exploration of material compositions and advanced use cases in printed soft electronics and thermal management.

REFERENCES

[1] P. Won, S. Jeong, C. Majidi, S.H. Ko, " Recent advances in liquid-metal-based wearable electronics and materials," iScience, vol. 24, no. 7, p. 1092698, 2021.

[2] N. Kazem, T. Hellebrekers, C. Majidi, "Soft multifunctional composites and emulsions with liquid metals," *Advanced Materials*, vol. 29, no. 27, p.1605985, 2017.

[3] M.H. Malakooti M.R. Bockstaller, K. Matyjaszewski, C. Majidi, "Liquid metal nanocomposites," *Nano. Adv.*, vol. 2, no. 7, pp.2668-2677, 2020.

[4] P.A. Lopes et al. "Bi-phasic Ag–In–Ga-embedded elastomer inks for digitally printed ultra-stretchable multi-layer electronics," *ACS Appl. Mater. Interf.*, vol. 13, no. 12, pp.14552-14561, 2021.

[5] W. Zu et al. "A comparative study of silver microflakes in digitally printable liquid metal embedded elastomer inks for stretchable electronics," *Adv. Mater. Tech.*, vol. 7, no. 12, p.2200534, 2022.

[6] Y. Hu, C. Majidi "Dielectric Elastomers with Liquid Metal and Polydopamine-Coated Graphene Oxide Inclusions," *ACS Appl. Mater. Interf.*, vol. 15, no. 20, pp.24769-24776, 2023.

[7] C. Majidi, K. Alizadeh, Y. Ohm, A. Silva, M. Tavakoli, "Liquid metal polymer composites: From printed stretchable circuits to soft actuators," *Flex. Printed Elect.*, vol. 7, no. 1, p. 013002, 2022.

[8] M. Reis Carneiro, C. Majidi, M. Tavakoli, "Gallium-Based Liquid–Solid Biphasic Conductors for Soft Electronics," *Advanced Functional Materials*, vol. 33, no. 41, p.2306453 (2023).

Developments with Indium-Gallium Thermal Interface Materials
for TIM1 Bare-die Server Processors

Tim Jensen, Bob Jarrett, Ricky McDonough
Indium Corporation, Clinton NY USA
E: Tjensen@indium.com

Dave Saums
DS&A LLC, Amesbury MA USA
E: dsaums@dsa-thermal.com

Extended Abstract

SUMMARY

Rapid increases in heat dissipation for server and AI server processors combined with increased total die area for these processor modules is challenging for development of improved thermal interface materials (TIMs). It is important to note that these challenges are not limited to maximizing thermal performance for a given material type. Practical application processes, retention requirements, and long-term reliability are principal challenges for materials in development. Traditional polymeric TIMs are not able to dissipate sufficient heat, given the heat dissipation levels and complexities related to die size and format. Further, heterogeneous integration brings memory stacks (HBM, High Bandwidth Memory) sensitive to temperature, placed in close proximity to processor die within the module; processor and system designers require the processor die TIM to also provide necessary coverage and heat dissipation for the HBM stacks.

These technology and packaging drivers have directed significant focus on developing different types of innovative metallic TIMs capable of meeting new requirements, as detailed in this presentation. An important basic categorization of metallic TIMs is the identification of applications for flat or patterned and compressible foils for one application category, generally as TIM2; and the development of different forms of liquid and hybrid/liquid metal TIMs that are specifically intended for use as TIM0 and/or TIM1, wherein the metallic TIM is to be in direct contact with the die surface. This presentation is focused on development of new liquid and hybrid liquid metal TIM1 solutions for extremely high heat dissipation for bare-die processors

1. INTRODUCTION

Significant current and forecast maximum power dissipation per processor for high-performance AI/GPU and general server processors illustrates the sharp recent increases in thermal design power (TDP). The total thermal resistances through the thermal and packaging materials stack is increasingly therefore seen as a primary focus for new materials developments.

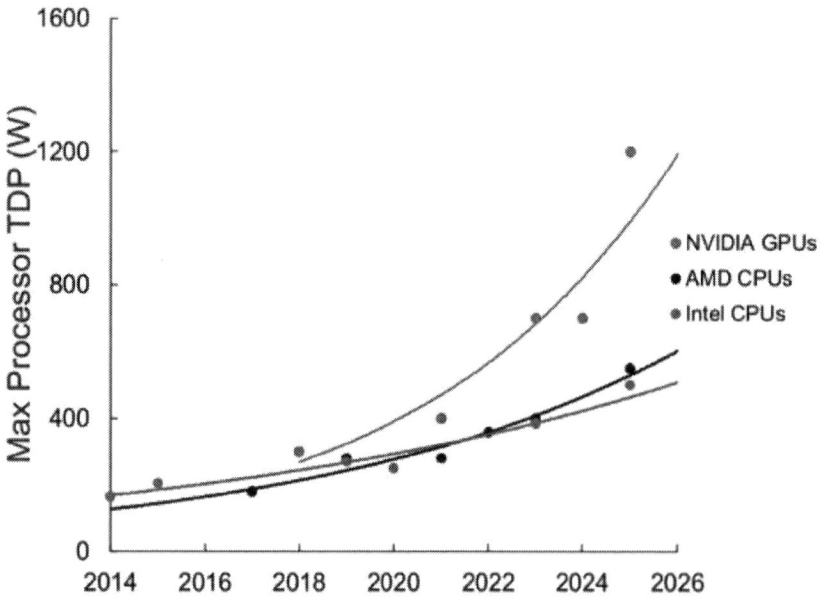

Figure 1. Recent and forecast maximum processor TDP for selected processor families. (Source: Dell'Oro Group.

This presentation is intended to focus on developments with TIMs used primarily in processor packaging and heat dissipation.

Metal alloys with thermal conductivities significantly higher than polymeric TIMs must be developed to meet process and application needs and to perform successfully when tested for reliability requirements. The range of requirements include addressing die warpage, uniform dispensing and thickness control, and successful interaction with the semiconductor die surface and the copper or metallized surface of a liquid cold plate, heat pipe heat sink assembly, or other thermal solution that serves as the processor's primary heat removal mechanism. There are numerous proposals for hybrid liquid metal (LM) formulations, metallic pastes, gallium-containing materials, phase change metal alloys (PCMAs), and graphene-enhanced graphitic TIM types for high-performance server processor applications where expected power dissipation is 850W or greater for a single die.

The use of gallium metal presents gallium-specific challenges, specifically its corrosive nature when in contact with other metals and the resulting surface degradation and intermetallic formation. This affects uniquely modified gallium alloys, non-gallium-containing mixtures including PCMAs, and lower-melt alloys not designed to change phase. PCMAs are typically a mixture of indium, bismuth, and tin. Using PCMAs as TIMs explores challenges, including processing methods for outsource assembly and tests (OSATs), packaging operations to apply, and protection from moisture and material movement (pump-out or squeeze-out). Thermal performance, processing, and protection methods will be discussed.

Key words
Artificial intelligence (AI), High bandwidth memory (HBM), High-performance computing (HPC), Hybrid liquid metal (HLM), Liquid metal (LM), Phase-change metal alloy (PCMA), thermal design power (TDP), thermal interface material (TIM), thermal resistance.

2. EXPERIMENTAL/NUMERICAL METHODS

Basics of development of metallic TIMs will be reviewed briefly, categorizing different materials types within this expanding area as more materials and more manufacturers continue to determine new mixtures and material concepts to address specific uses (i.e., TIM0, TIM1, TIM2) and application requirements.

Solder	Compressible	Liquid	Hybrid LM	PCMA
Soldered interface maximizing heat transfer	Metal TIMs compressed under pressure to ensure low interfacial resistance	Traditional LMs	LM base with solid metal fillers	Solid metal TIMs that melt during power cycling
In-based	In- or Sn-based	Ga Liquid	Ga Paste	Low Melt Alloy
sTIM	Heat-Spring®	51E	LMP007	19
Fluxed sTIM	High-profile Heat-Spring	300E	M2TIM®	PCMA2
InX Alloys	InTACK®	60	Barrier Technology	Oxidation Tech

Table 1. Types of metallic TIMs, identified in five material categories with examples of current commercial products available. (Note to reviewers: Table will be rewritten to remove commercial product names, if required, and insert additional characteristics.)

A second table listing phase-change metal alloys in development with corresponding phase-change temperature and bulk thermal conductivity will follow, to illustrate a range of potential materials suitable for use as thermal interface materils.

A section describing application of thermal test vehicles to metallic TIM development will be included. A section describing material testing and impact of die warpage and measurements of material compliancy will follow. TTV testing is important in order to address surface adhesion behavior of a given metallic TIM mixture, for liquid metals and hybrid liquid metals in direct contact with the die surface.

Consideration of different metal alloy and mixtures allows selection of the metal alloy to be used to meet the design requirements of a given set of package characteristics. Increasing specialization of TIM developments is required to meet rapidly changing design critera, as processor package requirements change, maximum TDP values increase rapidly, maximum hot spot heat flux increases, and previously less-significant factors such as die warpage and multiple large-area die introduce new challenges. One important advantage of the use of metal alloy TIMs as TIM0 or TIM1 is that small changes may be made in alloy composition to alter alloy characteristics, such as a change in the targeted melting point to match downstream module processing temperatures (such as secondary or tertiary solder reflow processes).

3. RESULTS

Illustrations of die temperature gradients achieved with an example phase-change metal alloy will be included, to illustrate maximized die surface wetting achieved, even with significant die warpage. Examination to determine that maximum uniform surface wetting has been achieved is an important characteristic of in-situ testing. Figure 2 is an example.

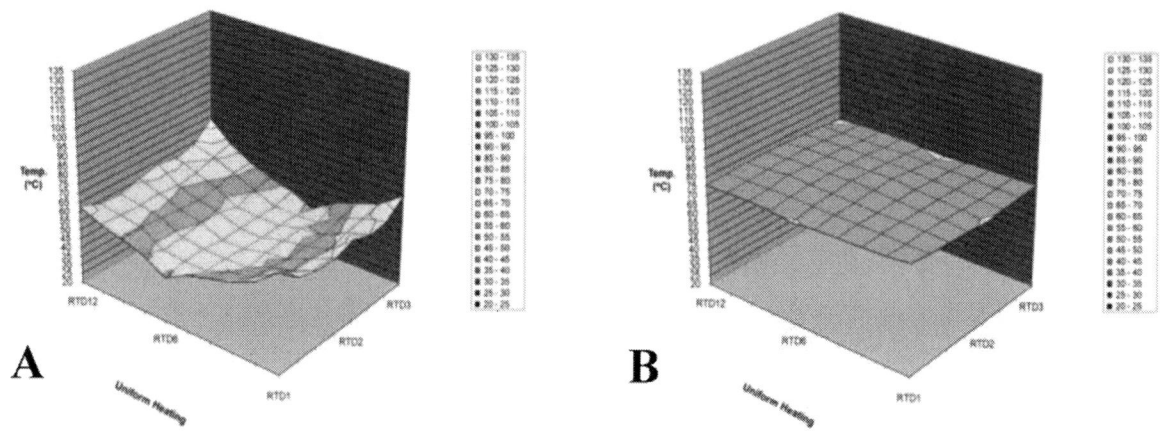

Figure 2. Temperature measurements achieved with TTV and development PCMA as TIM0: (A) Temperature gradients across the die are seen prior to initial heating to bring the die to the targeted PCMA2 phase-change temperature; (B) After phase-change temperature has been achieved, the sharply reduced temperature gradients across the die surface are apparent as the metallic PCMA development material has achieved 100% surface wetting on a die with significant warpage.

Power cycling of a completed TTV with development materials such as the PCMA illustrated in Figure 2 has been undertaken, cycling from 0W to 300W. This study was run for 1,800 cycles. The goal was to examine die surface temperature consistency and monitor thermal resistance values from start to finish. Tested performance for this power cycling regime was also compared to identical testing for a graphite film TIM (from another supplier) and a compressible metal foil TIM. Test data will be shown with figures similar to Figure 2 (above); test data showed no change in measured die temperature throughout the cycling period.

An identical power cycling test was also conducted for the test assembly with the cycling regime set for 0W and 600W; the development PCMA material was expected to change phase during each of these cycles. Performance results will be included in an additional figure. Degradation was seen after 300 cycles and testing halted. Analysis indicated that oxidation was occurring at the surfaces, impeding heat transfer slightly, yielding increased die operating temperature. A commercial material of this type would require use of a polymer barrier seal at the die perimeter to mitigate oxidation; this test was run purposely to determine the impact of oxidation as this is an important question. For this level of power dissipation, a commercial PCMA would be designed with a higher phase-change temperature as well as the use of the die perimeter barrier seal.

4. CONCLUSIONS

Explanation of the different categories of metallic thermal interface materials under development, in addition to those previously commercialized, is useful to build a picture of the increasing range of such metallic TIMs available in the thermal engineering tool box, in an era of rapidly increasing TIM specialization to meet newly-developing performance, assembly, and reliability criteria. The limitless number of metal alloys possible allows even further specialization of such TIMs to meet individual design requirements for different processor package characteristics. Minimized thermal resistance is only one criteria for evaluation and selection of such TIMs, especially for the TIM0 and TIM1 processor package requirements where new criteria such as application over multiple die with sharply higher power dissipation, die size, and die warpage are now in place.

This presentation illustrates developments with novel PCMAs for TIM1, as an example of new TIM developments targeting high performance AI/GPU server processors, with consideration for the material characteristics, application process, performance measurements, assessment of degree of surface wetting and adhesion achieved, and reliability over time and power and temperature cycling.

5. REFERENCES

C. Grossner, I. Beran, R. Bonner, A. Bryant, *"Direct-to-Chip Cooling for the AI Data Center,"* Open Compute Project, Educational Webinar Series, May 2024.

D-Central Technologies, *"ASIC vs. FPGA vs. GPU vs. CPU: Understanding the Differences,"* D-Central Technologies (on-line): https://d-central.tech/asic-vs-gpu-vs-cpu-understanding-the-differences/ (February 10, 2024).

Wang, Zhengfang; Wu Zijian, et al., *"A Roadmap Review of Thermally Conductive Poymer Composites: Critical Factors, Progress, and Prospects."* Advanced Functional Materials, 22.10.1002/adfm.202301549.

Saums, D., *"Thermal Interface Materials and Testing Methods for Semiconductor Test,"* TestConX 40 Workshop, Mesa AZ USA, March 3-6, 2024.

Thermotest NT20-3k-FC Data Sheet. Berliner Nanotest und Design GmbH, Berlin, Germany: http://nanotest.eu/fileadmin/downloads;datasheets;NT20-3k-FC_datasheet.pdf.

Thermotest Thermal Test Vehicle Solutions, Berliner Nanotest und Design GmbH, Berlin, Germany. 2024.

Abo Ras, M, *"Thermal Test Vehicle for Investigation of Thermal Path in Large Die Area Packages by Thermal Transient Impedance Analysis,"* Semiconductor Temperature Measurement and Thermal Management Symposium, SEMI-THERM 40, San Jose CA USA, March 25-28, 2024.

Evaluation of Graphene-Enhanced Thermal Interface Material in Air and Immersion Cooling Systems

Henrik Barestrand*‡, Markus Enmark†, Jonas Gustafsson*, Tina Stark*, Håkan Fredriksson*, Johan Liu†¶
Jon Summers*,
*RISE ICE Data Center, Luleå, Sweden
†Chalmers University of Technology, Gothenburg, Sweden
Corresponding authors: ‡henrik.barestrand@ri.se ¶johan.liu@chalmers.se

Abstract—**This study presents a detailed performance evaluation of a graphene-enhanced thermal interface material (TIM) conducted at RISE ICE Data Center. Tests were performed on Open Compute Project (OCP) Leopard servers using three different TIMs: conventional thermal paste, graphene-enhanced thermal pad GT90 from SHT Smart High Tech AB (SHT), and indium foil. Three sets of experiments were conducted: (1) air cooling with default chassis fan control in a bespoke server wind tunnel, (2) air cooling with controlled, fixed fan speeds and different heatsink mounting pressures operating in the wind tunnel and (3) immersion cooling tests with two coolant flow rates at fixed inlet temperatures. Results indicate that graphene-enhanced TIM and thermal paste exhibit similar performance in experiment (1), whereas indium foil TIM tests showed the undesired effect of increased CPU temperatures. In experiment (2), servers equipped with graphene-enhanced TIM showed lower CPU temperatures in comparison to the servers equipped with Indium foil TIM. In experiment (3), immersion cooling resulted in lower CPU temperatures overall, with the graphene-enhanced TIM again providing lower temperatures than indium foil at a similar mounting pressure. The findings suggest that the interfacial thermal conductivity and material compatibility of the GT90 TIM contribute to an improved performance in the tested immersion cooling system as well as the importance of mounting pressure.**

Index Terms—**Thermal Interface Material, Graphene, Indium, Air Cooling, Immersion Cooling, Data Center Thermal Management**

I. INTRODUCTION

Thermal management of IT systems is becoming increasingly challenging as the power densities continue to increase, lately driven by the surging need for AI services. This is provisioned by large installations of high-power graphics processing units (GPUs). With an increasing share of power going to IT systems and the supporting infrastructure, the need for efficient and rapid heat removal has become a necessity to continue the digitalization growth. Reducing thermal resistance in the heat rejection chain, from microprocessor junction to the point of heat

rejection/re-use is key, leading to heat recovery at high temperatures and increased energy efficiency. The thermal resistance between the microprocessor case, known as the integrated heat spreader (IHS), and the heatsink is the first thermal resistance faced at the system level in the heat transport chain [1]. To maintain the thermal conductivity between the IHS and the heat sink, a thermal interface material (TIM) is normally used. As the thermal power generated by the microprocessors is continuously increasing, the quality of the TIM and its material is crucial to maintain the microprocessor from overheating [2]. The effect of increasing the power of the CPUs is that the temperature gradient across the TIM increases, requiring heat sink temperatures to be kept cooler, which in turn requires more power in the cooling system, and reduces the possibility for recovering the heat in various heat recovery applications. [3]

In the data center domain, an increase of only a few degrees Celsius at the heat sink, could have a large impact on the cooling system cost of operation, and limit the amount of hostable servers.

As thermal densities in data centers keep on increasing, using air as the cooling medium is no longer sufficient to cool the power demanding high-end devices used for artificial intelligence (AI). Liquid cooling of servers can be achieved in different ways, using direct-to-chip solutions where a liquid is directed to a cooling head (heat sink) mounted on top of the high-end microprocessor through a set of pipes or hoses [4]. The technology concept has existed for many years, and is widely adopted in the high-performance computing (HPC) community. A drawback of the solution is that heat generated by other server components, requires air cooling, so a combination of liquid and air is required to operate direct-to-chip cooled servers in AI/HPC data centers. The system complexity of direct-to-chip solutions also tend to be high with the need for pipework to distribute liquid coolant to every microprocessor. The concept can however support very high powered GPUs, when the liquid coolant is kept cool and its flow rate high. Since the TIM will be placed in the dry space between the microprocessor IHS and the cooling head, no material liquid compatibility issues are expected for the TIM-materials used.

Another liquid cooling approach is to immerse the complete system (server) into a non-conductive liquid. This concept brings the advantage of transferring all the heat generated by the system (server), so no air-cooling is required. A drawback often expressed is service-ability, as the servers are most often mounted vertically, it is heavy to lift them up of the basin, and when out of the immersion-bath the servers are wet from the immersion liquid, which is commonly a mineral oil. [5]

In an immersion cooled system, the TIM will be exposed to the immersion fluid, and depending on the materials used in the TIM and the immersion fluid, material compatibility issues may occur. Using cooling paste in immersion environments is generally avoided, as the paste might dissolve in the fluid. To avoid this kind of compatibility issue, Indium is frequently used in immersion cooled systems [6], and Indium does not have any known material compatibility issues with the frequently used immersion fluids[1].

As the surface of modern GPUs and CPUs are growing, and Indium requires a relatively high mounting pressure to achieve a good thermal transfer, the risk of curvature of the heat sink and/or the microprocessor becomes apparent. If the structure is bending, the contact between the CPU casing and the heat sink, via the TIM, will become uneven, and the heat transfer capacity will be negatively affected. Graphene-enhanced TIM can be produced with various stiffness, reducing the force required to achieve good thermal contact with the chip case surface showing good performance and reliability [7].

II. MATERIALS & METHODOLOGY

The experiments were conducted using OCP Leopard servers[2]. Three identical servers were used in all tests to ensure statistical significance. The Leopard server was chosen as they provide a good cooling media flow across the two CPUs that are positioned in series, as seen in Fig. 1. The cooling media (normally air) is concentrated to flow over the CPUs and memory units using a "airflow housing" which can be seen in Fig. 6. Compared to most traditional 19" servers the Leopard has a lower flow resistance across the server. There are two Intel Xeon E5-2678v3 CPUs with a Thermal Design Power (TDP) of 120 Watts each. The Stress-ng[3] code base was

Fig. 1: OCP Leopard server used in the experiments. The airflow housing is removed in the picture to show the placement of the CPUs and Memory.

used to load the servers, excluding loading the memory modules in these tests. As an example, the command to launch stress-ng for 12 hours using 24 CPU cores is `stress-ng --cpu 24 --timeout 43200s`.

A. Thermal Interface Materials

The TIMs tested were thermal paste[4], Smart High Tech graphene-enhanced TIM (GT90) of 0.2 mm thickness, and 0.2 mm Indium foil[5].

In this paper, we are evaluating how well the graphene-enhanced TIM, GT-90[6] produced by Smart High Tech performs, both an air cooled environment and in immersion. The GT-90 does not require as high mounting pressure as Indium, but still higher than what is commonly used with cooling paste. According to Smart High Tech, the graphene-enhanced TIM is produced with a vertically aligned structure embedded in a polymer matrix. A schematic drawing of the heat dissipation path in such a structure is shown in Fig. 2. To evaluate the performance of the SHT GT-90, comparative studies were conducted both using air as cooling medium and in an immersion cooled setup.

B. Test environments

To achieve reliable results, a controlled environment is required. In the RICE ICE test and demo facility there are several dedicated test environments to study server performance. For this study, two test environments were used. First, the thermal interface material was to be tested and evaluated in an air cooled environment, and second in an immersion cooled environment, a short introduction to the test rigs follows.

---

[1]Material Compatibility Guide.https://www.engineeredfluids.com/our-resources/compatibility-guides/

[2]For more information, visit the Open Compute Project website at https://www.opencompute.org/

[3]stress-ng (stress next generation) stress test code. https://github.com/ColinIanKing/stress-ng.

[4]Thermal Grizzly:Kryonaut thermal paste. https://www.thermal-grizzly.com/en/Kryonaut/S-TG-K.

[5]Haines-Maassen indium foils. https://www.haines-maassen.com/en/indium_seals.

[6]GT90 by Smart High Tech AB. https://smarthightech.com/product/gt90/.

TABLE I: Thermal interface materials, specifications.

Product	Thermal Grizzly:Kryonaut	Indium Corp.	SHT GT 90
TIM type	Paste	Foil	Sheet/Pad
Thermal Conductivity k [W/mK]	12.5	86	90
Thickness [mm]	Pressure dependent	0.2	0.2
Mounting pressure [kPa]	117	690	275

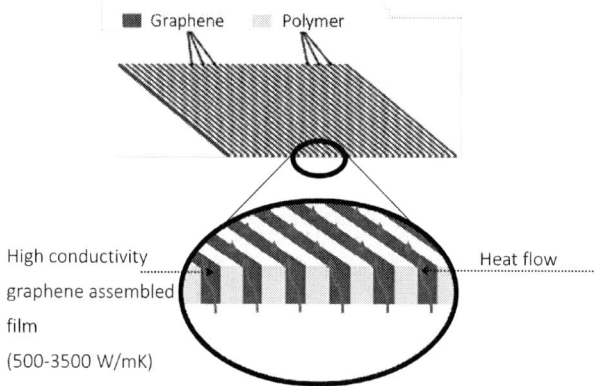

Fig. 2: Schematic drawing of the concept of the graphene-enhanced TIM (courtesy of Smart High Tech). The graphene film itself has a thermal conductivity between 500-3500W/mK in a highly pressed state.

Fig. 3: Overview of the wind tunnel setup for controlled air environment tests.

Fig. 4: Placement of the servers in the wind tunnel test section.

1) Air Cooled Test Environment: To achieve repetitive, controllable test iterations for the air-cooled case, the servers were installed in an semi-open ended server wind-tunnel. The server wind tunnel is depicted in Fig. 3 and a schematic is shown in Fig. 5. Starting from the left in the Fig. 5, the curved arrow indicate the air-mixture chamber which is used as an inlet air source, the air is sucked from the air-mixture chamber by the fan denoted as WT3-FA1, the temperature of the air is controlled by the liquid-to-air heat exchanger (WT3-AC). To achieve a more laminar and steady air-flow, the air is then pushed through a "flow-straightener" consisting of several small-size pipes (straws). After the first section of flow-straightener, the mass air-flow is measured by a Venturi meter sensor, mounted and denoted as WT3-SA2. The air then passes two 90°turns, a second flow-straightener and enters a larger diameter rectangular shaped zone in front of the server area. This zone would represent a cold-aisle in a well-contained operational data center. In this zone, we measure temperature and pressure to be able to supply the servers with air at the right tempeature, pressure and humidity.

As air enters the servers, mounted in between the pressure differential sensors WT3-SA7 and passes over CPUs and other heat-generating components (see Fig. 4), its temperature increases and the relative humidity drops. Depending on the speed of the server internal fans, pressure can drop or increase across the server. The final piece of the wind-tunnel represents the hot-aisle of a real data center, where temperature, humidity and pressure is observed, before the air enters back into the air-mixture chamber. The servers positioned in the wind tunnel is shown in Fig. 4 where 3 out of a possible 9 OCP servers were placed for these tests.

Controllable parameters in the wind tunnel are temperature, humidity and pressure in the "cold isle" in front of the servers. The setup provides a good environment to test and evaluate the TIMs in a realistic and highly controlled environment.

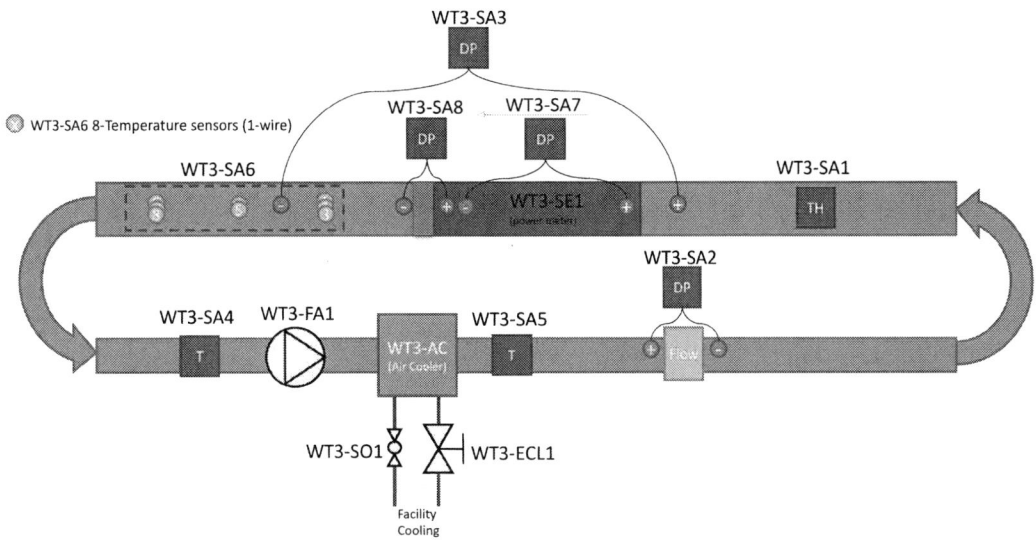

Fig. 5: Schematic of the wind tunnel setup from Fig. 3.

2) Immersion Cooling Test Environments: For the immersion tests, a modified Submer Micropod[7] immersion cooling system was used to perform the tests. To achieve a precisely controlled immersion fluid supply temperature, the immersion fluid default chiller was bypassed and connected to a liquid-to-liquid heat exchanger, where the primary side of the heat exchanger was connected to the RISE ICE facility cooling loop. The heat transfer rate, from the immersion liquid into the facility cooling loop, was controlled by a control valve mounted on the facility side of the heat exchanger. In this way, the temperature of the fluid entering the immersion basin is controlled with precision independent of the server load. Since the Submer micropod is designed to host 19" servers, the OCP servers used in the experiment did not fit without modification of either the Micropod or the servers. The solution for this study was to produce a fixture connecting the servers rotated at 90°in relation to how a 19" server would be mounted; see Fig. 6. This method created a gap between the servers and the Micropod basin edge, allowing immersion fluid to pass the server without passing the heat generating components (CPUs). To reduce this gap, a plastic blocking sheet was added to the fixture, guiding the immersion oil to pass through the servers.

To reduce any disturbance caused by fluctuations in ambient temperature, the immersion test setup is placed inside a controlled environment (room) where temperature and humidity are kept constant during the experiment. Note that immersion coolant properties can also affect the heat transfer characteristics [8], but in this case the same immersion coolant was used.

---

[7]Submer MicroPod edge data center immersion rig. https://submer.com/micropod/.

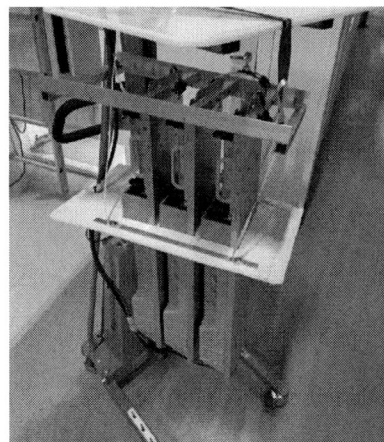

Fig. 6: Immersion fixture before submerging.

C. Experiments

1) Air Cooling with Default Fan Control: Initial tests were performed in one of RISE ICE Data Center wind tunnels depicted in Fig. 3. The servers were tested with default chassis fan speed control. The air supply temperature was set to 22°C for all test. To evaluate the impact of the CPU heat sinks heat transfer capacity, different airflows through the servers were tested. The change of airflow was achieved by adjusting the differential pressure across the servers. The differential pressures tested were increased in steps of 5Pa, ranging from 5 to 20Pa. The CPU temperatures were recorded when a steady state was achieved in the system.

During the initial tests, the factory default mounting pressure produced by the springs supplied with the servers holding the heatsink was used for all three TIMs. The average mounting pressure was measured to 117kPa.

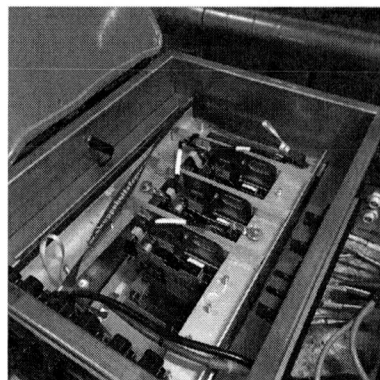

Fig. 7: Immersion tank setup with the three servers positioned vertically.

A problem with this setup is that the server internal control system adjusts the servers fan speed to achieve the temperature setpoint of the CPU for any particular CPU load, which might limit the TIMs impact on the CPU temperature.

2) Air Cooling Using Controlled, Fixed Fan Speed: In these tests, the server fan speeds were taken control of, and set to a fixed speed at approximately 10% of maximum RPM. The flow rate could then be fully controlled by the wind tunnel control system and fan. To ensure that enough cold air was provided to the servers, a differential pressure of 25Pa was dialed in.

To increase the performance of the GT90 and Indium tests, the mounting pressure was adjusted to meet the specifications of 276kPa for GT90 and 690kPa for Indium. To reduce the impact of bending the structure due to the high mounting pressure of 690kPa were the Indium tests were repeated with a mounting pressure of 240kPa.

3) Controlled Immersion Cooling Tests: In the immersion testing, only Indium and GT90 were considered for test, as the cooling paste could risk to dissolve in the immersion fluid. The mounting pressure used for the setup was 276kPa for the GT90 sheets, and 240kPa for the Indium. The reduced mounting pressure was chosen, as it showed better results in the air cooling tests, and we fear that the high mounting pressure of 690kPa might have damaged one of the servers during the air cooled tests (Server 2 was replaced with Server 7 for the immersion tests). The servers were immersed in a dielectric coolant oil hosted in the immersion test system explained in Section II-B2, see also Fig. 7. Tests were repeated at two different dielectric fluid flow rates, 1.8L/min and 15.6L/min to evaluate the TIMs performance with both high and low cooling capacity. The inlet temperature to the immersion tank was kept constant at 20°C across all tests.

TABLE II: CPU temperatures under varying airside differential pressure for different TIMs in the default fan control test with 100% CPU load.

CPU	ΔP [Pa]	CPU Temperature [°C]		
		Paste	Indium	GT90
1	20	58.4	67.2	58.8
1	15	59.9	68.7	60.1
1	10	62.0	70.2	62.2
1	5	65.0	70.7	65.0
2	20	65.1	76.3	66.0
2	15	67.8	79.1	68.5
2	10	71.2	82.2	71.8
2	5	75.4	83.1	75.9

III. RESULTS AND DISCUSSION

A. Experimental results

1) Air Cooling with Default Fan Control: The initial tests showed that thermal paste and GT90 resulted in similar CPU temperatures while indium foil gave a higher CPU temperatures. As seen in Table II, Indium resulted in an approximately 8°C higher CPU temperature on average. Increasing the cold aisle overpressure from 5 Pa to 20 Pa reduced CPU temperatures by up to 10°C due to increased airflow. This demonstrates the impact of airflow on cooling performance, regardless of the internal server fan response.

2) Air Cooling Using Controlled, Fixed Fan Speed: Under controlled airflow and fixed fan speeds, GT90 again resulted in a lower CPU temperature compared to the Indium, as presented in Table III.

The very high mounting pressure for Indium indicatively caused bending of the heat sink and/or the IHS as the results from the reduced mounting pressure of 240kPa resulted in lower CPU temperature. The assumed bending has possibly caused an uneven pressure over the Indium foil, which can be seen in Fig. 8. This might explain why the CPU temperature for the tests conducted at 690kPa mounting pressure, came out higher.

3) Immersion Cooling Tests: From the immersion tests we can see a similar difference in performance between indium and the GT90 TIM compared to the air cooled tests whilst the overall operating temperatures are lower for both CPUs, see Fig. 9. The graphene pad consistently resulted in lower CPU temperatures than indium foil at both low and high flow rates (see Table IV). Specifically, there was an average temperature reduction of about 10°C for CPU 1 and 12°C for the CPU 2 when using GT90.

4) Immersion Cooling Long Term Tests: The used and removed GT90 sheets from both the immersion tests and the air cooled tests as reported in this study are shown in Fig. 10. Visual inspection of graphene-enhanced thermal pads (GT90 sheets) shows that the thermal pads are intact.

TABLE III: Mean CPU temperatures and mounting pressure under controlled conditions for both CPU 1 and CPU 2 (25 Pa airside pressure drop across the server).

| | | CPU Temperature [°C] | | | | | |
| | | Indium | | | GT90 | | |
CPU	Mounting Pressure [kPa]	Srv. 2	Srv. 4	Srv. 6	Srv. 2	Srv. 4	Srv. 6
1	240	61	66	68	-	-	-
2	240	74	76	67	-	-	-
1	276	-	-	-	55	53	51
2	276	-	-	-	61	60	59
1	690	72	73	74	-	-	-
2	690	80	81	77	-	-	-

TABLE IV: Mean CPU temperature at 1.8 and 15.6 L/min volumetric flow rate through the immersion tank. The mounting pressure used were 240kPa for Indium and 276kPa for GT90.

| | | CPU Temperature [°C] | | | | | |
| | | Indium | | | GT90 | | |
CPU	Volumetric flow rate $\dot{V}$[L/min]	Srv. 4	Srv. 6	Srv. 7	Srv. 4	Srv. 6	Srv. 7
1	1.8	56.0	49.9	53.9	44.6	40.9	45.5
2	1.8	60.2	61.3	62.3	48.1	50.7	50.4
1	15.6	52.7	46.2	50.7	43.5	38.0	42.8
2	15.6	55.6	56.5	56.2	42.1	50.7	50.4

Fig. 8: Indium contact surface illustrated after removal of heatsink with 690kPa mounting pressure.

It is also seen that the GT90 sheets are fully exposed and wetted by the immersion oil. Possible changes in chemical composition due to immersion coolant exposure is currently being analysed.

Within the scope of another long term project conducted at RISE [9], a set of graphene pads in a commercial environment using a Submer System with Smartcoolant was tested during three months.[8] Albeit the default mounting pressure

[8]Submer SmartPod product website: https://submer.com/smartpod/.

for the servers was used, no signs of pad degradation was observed in terms of CPU temperatures.

IV. CONCLUSIONS

The study concludes that the graphene-enhanced TIM (GT90) provides an improved thermal performance over the indium foil with the given conditions of each experiment, cf. Fig. 9. In addition, the GT90 did not show any degradation of performance over a period of approximately 100 days in a fully operational immersion environment. The results also indicate that GT90 performs similar to thermal paste in air cooled environments. Indium foil and GT90 have comparable thermal properties, but their performance is seen to be affected by factors such as interfacial thermal contact and the need for a high and evenly distributed mounting pressure, which will become more important for microprocessors with larger form factors. The GT90 has a sponginess and ability to fill microscopic cavities at a lower mounting pressure and this is assumed to be the key factor contributing to its improved performance in experimental tests performed in this study.

ACKNOWLEDGMENT

This work has been performed with financial support from Vinnova SIO Grafen. M. Enmark and J. Liu also acknowledge the financial support from the Production Area of Advance at Chalmers University of Technology. We also acknowledge valuable contributions of the RISE ICE Datacenter team.

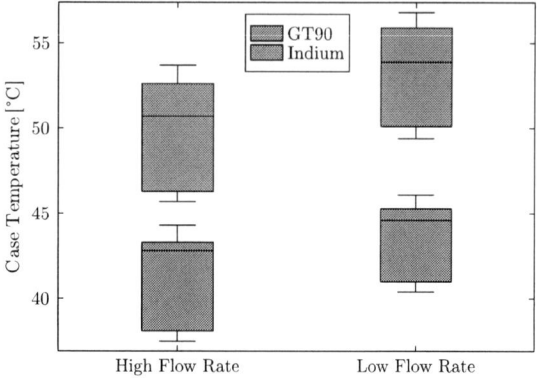

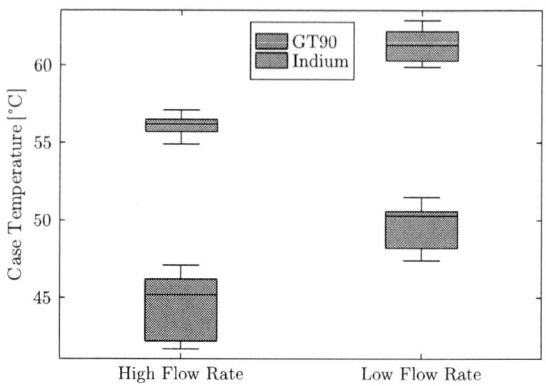

(a) CPU 1 case temperatures in the immersion tests at different flow rates.

(b) CPU 2 case temperatures in the immersion tests at different flow rates.

Fig. 9: Comparison of CPU 1 (a) and CPU 2 (b) case temperatures in the immersion tests at different flow rates.

Fig. 10: Graphene pads after immersion cooling tests.

REFERENCES

[1] K. M. Razeeb, E. Dalton, G. L. W. Cross, and A. J. Robinson, "Present and future thermal interface materials for electronic devices," *International Materials Reviews*, vol. 63, no. 1, pp. 1–21, 2018.

[2] J. S. Lewis, T. Perrier, Z. Barani, F. Kargar, and A. A. Balandin, "Thermal interface materials with graphene fillers: review of the state of the art and outlook for future applications," *Nanotechnology*, vol. 32, no. 14, p. 142003, 2021.

[3] R. Viswanath, V. Wakharkar, A. Watwe, V. Lebonheur *et al.*, "Thermal performance challenges from silicon to systems," *Intel Technology Journal*, vol. 4, no. 3, pp. 1–16, 2000.

[4] A. Heydari, A. R. Gharaibeh, M. Tradat, Y. Manaserh, V. Radmard, B. Eslami, J. Rodriguez, B. Sammakia *et al.*, "Experimental evaluation of direct-to-chip cold plate liquid cooling for high-heat-density data centers," *Applied Thermal Engineering*, vol. 239, p. 122122, 2024.

[5] Y. Chi, J. Summers, P. Hopton, K. Deakin, A. Real, N. Kapur, and H. Thompson, "Case study of a data centre using enclosed, immersed, direct liquid-cooled servers," in *Proceedings of the 2014 Semiconductor Thermal Measurement and Management Symposium (SEMI-THERM)*. IEEE, 2014, pp. 164–173.

[6] S. Sarangi, E. D. McAfee, D. G. Damm, and J. Gullbrand, "Single-phase immersion cooling performance in intel servers with immersion influenced heatsink design," in *2022 38th Semiconductor Thermal Measurement, Modeling & Management Symposium (SEMI-THERM)*. IEEE, 2022, pp. 1–5.

[7] M. Enmark, M. Murugesan, A. Nkansah, Y. Fu, T. M. Nilsson, and J. Liu, "Reliability characterization of graphene enhanced thermal interface material for electronics cooling applications," in *2022 IMAPS Nordic Conference on Microelectronics Packaging (NordPac)*, 2022, pp. 1–6.

[8] M. Hnayno, A. Chehade, H. Klaba, G. Polidori, and C. Maalouf, "Experimental investigation of a data-centre cooling system using a new single-phase immersion/liquid technique," *Case Studies in Thermal Engineering*, vol. 45, p. 102925, 2023.

[9] P. Taddeo, J. Romaní, J. Summers, J. Gustafsson, I. Martorell, and J. Salom, "Experimental and numerical analysis of the thermal behaviour of a single-phase immersion-cooled data centre," *Applied Thermal Engineering*, vol. 234, p. 121260, 2023.

Die Warpage as a Critical Factor for Thermal Interface Material (TIM0) Development and Selection

Dave Saums
DS&A LLC, Amesbury MA USA
E: dsaums@dsa-thermal.com

Extended Abstract

Die warpage for high-performance AI and server processors is an increasingly critical challenge for practical thermal interface material development, as power dissipation, die area, and warpage values are each rapidly increasing in new processor module designs. Increasing warpage values over larger die footprints present serious challenges for developments of any thermal interface material implemented as a TIM0 or TIM1 in system design. This is of particular importance for very high performance liquid/hybrid metal and polymeric TIM0 materials, where the TIM is in direct contact with the (typically) backside of the semiconductor silicon die and material retention within the interface is critical. The retention and maintenance of the designed thermal resistance value through this interface is critical for liquid immersion system applications. A review of processor design and explanation of certain performance drivers, specifically die warpage, is important but has not been documented for the general thermal design engineering community.

SUMMARY

Die warpage is a semiconductor industry phenomenon that is distinct from substrate warpage for IC packaging, and rapidly increasing measured die warpage is adding challenges to practical thermal interface material development, as warpage is seen as potentially combining with other physical and mechanical factors in how new high-performance TIMs perform and potential for reliability issues over time.

Definitions of what is meant with industry use of terms such as "die area" and "single die" are important to this discussion. Previously, a "single die" processor was indicated as meaning a single silicon die manufactured within the silicon wafer manufacturing industry, meeting the maximum reticle size limitations for fabrication equipment. Today, references are made to as much as "5x reticle" and suggests that the definition of what constitutes a "single die processor" has changed. This reinterpretation impacts the total surface area for how a TIM is to be applied (especially for what is termed as TIM0) and that, in turn, increases the total surface contact area over which die warpage is to be measured.

Assessing the impact of these changes is important for practical discussion of development targets for new thermal interface materials. It is most important in module and system design to evaluate each of the several factors that impact TIM0 materials, their function, and reliability over time and temperature.

Key words
AI, coefficient of thermal expansion (CTE), die area, die warpage, high bandwidth memory (HBM), high performance computing, metallic TIM, phase-change metal alloy (PCMA), thermal interface material (TIM), thermal interface material level zero (TIM0), thermal resistance

1. INTRODUCTION

It is the integration of a thermal interface of a specific material type into a processor module containing one or more large area silicon die that presents the greatest challenge for current TIM development and commercialization for IC applications. Definition of die size is critical for current and forecast die area.

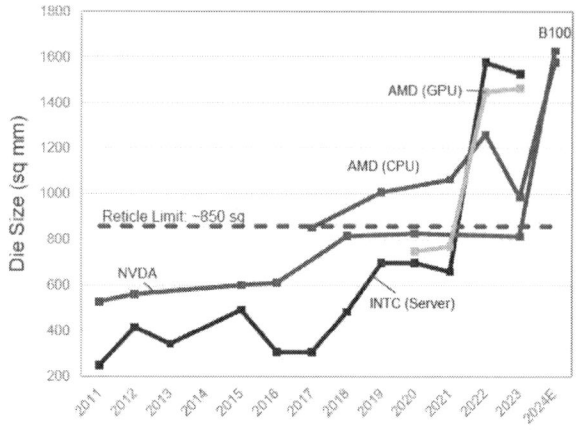 

Figure 1 (Left). Current and forecast die area for high-performance server processor modules. (Source: Kini, G., AMD, 2024.)

Figure 2 (Right). AMD Instinct MI300X. An example of a current high performance heterogeneous integration server processor with large die area. Source: AMD.)

Factors which have potential to increase difficulties for TIM0 development (for the TIM manufacturer) and TIM0 selection and implementation (for the module and systems OEM thermal engineer) include but are not limited to:

A. Large area silicon die AI server processor modules may contain multiple large die placed together with minimal separation of the multiple die (e.g., 10µ spacing, edge to   lthough termed as a "single die" by the semiconductor manufacturer for the purposes of internal module definitions. Id   on of what constitutes die area in a heterogeneous integration module, such as a current AI server processor, when in fact separate silicon die may be placed in close proximity, is creating new definition and application challenges as total warpage may be measured over a significantly larger contact area.

B. Current processors utilize polymeric substrates typical of FR4 printed circuit board manufacturing. Semiconductor die of increasing size are subject to increasing warpage during assembly and operation and the substrate is also subject to differing degrees of warpage, which is typically opposite to that of the die.

C. Silicon die warpage and substrate warpage occur for several reasons. Materials under compression during manufacturing processes may create some degree of warpage, while mismatched coefficient of thermal expansion for semiconductor die with substrate and module materials create warpage die, also.

D. Rapidly increasing power dissipation for succeeding generations of enterprise server, AI server, GPU and server processors, and ASICs increases stresses incurred due to material CTE mismatch.

E. Increased die operating and/or die junction temperature impact how specific types of liquid, gel, paste, and non-solid TIM formulations behave. Relative melting point (if relevant), thixitropicity and surface adhesion characteristics are important determinants in higher (and lower) temperatures during operation, shipping, and storage.

F. Implementation of such high-performance processor modules in single-phase and two-phase liquid immersion systems requires analysis of TIM behavior and the potential for contamination of the immersion fluids. Implementation in liquid immersion systems also requires analysis of how fluids may affect TIM materials, either through dissolution, extraction, or other actions that affect the interface material performance and reliability over time. Change is often measured by evaluating potential for deteriorating (increasing) operating temperatures via on-die sensors. Different TIM0 (and TIM1/TIM2) materials are subject to deterioration that impacts thermal performance in different ways, and different degrees, for differing reasons. This is important to recognize for material selection.

G. Placement of high bandwidth memory stacks that are typically highly temperature sensitive in close proximity (e.g., 10µ spacing) to main processor die that are the primary heat-generating sources, typically also is seen as subject to coverage with the single large TIM0 that is to protect the main processor die within the module.

These issues will be explained and evaluated and offer potential to impact TIM0 performance and reliability.

2. Experimentation Methods and Discussion

These issues will be explained and evaluated and offer potential to impact TIM0 performance and reliability.

Thermal test vehicles (TTVs) have been developed and implemented for in-situ testing of TIMs for more than thirty years and provide a useful tool for measuring performance of a given TIM0 installed on the die backside with a liquid cold plate or other hardware solution and mechanical fasteners. Output provides on-die readings for thermal resistance values in each sector of the die, where an RTD is located in each cell. Depending on the number of cells in the TTV die, a detailed mapping of temperature value is used to indicate actual in-situ performance of the TIM0 under test and the surface conditions for the die.

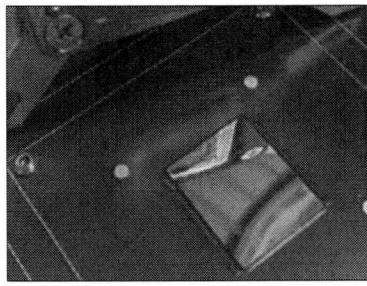 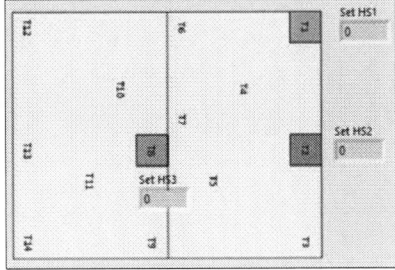

Figure 2. Example, thermal test vehicle (TTV) with cell identification map and RTD placement. (Source: Indium Corporation, October 2024.)

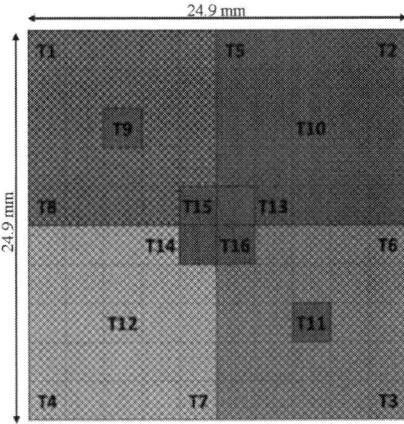

Figure 3. Example of an alternative TTV arrangement with four quadrant power zones; total die area of 620mm², 16 RTDs placed at corner edge center, as necessary. (Source: Berliner Nanotest und Design Gmbh, March 2024.)

An illustration of a test system designed to adjust to different warpage values was presented previously, by Laird Performance Materials.

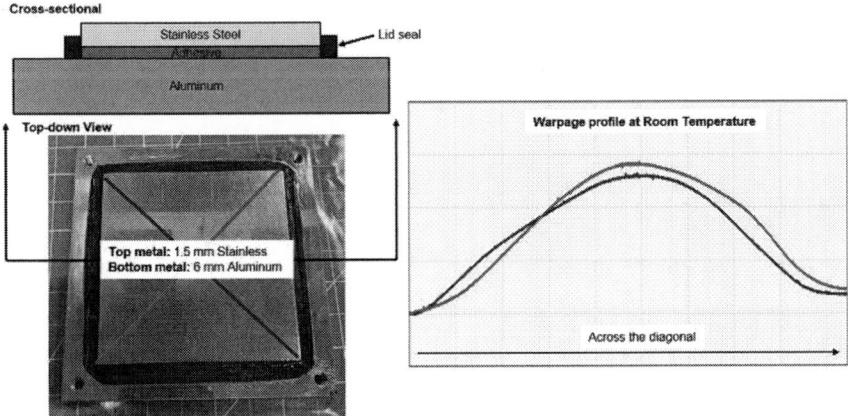

Figure 4. Example of a TTV system designed to allow control of surface warpage for measurement purposes. (Source: Laird Performance Materials, March 2024.)

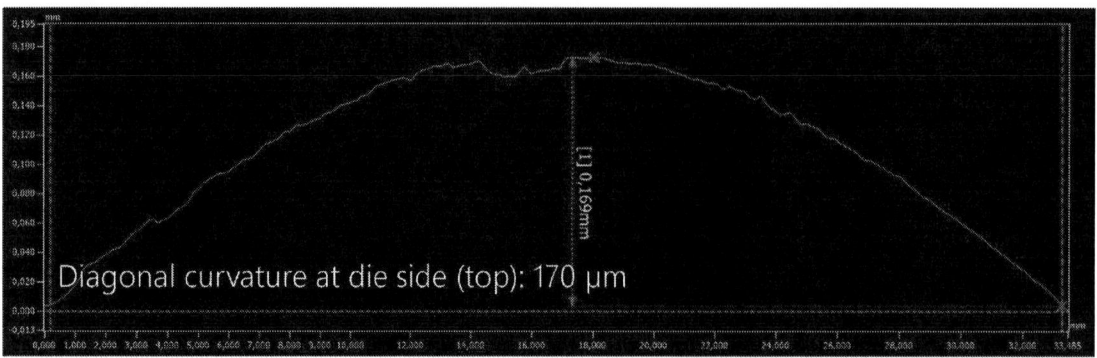

Figure 5. Example of warpage measurement across 25x25mm die. (Source: Berliner Nanotest und Design Gmbh, March 2024.)

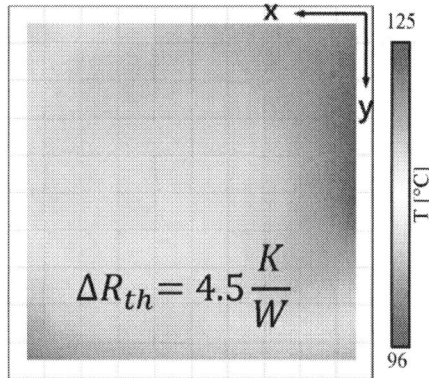

Figure 6. Example of temperature measurement across 25x25mm die, showing impact of die warpage. Note that temperature measurement is not centered and uniform, as warpage is not centered due to location of hot spots. (Source: Berliner Nanotest und Design Gmbh, March 2024.)

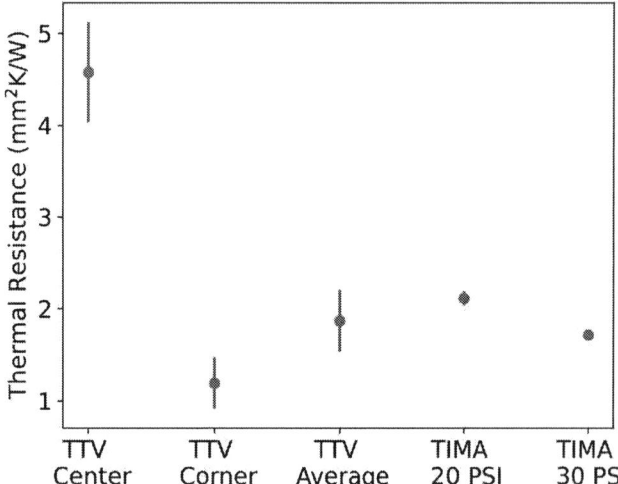

Figure 7. Example of thermal resistance data for TIM development taken with TTV, as compared to thermal resistance values taken with an ASTM D 5470-17 test stand with uniform surface finish, roughness, co-planarity, and specific clamping force applied. (Source: Arieca Inc., March 2024.)

Note that discussion of die measurement is separate from discussion of co-planarity of mating surfaces (die backside and cold plate or heat spreader or heat sink) and die surface roughness. Surface roughness of thermal solution hardware contact surface is also a separate issue that affects performance of thermal interface materials.

3. SYSTEM DESIGN IMPLICATIONS

Large-area die and concomitant die warpage are factors affecting system design, also, as significant die warpage may require a thicker TIM0 application than would otherwise be necessary. System design implications include behavior of certain types of polymeric TIMs in an immersion fluid, for either single- or two-phase system immersion cooling. Changes with fluid chemistries, if a change in fluid vendor is required for other reasons, and increased thickness of a polymeric TIM to adapt and provide adequate die surface coverage for a module with large measured die warpage, may result in erosion and deterioration of the TIM0, especially with rapid boiling in a two-phase system.

Such system-level design considerations have in certain designs required establishment of new TIM0 selection prioritization. If a vapor chamber or boiling enhancement plate is attached to the processor module with a TIM0 and mechanical fasteners and the designed structure is very thin, a mechanical pumping action may be induced during operation by changes in temperature and CTE mismatch between materials. In reported instances of polymeric TIM0 (silicone oil thermal greases) erosion in a two-phase immersion system, for example, solution to the TIM0 degradation problem has led to development of this material selection criteria, prioritized:

1. Select a solid material preform with no silicone oil (or other) carrier that will be subject to erosion and potential dissolution in the two-phase fluid with boiling action;
2. Select a solid TIM0 material that will not be subject to mechanical pumpout over time, during system operation;
3. Determine if sufficient ("good enough") thermal resistance values are achieved to meet system design goals.

Other combinations of system design may require consideration of different TIM0 material selection criteria, such as testing for potential contamination of a single-phase immersion fluid for a polymeric TIM, prior to implementation.

4. CONCLUSIONS

Large-area die warpage is adding significant challenges for thermal interface material development at the TIM0 level, as well as other TIM types. High performance AI/GPU server modules with high power dissipation, constructed in heterogeneous integration formats with large die area, made up of multiple silicon die, have measured warpage across the entire area for some processors that is greater than 200μ. Significant warpage increases TIM thickness requirement, in order to address center to die corner coverage requirements. Increased TIM thickness for certain material types requires careful consideration of thixitropicity for polymeric materials and retention mechanisms for TIM0 types such as liquid/hybrid metallic mixtures containing gallium.

Implementation of TIM0 for these large- area die with high power dissipation in system design must also take into account aspects of system thermal management that may impact TIM performance and long-term reliability. An example is the use of silicone thermal greases, as an example, in two-phase liquid immersion server systems where the boiling fluid action may cause erosion of the TIM0 from the die interface, causing deterioration of overall thermal performance.

Consideration of these die and TIM0 characteristics and system performance impacts of thermal solutions employed are important factors to be considered for system long-term reliability.

5. REFERENCES

Abo Ras, M., "Thermal Test Vehicle for Investigation of Thermal Path in Large Die Area Packages by Thermal Transient Impedance Analysis," Semi-Therm 40 Symposium, San Jose CA USA, March 25-28, 2024.

Feller, K., "Liquid Metal Embedded Elastomers (LMEE) as Low-BLT Thermal Interface Materials," Semi-Therm 40 Symposium, San Jose CA USA, March 25-28, 2024.

Jain, S.; Johnson, R.; Foltz, B.; Duffy, J.; Strader, J., "Development of a Controlled Warpage Thermal Test Vehicle," Semi-Therm 40 Symposium, San Jose CA USA, March 25-28, 2024.

Kini, G., "Thermal Landscape for Data Center GPUs," Binghamton University/IBM Research/GE Global Research Electronics Packaging Symposium 25, Vestal NY USA,

Warzoha, R.J.; Donovan, B.F., "High Resolution Steady-State Measurements of Thermal Contact Resistance Across Thermal Interface Material Junctions," Rec. Sci. Instrumn. 88, 094901 (2017); https://doi.org/10.1063/1.5001835. (Published on-line, September 19, 2017).

Novel Metal Compressible Thermal Interface Materials (TIMs) for non-planar surfaces

Miloš Lazić,
Bob Jarrett,
Dr. Ricky McDonough,
Indium Corporation, 34 Robinson Rd., Clinton, N.Y., 13323
+1 (315) 381-2037, mlazic@indium.com

SUMMARY

Metals have been used as thermal interface materials (TIMs) for many years. Even though they have a higher thermal conductivity than thermal pastes or polymeric phase change materials (PCMs), there is a need for novel metal TIMs with better thermal performance that can be used in high-power applications. In this work, we will present a new type of compressible metal TIMs.

1. INTRODUCTION

More than twenty years ago, solder TIMs (sTIMs) were introduced to the semiconductor industry. Solder preforms made of pure indium or indium alloys would be soldered between the semiconductor die and the lid (heat spreader). Thus, the solder TIM would not just play a role in taking away heat from the heat source, but it would also form a mechanical bond between those two surfaces. The downside of this approach was that metallization is needed on both surfaces (die and lid), which would add an additional step in the manufacturing process and create an extra cost.

Compressible TIMs came out as the next generation of metal TIMs to solve that problem. These compressible TIMs, primarily made from indium and indium alloys, have a special pattern to address any non-planarities between surfaces that are connecting, as well as CTE mismatch when the material is used as a TIM1 or TIM0 (TIM1.5). There is no need for any metallization on the surfaces, but pressure is needed. Compressible TIMs can achieve effective thermal conductivity of more than 20 W/m*K when they are compressed with pressures of 30 psi. While sTIMs are limited to TIM1 applications, compressible TIMS could be used as TIM1, TIM2, or TIM1.5. These metal TIMs were invented more than 15 years ago, when dies were much smaller and semiconductor applications used much less.

Pattern Type	Appearance	Thickness Expansion
HSD	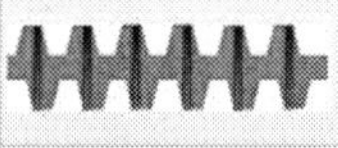	3-mils
HSHP	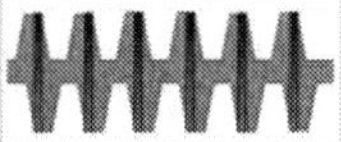	6-mils
HSK		3-mils

Figure 1. Three different types of compressible TIMs

There are three different types of compressible TIMs: HSD, HSHP, and HSK. HSK is specially made for burn-in and testing applications, hence the aluminum cladding on the bottom. Indium tends to create intermetallics with certain metals, such as copper; the aluminum layer prevents the indium from leaving any marks on the device under testing.

These three compressible TIMs have a special pattern that makes extensions of 3 mils or 75 µm (HSD) and 6 mils or 150 µm (HSHP and HSK). With that said, HSD can address up to 75 µm of warping (or up to 75 µm non-planar), while HSHP and HSK can address warping of 150 µm. Many of the semiconductor applications have large and curved dies (typically above 150 µm), and with warping and CTE mismatch issues, HSD and HSHP compressible TIMs don't have enough compliance to address those issues. That's why we invented a novel type of compressible TIM called "HSx", specially-made for the most challenging applications.

2. EXPERIMENTAL/NUMERICAL METHODS

We use the standard ASTM D-5470 system for the evaluation of TIMs (Figure 2).

Figure 2. ASTM D-5470 system

These systems can measure the effective thermal conductivity and effective thermal resistance of TIMs. However, they test TIMs under perfect conditions. Test heads are perfectly flat and parallel, and pressure is spread evenly on the testing TIM. In Figure 3, we can see how thermal resistance changes over time (48 hours) when the material is tested in the ASTM D-5470 system. The thermal performance of compressible TIMs is lower quality at the beginning because these materials plastically deform when pressurized. The plastic deformation process is a function of pressure and time. As seen in Figure 3, the effective thermal resistance of a 100 μm-thick HSD compressible TIM is two-times lower after 48 hours.

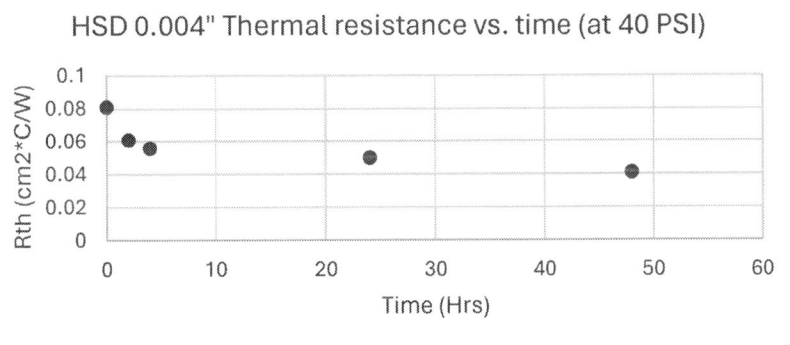

Figure 3. Thermal resistance vs. Time for an HSD compressible TIM

TIMs in a power cycling system were also tested using thermal test vehicles (TTVs), shown in Figure 4.

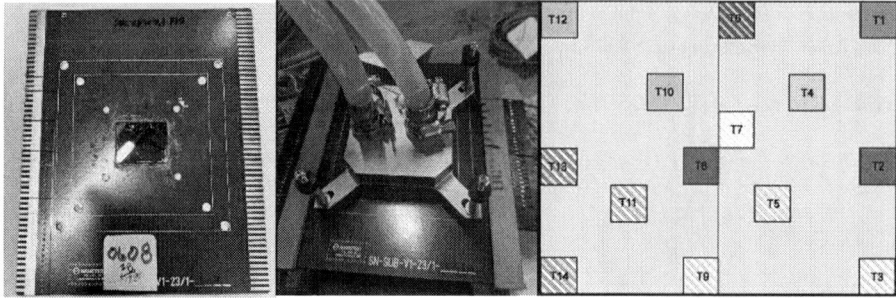

Figure 4. Power cycling TTV

TTV uses integrated circuit (IC) materials and assembly methods to create simulators. The die is configured with resistive heaters and RTD circuits to apply heat and measure temperatures at the active surface. These TTVs have 20 x 25 mm, 125 μm crowned die, shown in Figure 5, which is very similar to standard semiconductor applications and allows the testing of TIM compliance.

3. RESULTS

The compressible TIM works on the principle that two surfaces joined with metal will conduct heat very well, and as previously described with HSD, that pattern extends for 75 μm on a compressible TIM and for 150 μm on HSHP. That special pattern is applied on the flat metal shim and acts as an array of individual compressible columns. Since metal TIM materials have high thermal conductivity, interface resistance is very weakly affected by TIM thickness. The resistance decreases as the metal plastically forms to the interface surfaces either due to creep or with an initial preload. When we tested HSD in our power cycling assembly, it was obvious that this type of compressible TIM doesn't have enough compliance for a 125 μm crowned die (Figure 5.). HSD was pressurized with 50 psi and tested with 300W power.

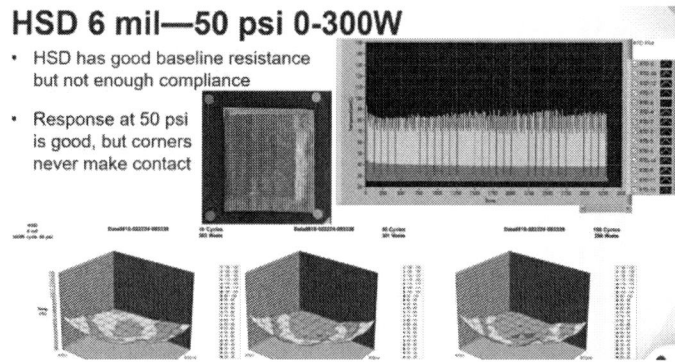

Figure 5. HSD compressible TIM power cycling

After HSD, we tested HSHP compressible TIM in the same system (Figure 6.).

HSHP 6 mil—50 psi 0-300W

- HSHP has good baseline resistance and slightly better compliance at 50 psi
- Response at 50 psi is good, but corners never make contact

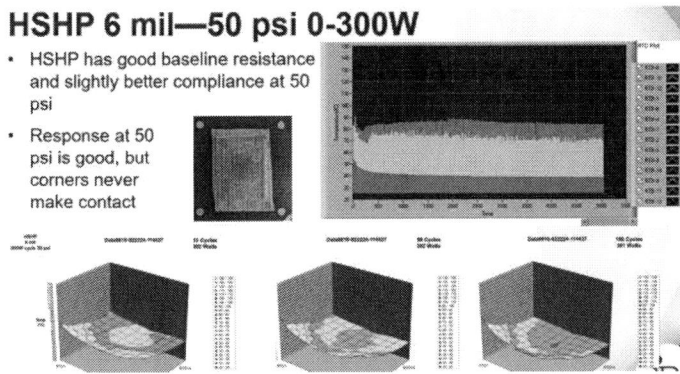

Figure 6. HSHP compressible TIM power cycling

HSHP was tested with the same pressure (50 psi) and the same power (300 W) as HSD. The results were better than HSD, but HSHP still didn't have enough compliance and the corners were unable to make contact between the die and cold plate. That was an obvious sign that a compressible TIM with higher columns that could be used for an extremely curved heat source, or in applications with warping issues, was needed. Another idea was to make a different compressible TIM that would have better performance with lower pressures of 20-30 psi, which HSx became. For HSD and HSHP, the recommendation was to compress them with pressures of 30-40 psi.

HSx, the new type of compressible TIM, has columns that are 250 μm tall (Figure 7.)

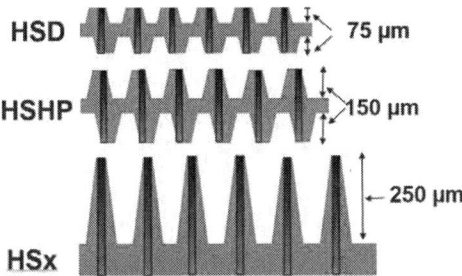

Figure 7. Difference between HSx, HSHP, and HSD compressible TIMs

HSx has a pattern applied only on one side. The flat side of HSx should be applied on the heat source, while the pattern side should meet the heat-sink or a heat pipe (in TIM1 or TIM 1.5 applications). This is an advantage compared to HSD and HSHP. Since metal and alloys will transfer heat isotropically when a flat side is covering the full surface of the heat source, this maximizes heat dissipation. To make HSx more compressible, we tried different approaches in the design of the columns. It was crucial to define an adequate number of those columns per mm^2. If we have too many of those columns, we will have more contact points, and with that, heat transfer will improve. On the other hand, with more columns, more pressure is needed, and that can be a very challenging request for some applications. Also, the angle of the columns can influence compressibility. Before choosing HSx as the final design of the new compressible TIM, more than 40 different patterns were tested. HSx can provide good thermal performance with lower pressures. Effective thermal conductivity can get up to 18 W/m*K, with pressures of 20 or 25 psi. Since this type of compressible TIM has the tallest columns, it's recommended for extremely non-planar applications, curved surfaces, and applications with a high CTE mismatch. HSD and HSHP types have a limitation in maximum thickness. Both of them can be made up to 400 μm thick. HSx can be made in the thickness of 1 mm or even higher. On the other hand, the minimum thickness of HSx is 300 μm, which is higher than both HSD (100 μm) and HSHP (150 μm). With 250 μm high columns, HSx has enough compliance to be used in more wrapped and irregular surfaces and will provide great performance with lower pressures. Figure 6 shows how HSx performed when it is power-cycled in our TTV.

HSx at 50 psi – 300W

- Burn-in <20 cycles
- Much faster than the other Heat-Springs®

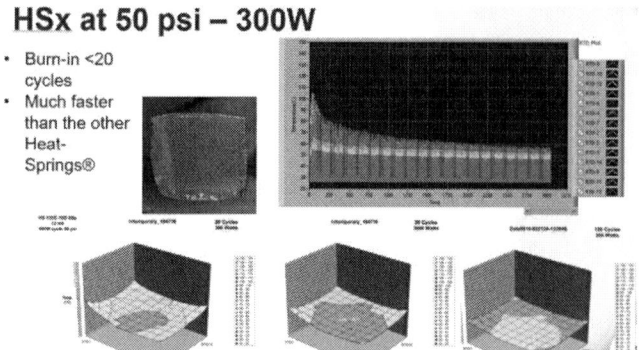

Figure 8. HSx compressible TIM power cycling

4. CONCLUSIONS

HSD and HSHP compressible TIMs are still depended on. With their lower minimum thickness, they can consistently be used between flat and parallel surfaces. The introduction of HSx, however, is a game-changer, as it can be used in the more challenging applications between curved surfaces, or in applications with high warping or CTE mismatch issues.

5. REFERENCES

1. Robert N. Jarrett, Jordan P. Ross, and Ross Berntson – "Full Metal TIMs", Power Systems Design Europe, September 2007.
2. Miloš Lazić – "Metal Thermal Interface Materials for Electric Vehicles" – E-mobility, IPC Webinar, November 2023
3. Miloš Lazić – "Thermally conductive agent for integration with Heat-Spring®" – SEMI-THERM, San Jose CA USA, March 2024
4. Tim Jensen – "Metal TIMs for Bare Die Applications" iMAPS, Boston, MA, USA, October 2024

MXene-Liquid Metal Embedded Elastomer Aggregate Composites

Mason Zadan[1], Yafeng Hu[1], Jeremiah Lipp[2], and Carmel Majidi[1*]

[1] Mechanical Engineering Department, Carnegie Mellon University, 5000 Forbes Ave, Pittsburgh,
PA 15213, USA
[2] Materials and Manufacturing Directorate, Air Force Research Laboratory, 5135 Pearson Road,
Wrigh-Patterson AFB, OH 45433, USA
* Corresponding Author, email: cmajidi@andrew.cmu.edu

Extended Abstract

SUMMARY

Conductive fillers embedded within curing on non-curing elastomers or greases have been used extensively as thermal interface materials (TIM) between power modulus, processors, etc. and heat sinks. Traditionally, these filler materials consist of rigid materials such as aluminum or copper particles. More recent work has introduced soft composite materials such as liquid metal (LM) to improve thermal and electrical performance. This presentation introduces $Ti_3C_2T_x$ MXene sheets as a coating onto LM micro-droplets to improve thermal and electrical performance at low bond line thicknesses (BLT) through induced aggregation. This presentation introduces the fabrication method, microscopy, microCT imaging, and thermal and electrical performance of these composite materials embedded within a silicone oil matrix. Results indicated that using tip sonication, MXene sheets can be exfoliated and wrapped around LM droplets through Van der Waals forces causing large clusters 3 orders of magnitude larger then without the introduction of MXenes. When the clusters are combined in a silicone oil even at low volume fills of 25 % LM and 1 % MXene there was a 4.6 times improvement over the 25 % by volume LM control at a 50 μm BLT. This work highlights how small amounts of 2D materials can be added to design nano-composites with vastly different properties and morphologies.

1. INTRODUCTION

In this talk, 2D MXene nanosheets are introduced into liquid metal embedded elastomer (LMEE) thermal interface materials (TIM) to wrap around the LM droplets to act as a coating and aggregation mechanism for the LM droplets. This talk explores the fabrication method for these composites along with how these aggregated MXene-LM clusters improve thermal and electrical conductivity at low bond line thicknesses (BLT). More generally, the need for these improved TIMs comes the high thermal resistance at the interface between heat sinks and electronic components due to surface roughness and poor contact causing poor heat transfer and decreasing device performance. TIMs act as a solution to this issue, bridging the poor interface with high thermal conductivity deformable materials such as thermal pads or greases. In particular, curable and non-curable thermal elastomers or greases are exceedingly popular for their ability to be deposited and compressed to small BLTs while adhering the components together. After depositing the paste, pressure is applied to decrease the BLT improving contact and heat transfer. As the BLT decreases, rigid conductive particles such as copper or aluminum embedded within curing on non-curing elastomers begin to act as bridges between the two contacts increasing heat transfer. One issue is that these rigid materials cannot deform as the TIM is compressed. To address this, soft conductive fillers such as LMs including EGaIn and Galinstan have been introduced.

The addition of LM into soft silicone composites removes the rigid-soft interface that exists in most soft composite materials. These materials often have stress concentrations occur at this interface. Recent work has

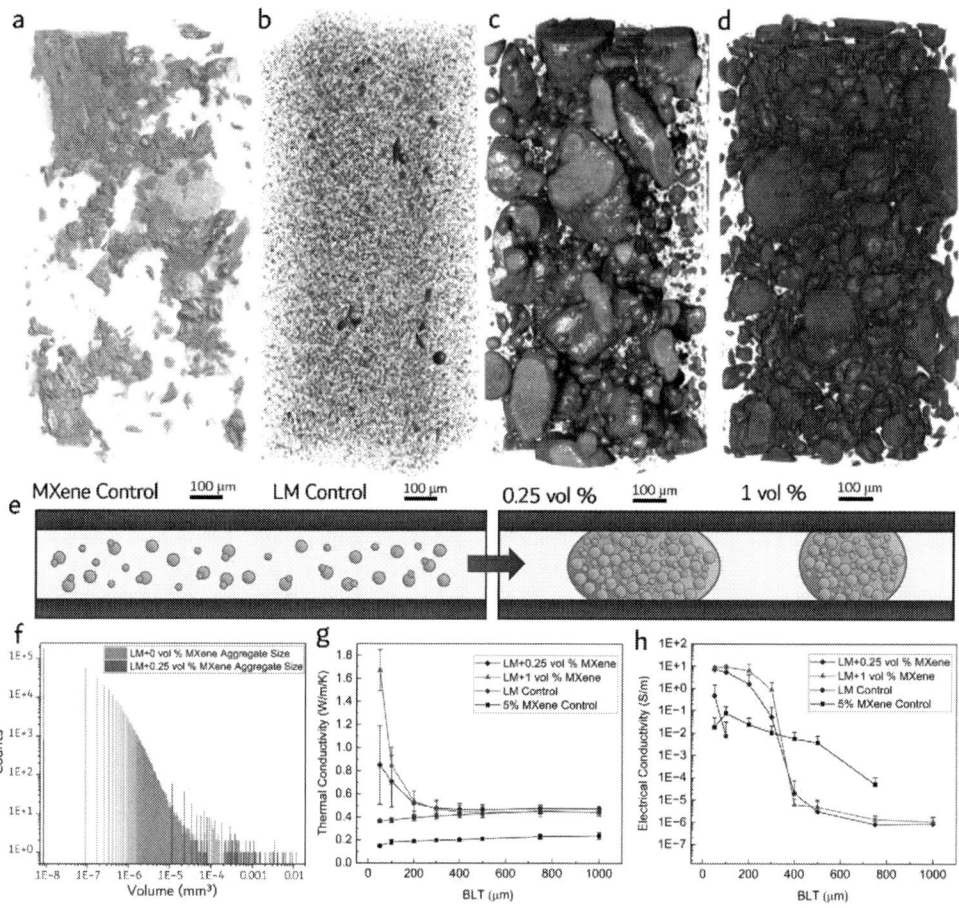

Figure 1: a-d, MicroCT reconstructions of samples for MXene control, LM control, 0.25 vol %, and 1 vol % MXene content. **e,** Diagram highlighting the influence of aggregate size on percolation threshold. **f,** Graph highlighting the difference between MXene volume content and aggregate size. **g,** Thermal conductivity vs BLT data for various samples. **h,** Electrical conductivity vs BLT data for various samples.

shown that LM fillers are able to deform along the axis of strain with the bulk material exhibiting no degradation in mechanical performance[1]. This material architecture has become useful as the deforming LM droplets create excellent contact between the metal pads when compressed. Current work in the field produces composite materials with evenly distributed fairly monodisperse droplets that are statistically uniform and cannot aggregate or self-assemble[2].

To increase thermal and electrical performance of these materials, one potential future direction is to increase heat transfer by forming the individual droplets into self-assembled connective aggregates. To explore this concept, this talk introduces a method to wrap 2D MXenes sheets around the surface of LM droplets to functionalize them and make them sticky. This allows these droplets to aggregate and form large clusters three orders of magnitude larger than the constituent droplets for improved thermal and electrical performance when added to a silicone matrix.

2. EXPERIMENTAL METHODS

MXene-LM composite synthesis was begun with the addition of EGaIn into a DMSO:DI water solution. The LM was broken up into droplets on the order of 1 micron through focused tip sonication for 1 hour at a high 40

% amplitude. Once the LM droplets were fully sonicated, $Ti_3C_2T_x$ MXenes suspended in a DMSO: water solution were then added corresponding to the desired ratio. This mixture was then sonicated a second time at 20 % amplitude for 30 minutes to maximize the Van der Waals interactions and allows for coatings of the MXenes onto the LM. Next, the contents were vacuum filtrated to remove the solvent. The material in a paste form is then transferred to a mixing cup, where high viscosity silicone oil is added (60,000 cSt @ 25 °C). The contents were then planetary mixed for 5 minutes at 2000 rpm for 3 cycles inducing the large aggregates on the order of 300 microns. For microCT measurements, material was injected into a micropipette tube and imaged with a pixel resolution ~ 1.4 µm. Manual reconstructions and intensity calibrations were conducted. Thermal and electrical measurements were conducted on a Nanotest TIMA5, with BLT dependent data being recorded for thermal and electrical performance.

3. RESULTS

MXenes are a 2D electrically conductive material that include transition metal carbides, nitrides, and carbonitrides with alternating layers capped with a termination layer. One of the most common forms of MXenes are $Ti_3C_2T_x$ MXenes with intercalated Ti and C layers and -F, -OH, and -O acting as the termination layers[3]. MXenes are prepared from a MAX precursor (Ti_3AlC_2 in this case) with HF used for etching followed by sonication for exfoliation. MXenes exhibit high electrical and thermal conductivities along with large negative zeta potentials when in proper solvents, making them able to bond well through electrostatic interactions[4]–[6]. In this presentation we introduce a method for these high aspect ratio MXenes sheets to coat and wrap around LM droplets at the micron scale. These sticky functionalized particles are then shown to self-assemble into larger aggregates inducing macroscopic changes in the material morphology. Thermal and electrical results will be discussed highlighting the influence of MXene filler on performance.

Samples are prepared with 0.25 vol % and 1 vol % MXene along with LM and MXene control samples (Figure 1a-d). All samples with LM were prepared with 25 vol % LM. The LM control sample microCT images exhibited small uniform droplets on the order of 10^{-7} mm^3. With the addition of only 0.25 vol % MXene during synthesis, the aggregates were then shown to be on the order of 10^{-4} mm^3, a 3 order of magnitude increase in size (Figure 1e). The particle size distribution graph is shown in Figure 1f. The MXene control sample microCT images indicated the formation of large MXene sheets (Figure 1a). As the polar aprotic solvent (primarily DMSO) was removed, the MXenes lost surface stability causing uneven charge distributions and a crashing out of the MXenes out of the solution. LM and MXene control sample thermal conductivities stayed generally constant vs BLT down to 50 µm. With a small amount of MXenes added to the LM composite the thermal conductivity increased exponentially starting at around 300 µm, near the size of the aggregates themselves. The maximum recorded thermal conductivity was 1.67 ± 0.18 W/m/K. The electrical data showed similar exponential trends with a decrease in BLT for the MXene-LM composites, with a 7 order of magnitude increase in electrical conductivity between 1000 to 50 µm.

4. CONCLUSIONS

In conclusion, this presentation will introduce 2D MXenes into LMEE silicone based TIMs to introduce a novel method to induce the aggregation of LM droplets within an elastomer matrix. This talk explores a potential new direction for designing TIMs with self-assembled non-statistically uniform dispersions to decrease the filler material volume percentages needed for appropriate heat transfer. This talk introduces a reliable coating method of MXene sheets onto LM droplets comparing the influence of MXene volume content on the bulk material properties. Once planetary mixed, the filler material aggregates into large spheres on the order of a few hundred microns. Extensive microCT imaging is conducted under various MXene volume percentages with results indicating large aggregation formation with the inclusion of as little as 0.25 volume percent MXene. Thermal and electrical performance increases exponentially as BLT is decreased highlighting the role that the individual aggregates play in improving thermal and electrical performance.

5. REFERENCES

[1]   M. D. Bartlett *et al.*, "High thermal conductivity in soft elastomers with elongated liquid metal

inclusions," *Proc. Natl. Acad. Sci. U. S. A.*, vol. 114, no. 9, pp. 2143–2148, Feb. 2017, doi: 10.1073/pnas.1616377114.

[2]   C. Pan *et al.*, "A Liquid-Metal–Elastomer Nanocomposite for Stretchable Dielectric Materials," *Adv. Mater.*, vol. 31, no. 23, p. 1900663, Jun. 2019, doi: 10.1002/adma.201900663.

[3]   M. Naguib, M. W. Barsoum, and Y. Gogotsi, "Ten Years of Progress in the Synthesis and Development of MXenes," *Adv. Mater.*, vol. 33, no. 39, pp. 1–10, 2021, doi: 10.1002/adma.202103393.

[4]   Y. Liu, W. Zou, N. Zhao, and J. Xu, "Electrically insulating PBO/MXene film with superior thermal conductivity, mechanical properties, thermal stability, and flame retardancy," *Nat. Commun.*, vol. 14, no. 1, pp. 1–9, 2023, doi: 10.1038/s41467-023-40707-x.

[5]   M. Safarkhani, B. F. Far, Y. S. Huh, and N. Rabiee, "Thermally Conductive MXene," *ACS Biomater. Sci. Eng.*, vol. 9, no. 12, pp. 6516–6530, 2023, doi: 10.1021/acsbiomaterials.3c01420.

[6]   K. Maleski, V. N. Mochalin, and Y. Gogotsi, "Dispersions of Two-Dimensional Titanium Carbide MXene in Organic Solvents," *Chem. Mater.*, vol. 29, no. 4, pp. 1632–1640, 2017, doi: 10.1021/acs.chemmater.6b04830.

Overcoming Thermal and Electromagnetic Challenges with a Single Dispensable Multifunctional Material for Consumer Electronics

Arun Raghupathy[1], Robin Huang, Judy Guu, Sathya Kaliyamoorthy, Jay Lee, RZ Chiang, *Google LLC*

and

Himanshu Pokharna, *Deep Materials Inc,*

Abstract— This paper discusses exploratory work done to address challenges related to thermal and electromagnetic performance in consumer electronics by incorporating a new multifunctional material. Specifically, we describe multi-disciplinary qualification efforts of using a dispensable material that combines thermal and electrical conductivity to address thermal and desense challenges in these products.

Keywords—desense, thermal gel, electrically conductive, dispenseable

I.   INTRODUCTION

Consumer Electronics (CE) and smart electronics products are used for entertainment, communication, and information. They are used in many ways in the home including leisure and work. Some examples of these products include: Smart speakers, streaming devices, display speakers, Mobile phones, Game consoles, Computer hardware and software, Digital cameras and camcorders, and even home appliances.

CE is a multibillion-dollar industry that is constantly growing and innovating. One of the main trends is miniaturization, which presents several challenges:

Heat dissipation: Miniaturized components are often tightly packed, which can increase heat generation. This can affect device reliability and performance if not properly managed.

Signal integrity: As components are placed closer together, the risk of signal crosstalk and electromagnetic interference (EMI) increases, affecting device performance, especially in critical electronics.

Durability and reliability: Smaller components may be more susceptible to physical damage, such as mechanical stress or vibrations. In an era of ever-lower cost expectations from consumers, mechanical tolerances are trending higher, putting even more pressure on reliability.

Manufacturing complexity: Miniaturization can increase the complexity of manufacturing processes, making quality control and reliability more challenging.

Assembly: Miniaturization increases the demand for precise assembly of thermal and grounding solutions. Currently, much of this is done manually, which can lead to unacceptable yield issues due to operator error.

II.   BENEFITS OF DISPENSABLE ELECTRICALLY CONDUCTIVE THERMAL GEL

In electronic products, often a variety of components and solutions are deployed to solve various problems. For example, Desense, or radio frequency (RF) desensitization, is a phenomenon that can affect electronic products, particularly wireless receivers. It occurs when the sensitivity of a wireless receiver is reduced by electromagnetic interference (EMI), which can come from external sources or from within the product itself. Desense is typically managed by utilizing products like fabric over foam (FoF) or other conducting gasket materials. In fabric over foam, an electrically conducting cloth is wrapped over a soft foam and then cut to the right size and incorporated in a product where one needs electrical contact between two surfaces to minimize desense. This FoF or gasket is specifically designed to ensure electrical conductivity between two surfaces, commonly used in electronic devices or equipment where minimizing electromagnetic interference (EMI) is essential.

Although these FoF products are the workhorse of the industry, there are several challenges that they pose. Oftentimes they are assembled by hand and require precise placement. Vendors have developed specialized automation equipment to apply these but they remain expensive. Also, their handling requirements drive certain physical form-factors of these components and they are often 2 mm or more in thickness. This limits design flexibility to achieve tight integration. Furthermore, these products have very low thermal conductivity and do not help with transfer of heat between the two components they straddle. Also, in some specific areas using FoF adjacent to silicone-based thermal material can potentially cause desense issues due to oil migration from the thermal material.

These days many of the consumer electronic systems are moving to dispensable TIMs. That is because of several advantages they afford over traditional pre-formed thermal pads, especially in electronics and high-performance, and high volume cost sensitive consumer electronics applications. Some key benefits include:

Reduced Risk of Damage: Dispensable materials conform to the shape of the gap in the product, reducing stress on fragile components and packages.

Customization: Dispensable TIMs can be applied in precise amounts and shapes tailored to unique designs, improving contact and thermal transfer efficiency.

---

[1]Arun Raghupathy is with Google LLC in the Platform and Devices organization (e-mail: araghupathy@google.com).

Automated Application: Automated dispensing equipment improves consistency, reduces human error, and speeds up production, crucial for mass production. This is increasingly important as labor costs rise globally.

Filling Gaps and Irregular Surfaces: Dispensable materials conform better to irregular surfaces and fill gaps compared to solid pads, ensuring optimal contact for improved heat dissipation. Pads are fundamentally two-dimensional, whereas dispensed materials easily take 3D shapes.

Reduced Waste: Applying material directly to the needed area reduces waste compared to pre-cut pads or over-application of grease.

Improved Thermal Performance: Many dispensable TIMs have excellent thermal conductivity, improving heat transfer and enhancing component performance and longevity.

Reworkability: Many dispensable TIMs allow for easy rework without damaging components.

Compatibility with Complex Geometries: Dispensable TIMs can be used in complex assemblies where traditional TIMs may not be practical.

No Compression Rate Requirement: Fabric-over-foam requires a specific compression rate for minimum impedance, introducing force that can complicate designs. Dispensable TIMs do not have this requirement.

Overall, dispensable TIMs offer greater flexibility, efficiency, and performance in thermal management for electronics and high-heat applications. The application of this material spans various industries, including aerospace, defense, medical devices, and space electronics.

III. EXPERIMENTAL METHODS - DOWNSELECTION

A number of potential materials were investigated for electrical impedance prior to downselection.

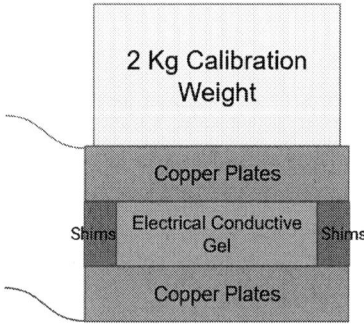

Fig.1. Schematic representation of setup used for testing electrical impedance

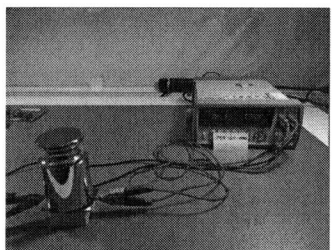

Fig.2. Actual of setup used for testing electrical impedance with 4-wire measurement

The requirements for downselection were lowest electrical impedance (0 ohm) and reasonable thermal conductivity (2 W/mK). Instead of testing them directly in the product, a simplified approach was created. This simple setup shown in Fig.1 comprises two 50 sq.mm copper plates set at a fixed distance. Distance represents the distance between two surfaces in a product. The weight is representative of the clamping force on the product as well. Plastic shims are used for repeatable measurements of different candidate materials.

Conductive FoF was tested in the same setup as baseline and its impedance (0.004 Ohm) was set as the target to achieve with different material formulations. However, the best case that was achieved was 0.269 Ohms with the dispensable electrically conductive thermal gel (EC gel). While not as perfect as conductive foam, this material offered other benefits, such as higher thermal conductivity and better dispensability. This material was selected for system-level validation.

IV. SYSTEM-LEVEL SETUP

To verify the performance of the new gel material, a Google hardware product used for streaming online content is used as the test platform. This product demands both strong wireless performance for 4K streaming and efficient heat dissipation. Fig. 3 shows the exploded diagram of the product; as indicated in the figure, the conductive foam and thermal gel of POR (main) configuration are replaced by the EC Gel in the DOE (experiment) config. The EC gel is dispensed over the EMI shield can over the SOC package and it is also dispensed on the EMI shield in the area of the conductive foams. The dispensed EC Thermal gel can be seen in Fig 4 as black areas over the EMI shield can.

To get confidence in the performance of the material, about 29 units were built in the factory with processes used in mass production. The performance of the 29 devices was compared to a larger volume of 800 devices which used the POR configuration. Device performance was tested in terms of Thermal, Antenna and Desense. The system was also tested for reliability with multiple drops, heat soak and thermal cycling tests..

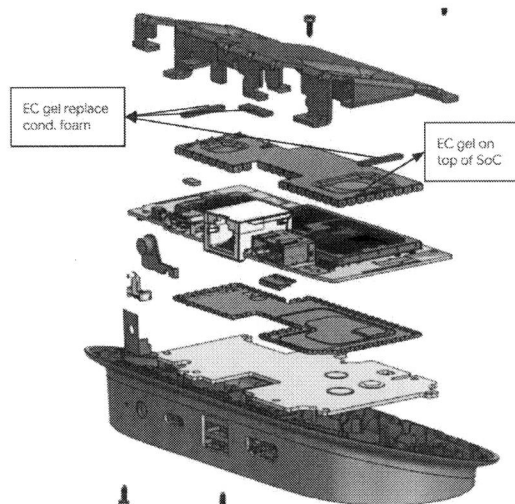

Fig.3. Exploded view showing EC Thermal Gel replacing both Thermal Gel and Conductive Foam in 4 locations.

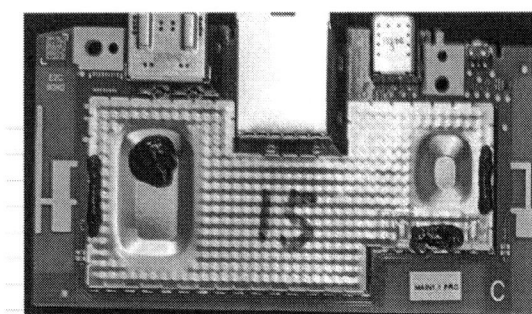

Fig.4. Dispensed EC Thermal Gel on EMI Shield Can

Thermal Performance

Table 1 shows the thermal performance collected from multiple devices in the factory running the same workload in a controlled testing environment with the only change being in the material used.

Table 1. Comparison of Thermal Performance of POR vs EC Gel Configuration

Configuration	POR	DOE
Thermal Solution	Thermal gel K=4 W/mK	EC gel K=1.8 W/mK
Quantity	40 devices	10 devices
Surface temperature (°C)	~54°C	~ 54°C
SoC temperature (°C)	~77°C	~82°C

As the results show, there is a slight increase in the SOC Tj and not much of a change in the surface temperature of the device. Since the higher SOC Tj is still within acceptable specifications, it was considered acceptable for thermal performance.

Desense Performance

Google TV streamer relies on antenna performance to have a reliable connection with the router. When an aggressor with high noise is turned on, that noise might interfere with wireless communication, which causes connection problems. Sensitivity search test is conducted to measure such interference level, and the desense level was defined as the difference between "baseline sensitivity when aggressors are turned off" and "sensitivity when an aggressor is turned on". Desense upper limit is defined as 6dB to ensure wireless performance can support smooth 4K 60fps video streaming.

As shown in Fig. 3, DOE config uses EC gel to replace the conductive foams, which are grounding and shielding solutions to ensure good wireless performances. This change was evaluated by measuring HDMI desense, a key performance indicator for our Google TV streamer product. As shown in Table 2 and Figure 5, the DOE thermal gel configuration (29 units) resulted in a 2-3 dB higher desense value compared to the original configuration (10 units). Despite this difference, both configurations meet the 6dB product requirement.

Table 2. De-sense performance comparison

Phase name (TH DOE1 29x)	Quantity	Min	Mean	Median	Max	STDDEV *
1080P59.94Hz10bit_Desense_2412_RX2	29	0.3	2.48	3.87	5.55	1.67
1080P59.94Hz12bit_Desense_5560_RX2	29	0.11	2.16	2.2	6.02	1.56
1080P60Hz10bit_Desense_2412_RX2	29	0.32	2.44	3.86	5.03	1.60
4k60Hz8bit_Desense_2412_RX2	29	0.41	2.57	4.03	6.31	1.96

Phase name (Main1 10x)	Quantity	Min	Mean	Median	Max	STDDEV *
1080P59.94Hz10bit_Desense_2412_RX2	10	0.82	1.61	1.7	3.98	0.92
1080P59.94Hz12bit_Desense_5560_RX2	10	-0.13	0.47	0.32	1.78	0.64
1080P60Hz10bit_Desense_2412_RX2	10	0.71	1.44	1.27	3.87	0.96
4k60Hz8bit_Desense_2412_RX2	10	0.25	1.00	1.11	2.25	0.58

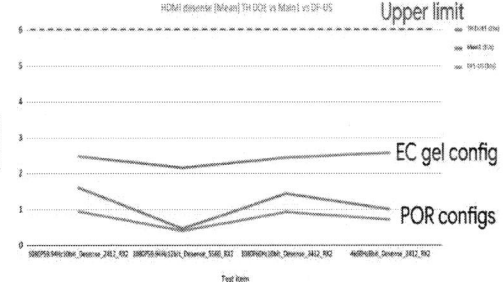

Fig. 5. Comparison of De-sense in POR Configuration vs EC Gel Configuration

Antenna Performance

Antenna performance is studied at both 2.4GHz and 5 GHz frequencies. Antenna performance is critical to user experience. Beyond the antenna's design, its performance is highly sensitive to environmental variations, particularly those related to grounding. Tests are typically done in shielding boxes. Given that assembly tolerances can introduce measurement deviations, acceptable upper and lower limits for antenna performance fluctuations within the shielding box are established. Acceptable performance fluctuation limits (10 dB) are based on anechoic chamber testing for this product. While meeting these limits, the goal is a Cpk of at least 1.33 or 1.66.

decrease, it continues to adhere to the performance limits. The Cpk values for both configurations remain comparable (POR: 2.55, DOE: 2.9).

Consequently, the DOE configuration not only fulfills the antenna performance specifications but also satisfies the Cpk requirements for mass production in a factory setting.

2442MHz Tx1

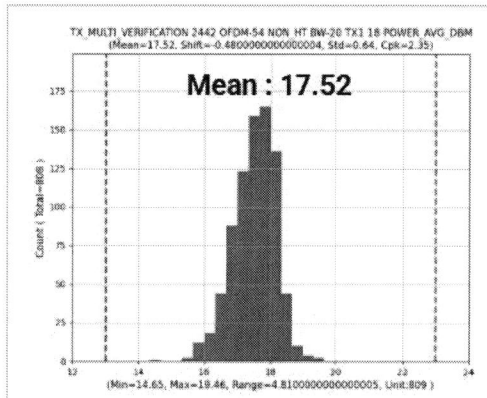

POR build (>800)

DOE build (=29)

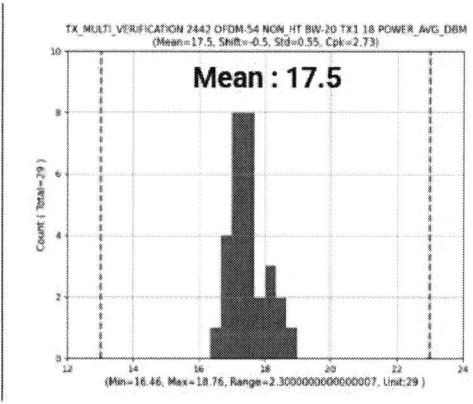

Fig.6 Antenna performance at 2.4 GHz compared 29 units with EC Gel vs 800 POR units

5500MHz Tx1

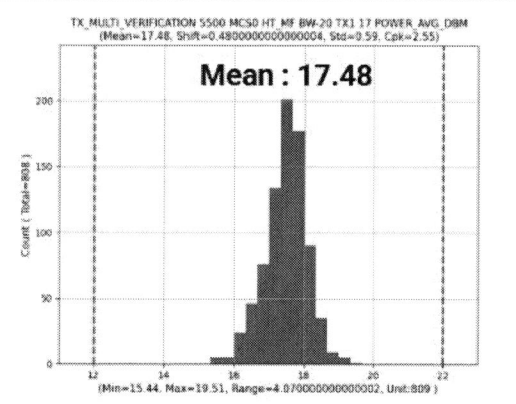

POR build (>800)

DOE build (=29)

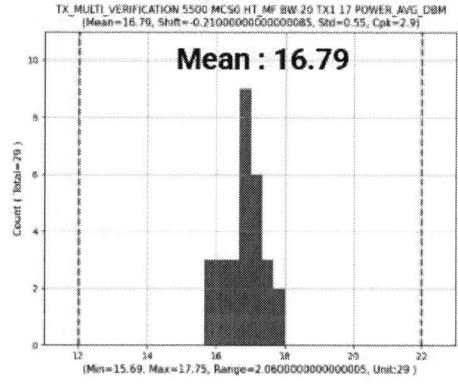

Fig.7 Antenna performance at 5 GHz compared 29 units with EC Gel vs 800 POR units

As illustrated in Figure 6, the DOE configuration, which substitutes conductive foam with EC gel, demonstrated that all 29 samples tested remained within the defined upper and lower limits in the 2.4 GHz band. Moreover, the Cpk performance is virtually indistinguishable from that of the POR configuration, which was based on a sample size exceeding 800 units (POR: 2.35, DOE: 2.73).

Similarly, as depicted in Figure 7, although the median value for the DOE configuration exhibited a slight

Drop Test Performance

One of the main concerns of using an electrically conductive thermal gel is migration from its original location of dispense and causing a short-circuit. This was studied by doing multiple drop tests (3 consequent drops) with functional and visual examinations after each drop test. The drop condition is with a height of 1m on wooden surface with 10 drops in each round. The gel after curing attains a thick paste like consistency that prevents it from migrating from its original location. Even after three rounds of drop, no migration of the EC gel was observed showing that the risk of short-circuit was very low. In spite of these results, certain design changes can also help in making sure there is no migration such as limiting gaps between mating surfaces to less than 0.5 m, using concave shapes on the shields locally to avoid gel migration and using foams as dams around the EC gel. Besides drop tests, other reliability tests such as heatsoak and thermal cycling were conducted. Desense and antenna performance remained within specified limits (less than 2 dB increase) after the reliability tests.

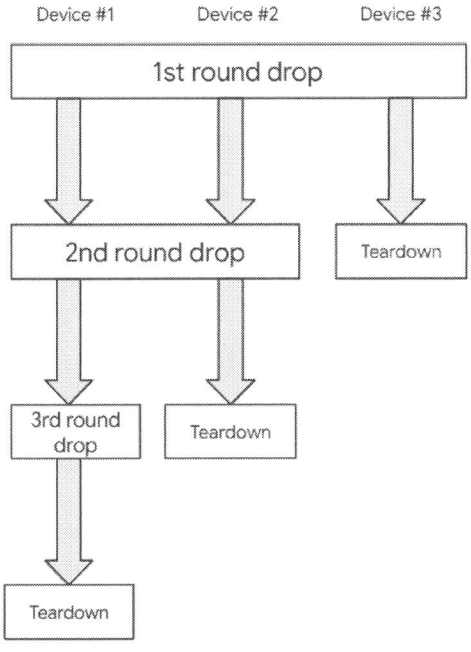

Fig. 8 Schematic showing the drop test study

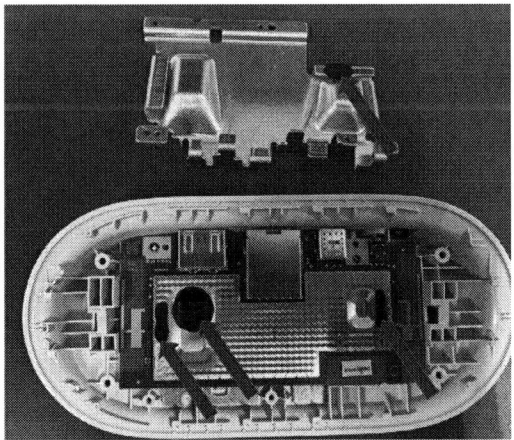

Fig. 9 Teardown after drop test showing that the gel shows no sign of migration from its original location.

V. CONCLUSION

The use of electrically conductive thermal gel provides many benefits and design flexibility for consumer products. This work was exploratory in nature and does not indicate future direction of design in the products. Controlled and statistical testing for thermal,de-sense and antenna performance shows that the material can serve as a good alternative to springs and conductive foams used for grounding purposes. The initial intent of the study was to replace the thermal gel with EC gel only in the locations of the thermal gel since the heat transfer path and the grounding path were the same. However, with some additional parametric testing, it was identified that antenna and desense performance gets impacted significantly if the EC gel is not used in the correct locations as needed for desense and antenna performance.

VI. ACKNOWLEDGEMENTS

The authors would like to thank the engineering leadership within Google Platform and Devices organization and other personnel including support from Deep Materials who made this controlled study possible to qualify the EC gel material for use in future products.

Thermal Interface Materials for Power Electronics Application

Debabrata Pal, Mark Severson

Collins Aerospace, Rockford, IL, US

debabrata.pal@collins.com

Abstract

Power electronics cooling has always been dependent on thermal interface materials (TIM) between semiconductor modules (Power modules, IGBT's, Diodes, etc) and cold plates. Increased product power densities and increasing component heat fluxes have magnified the importance of TIM materials. In high power liquid cooled applications, the TIM interface is often the highest thermal resistance that is controllable by the designer. Therefore, selection of an appropriate TIM is key to the success of the cooling design. Selection of an appropriate TIM generally begins by evaluating vendor literature which typically provides performance data collected using ASTM test methods. The ASTM methods provide a great starting point for performance comparisons but cannot compare all the 'real world' variations in mounting methods present in modern power electronics cooling designs. This paper helps the designer consider other non-ideal factors when selecting a preferred TIM. The ASTM method utilizes a small contact area and closely defines flatness and surface finish between the two mating heat transfer surfaces. This is a great starting point, but many applications have much larger contact surfaces, non-ideal flatness and non-uniform contact pressures. The designer generally has control of surface finish and flatness of the cold plate, but the designer has no control of the shape of the commercially available power electronics modules.

Introduction

For many applications the power electronics modules are large and can be moderately non-flat creating peaks and valleys in the mating surfaces. This defines a micro and macro thought process for TIM evaluation. The literature does a very good job of helping designers understand the micro imperfections where the TIM fully fills the small gaps between mating surfaces. The macro problem, especially on large modules is not commonly discussed in the literature. In real applications, modules can be non-flat as much as 0.005 in or more and thus the designer must perform due diligence to ensure the selected TIM

product is suitable for both micro and macro imperfections.

Background

This paper presents simple test and evaluation methodologies allowing a designer to both qualitatively and quantitatively evaluate a selected TIM on real applications involving large modules, imperfect flatness and imperfect contact pressures. The material presented is based on actual results with a commercially available module and a custom cold plate.

The first part of this methodology is to determine the flatness of the power electronics module and cold plate. The flatness and surface finish of the cold plate are controllable by the designer, but the module parameters can vary by supplier and even part to part. The flatness profile of a module will vary when measured in free (unbolted) and bolted conditions. Module baseplates can be convex, concave, or wavy and the magnitude of these variations can influence the TIM strategies. Clearly a 0.002-inch-thick TIM cannot fill a 0.005-inch gap between the module and cold plate so test data on TIM thickness and gap thickness is key to success.

The second part of the test methodology is to evaluate and test the robustness of the TIM interface. It is impossible to describe all the combinations of parameters leading to robustness and hence the literature is inadequate in this area. It is relatively easy to evaluate robustness of TIM performance between two well controlled blocks of aluminum but more difficult to evaluate robustness of an interface between a large power electronics module and a cold plate. An excellent TIM material can underperform if mated with a power module that does not remain flat. An excellent power module can underperform if mated with an inappropriate TIM.

Large power modules are constructed of multiple layers of dissimilar materials soldered together. Si or SiC dice are commonly soldered to the top of a direct bond copper stack up with thin copper attached to the top and bottom of an insulating material. The bottom direct bond copper layer is typically soldered to the top of base plate. Base plates can be copper, or CTE controlled materials. During manufacturing, the solder joints are stress free during the soldering process and become stressed when cooled to room temperature. The solder can creep over time and prior to use by the consumer. The shape of the baseplate changes during the manufacturing process due to solder creep and resulting baseplate flatness profiles can be a concave surface [1], convex surface, or wavy surface. Base plate contour and thermal interface visualizations are reported in previous studies [2,3].

Each material in the component stack up has a different coefficient of thermal expansion (CTE) hence the shape of the module baseplate can change in service at high and low temperatures as the solder stresses are relieved and reintroduced. This implies that module health is influenced by module temperature and thermal cycling. For a thermal interface to be robust, the thermal resistance should be almost constant at both high and low temperatures. Thermal cycling of the module/TIM/cold plate assembly can expose weaknesses in the thermal connection which typically manifests itself as a module overtemperature. This doesn't mean the TIM is bad, it might mean that the module has changed shape. Sometimes the TIM flows in a non-satisfactory manner so sometimes thermal cycling exposes an inappropriate TIM selection.

Evaluation

This paper helps designers evaluate the suitability of a TIM material using some simple evaluation methods. Methods are presented in qualitative summaries and data trends, rather than presenting performance data quantitively. Recommendations are presented allowing designers to wisely choose TIM materials appropriate to their application how to wisely apply the materials. Application methods add variability to the TIM performance, these methods include, stencils, trowels, freehand etc. For phase change TIM products, a burn-in may be necessary to melt and set the material to achieve desired results.

The structure of a typical power module is shown in Figure 1. Note the variety of materials hence a variety of CTE values.

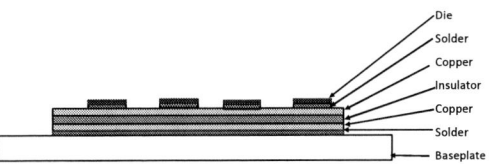

Fig.1: Typical Power Electronics Stack up Structure

Figure 1 is ideally presented because it shows the layers as perfectly flat. Real devices, especially large devices are rarely perfectly flat. The baseplate that interfaces with the cold plate can be flat, convex, concave, or wavy.

Thermal interface resistances for flat surfaces are shown schematically in Figure 2. This resistance path is for a situation where there is no metal-to-metal contact and heat is conducting across the thickness of the thermal interface material.

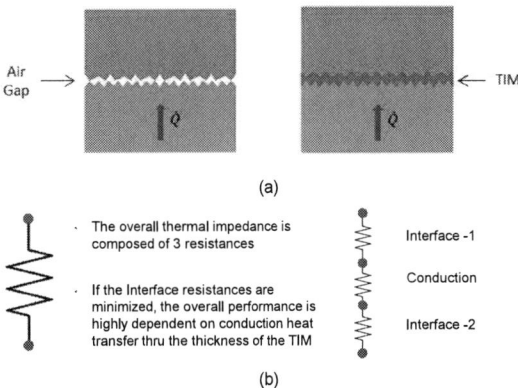

Fig.2: Thermal interface

Thermal impedance is a function of bond line thickness and thermal conductivity of the thermal interface material. This is shown on Figure 3. The slope of this curve is basically equal to $1/k$, where k is the thermal conductivity of the thermal interface material.

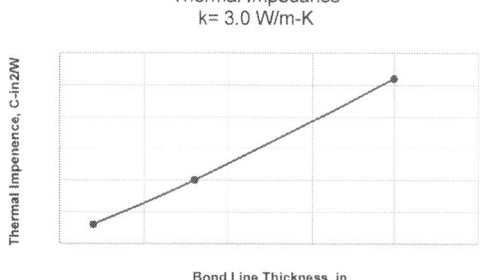

Thermal Impedance
k= 3.0 W/m-K

Thermal Impenence, C-in2/W

Bond Line Thickness, in

Fig.3: Thermal interface

The above figures are commonly described in the literature and illustrate the micro problem when a flat surface interfaces with another flat surface. This is the data commonly associated with the ASTM test.

In real life the power module baseplates are not necessarily flat and can be convex, concave or wavy. These imperfections are the macro problem that exists in addition to the micro surface roughness problem.

Figure 4 illustrates a convex power module baseplate mating with a flat cold plate. Mechanical fasteners illustrated by the red lines attach the component to the cold plate. This arrangement is preferred from a TIM perspective because tightening the fasteners flattens the module baseplate and ensures contact pressure in the middle of the module where predictable heat transfer is needed the most. The disadvantage of this arrangement is if the convex shape is too large, tightening the fasteners stresses the solder joints and reduces module life.

Fig.4: Convex Baseplate

Figure 5 illustrates a concave module baseplate. This arrangement creates a gap in the center of the interface that if too large will greatly degrade the performance of the TIM interface. Tightening the mechanical fasteners does not remove the center gap.

Fig.5: Concave Baseplate

Figure 6 illustrates a wavy baseplate which contains both concave and convex features. Tightening the fasteners might help some gaps but not all.

Fig.6: Wavy Baseplate

The concave and wavy baseplates require engineering evaluation because if the gap illustrated in Figure 7 is too big, it creates a potential performance issue. It is key to note that the gap (d) is not necessarily constant. In some modules, the gap might reduce at high temperatures due to CTE issues. In almost all cases the module baseplate flatness is not constant and the module baseplate shape changes with temperature and thermal cycling.

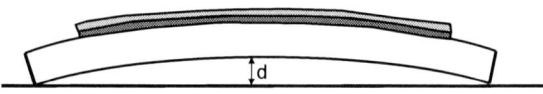

Fig.7: Creation of a gap (d)

Figure 8 illustrates the potential impact when the curvature of the module baseplate exceeds the thickness of the TIM. A representative TIM shape deposited by a stencil is shown. When the module is properly fastened to the cold plate, the TIM must reach across to contact both surfaces. Otherwise, an air gap is created, and the TIM performance is compromised. Airgaps are highly resistant to heat flow and will increase module operating temperatures.

134

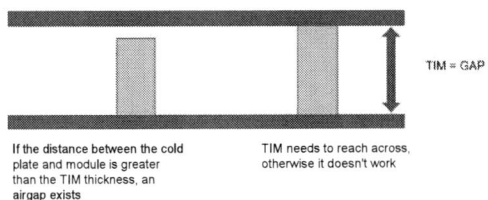

Fig.8: Creation of an Air Gap

Base plate contour characterization

It is important to characterize the base plate surface in both the free (unbolted) state and the assembled (bolted) state to determine the integrity of the thermal interface.

Figure 9 illustrates a common contact pressure issue when fastening a power module to a cold plate. The fasteners create a non-uniform contact pressure distribution with high pressures near the bolt location and low contact pressures away from the fasteners. This results in low contact resistances near the bolts and higher contact resistances away from the fasteners. Unfortunately, this is exactly the opposite of the optimum arrangement. Typically, the high flux die is not near the bolts and are in the middle where contact pressures are low, and TIM performance can be poor.

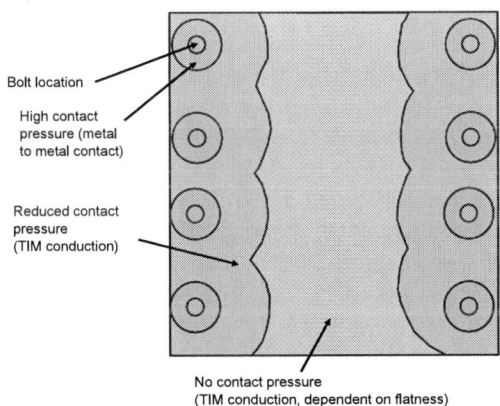

Fig.9: A large thermal interface and various areas of interest

Figure 10 illustrate examples of high contact pressures at the fasteners and virtually zero contact pressure near the center of the modules. These images were created with a 1-inch Lexan plate and a grease interface. The left image (a) had a 0.005 in thick grease application,

and the right image (b) had a 0.010 in grease application. It is commonly understood that the thinnest TIM layer that can accomplish the job is preferred. An interesting observation on both samples is the Lexan remained flat. The module deflected leaving the grease thicker in the middle than the edges. The shadow in both images is an air gap, and highly undesirable.

While these images were created with grease, this phenomena with high contact pressures at the bolts and low contact pressures occurs with almost all TIMs.

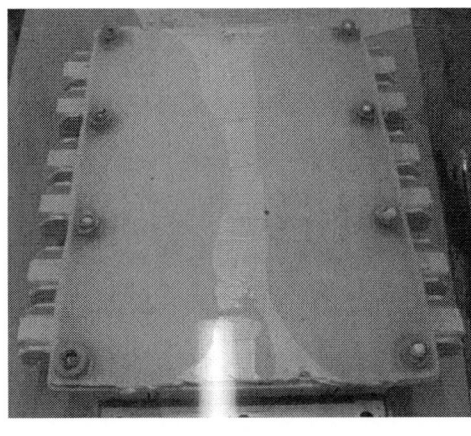

(a)

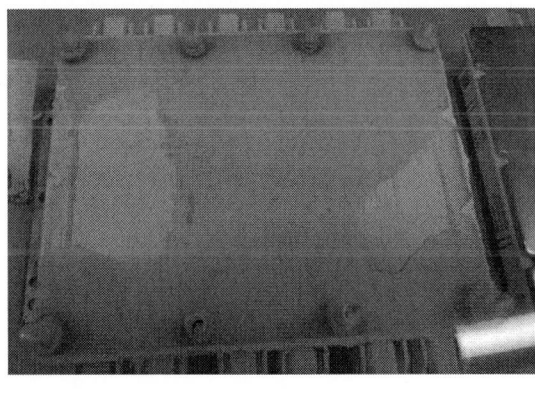

(b)

Fig.10: Example of high contact pressure near bolts and low pressure in the center

As illustrated by the grease example, base plate flatness can be inherent in the part or flatness can be influenced by mounting fasteners. Both are unattractive. For the heat sources that are in the center of the module where the concavity is maximum, there is additional thermal resistance between the base plate and the cold plate due to the conduction resistance through the thermal interface material.

Multiple methods of understanding module flatness can be utilized, each with advantages and disadvantages.

- Surface contour measurements can be measured using laser scanning or similar methods. This is accurate and insightful, but not all researchers have access to this equipment. The other issue (and perhaps more important) is the module can only be mapped in the free state (unbolted) and the profile in the bolted state is of higher interest.

- Measurements at bolted state is possible by similar methods and are not described here.

- If the TIM is applied with a stencil, the change in the stencil patterns after clamping up is a good indicator of whether the TIM extends across to both the cold plate surface and the module surface. Figure 11 illustrates a stencil pattern both before installation and after installation. Ideally after installation the TIM pattern looks like the bottom (b) photo. If the TIM pattern looks like the top photo (a) both before and after installation, then the TIM did not reach across to both surfaces and the TIM performance is not optimum.

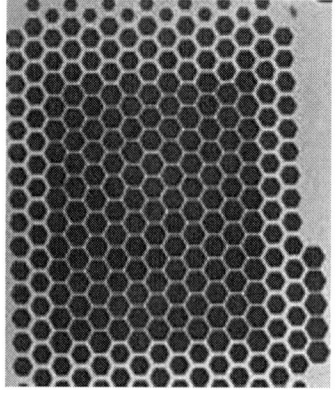

(a)

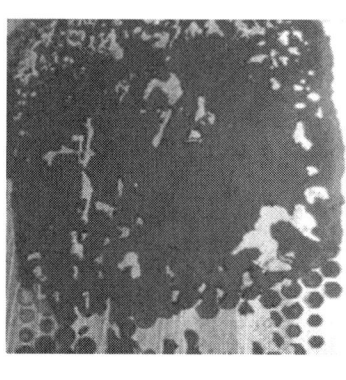

(b)

Fig. 11: Pattern before and after installation

Discussion

Large modules have base plates that are not often flat or in favorable shape to create a uniform contact with the cold plate. The characterization of base plate can vary from module to module and from manufacturer to manufacturer. Even for a module from one supplier, it is possible to see range of flatness variation. Thus, it is important to characterize module flatness by experimental methods as described to design the thermal interface.

Thermal interface materials have their own characteristics. Due to module flatness variation and temperature cycles, materials can migrate, creating airgaps under components and between the base plate and the cold plate. Such an airgap will result in increased component temperature.

Thus, a thermal cycle characterization test is required to realize the effect on module temperatures. Thermal interface characterization is thus an important part of thermal design of power electronic equipment.

Selection of proper thermal interface for large power module favors materials with higher thermal conductivity but also need to flow slightly to form to thicker and thinner gap areas. Phase change thermal interface materials have temperature dependent viscosity in addition to having high thermal conductivity (~ 2-5 W/m-K). Having temperature dependent viscosity results in reduced migration and higher thermal conductivity allows reduced thermal resistance. Rigid TIM materials should be avoided if the mating surfaces are not perfectly flat.

Assembly Considerations

The application or deposition of TIM is another important parameter. Thermal interface materials can be printed, applied by trowel, squeezed, come in sheet forms or other methods. Having a proper and repeatable deposition technique in manufacturing is required. The thickness of the material must be sufficient such that can fill the air voids, thus providing good thermal contact between the module base plate and cold plate over temperature range.

Effect of thermal cycles

Thermal cycling from hot/cold and cold/hot can change module flatness profiles due to CTE mismatch. This results in thickness change to TIM materials and subsequent movement of phase change materials and other soft TIMs including grease. The amount or severity of pumping characteristics changes with different types of TIMs.

The robustness of TIM materials with exposure to thermal cycling can be evaluated by accelerated testing. Thermal cycle tests have both hot and cold setpoints that normally exceed temperatures expected in service. The test assembly should include the cold plate/TIM/cold plate assembly and dwell at the hot and cold setpoints long enough for CTE movement to occur. As a rough estimate the metal in the assembly should dwell about 15 minutes at each setpoint. A sample thermal cycle is shown on Figure 12.

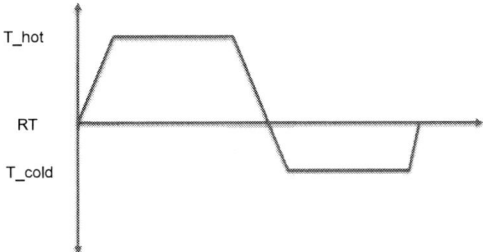

Fig.12: Thermal Cycle Profile

Evaluating Results

The most reliable way to evaluate the effectiveness of the TIM is to monitor temperatures both at beginning of life and during life testing. There are several locations and methods to monitor temperatures, some more involved than others.

1. Infra-red imaging: In these tests the module is partly disassembled by removing the plastic covers and exposing the die. Sometimes the electrical configuration of the devices is modified to operate in conduction only modes. This allows measurement of losses and simplifies testing. Infra-red imaging is done to measure temperatures directly. Caution will need to be taken to correct for emissivity of the die surfaces. In some instances, a thin coating of black paint is used to simulate emissivity close to 1.

2. Structure function: Temperature sensitive parameters of the power module transistors or diodes are calibrated to temperature. Modules are mounted to a cold plate and a transient heating profile is applied to the module. The transient temperature measure from the temperature sensitive parameters is provided to an equipment. Temperature data from these cycles are used to create structure functions. In addition to temperature of the junctions, structure function can provide thermal resistances in materials and interfaces in the heat flow direction.

3. Sensor temperature: Many power modules such as IGBT modules have internal negative temperature coefficient (NTC) sensors. The NTC sensor data is often used to indirectly provide die junction temperature. Temperature differential between die

junction and the sensor location from thermal model is used to calculate die junction temperature from NTC temperature.

Perhaps the simplest means to get a quick determination of the TIM performance is to combine thermal modeling with NTC measurements. The NTC temperature can be compared to expectations from the model, and changes to the NTC temperatures during testing indicate changes to the TIM performance.

Expectations for Thermal Cycling Tests

Sample results from thermal cycling tests is shown in Figure 13. The test data showed the deterioration of grease TIM performance on a non-flat module. The non-flat baseplate moved during hot/cold cycles causing baseplate movement and pumping action of the TIM. This grease example was applied at 5-10 mil. It is noted that the same test on a flat module did not experience a deterioration. The same module tested with a phase change TIM did not experience the temperature increase exhibited by grease. The phase change TIM behaved a bit worse with larger temperature cycles than smaller temperature cycles.

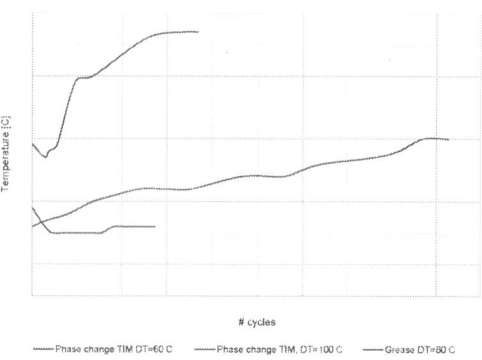

Fig.13: Changes in temperature performance during thermal cycling for different TIMs and temperature cycle ranges

The phase change TIM remained viscous at elevated temperatures and does not exhibit the pumping action exhibited by the non-flat module with grease. The preliminary conclusion is grease pumping action is related to viscosity and the CTE induced movement of module baseplate. More viscous TIMs tend to migrate less than less viscous TIMs especially at elevated temperatures. Inherently flat baseplates (that do not change shape) have reduced TIM pumping movement.

Conclusions

The ASTM methods provide a good means for designers to initially down select TIM materials. These tests only provide part of the selection criteria due to the small dimensions and favorable flatness called out for ASTM tests. Commercially available power modules can be 4-6 inches per side or larger hence considerably larger than the ASTM samples. It is very difficult to construct power modules that are ideally flat, which implies that many power modules are inherently non-flat. The internally stack-up on the modules makes the modules sensitive to CTE movement in thermal cycling tests and these movements can impact TIM performance.

The above issues are generally solvable, but the electronics packaging designer needs to conduct independent tests on the selected TIM using the actual selected module and selected cold plate. Typically, there are combinations of modules/TIM/cold plates that are acceptable and some unacceptable. The vendor published thermal performance of TIMs only applies to ASTM dimensions, flatness, etc. The TIM performance on large power modules is almost always different (worst) than the ASTM results.

The issues described in this paper will help designers ensure successful results and identify issue in the lab rather than in product application environment.

Key takeaways include

- Contact pressure between a power module and cold plate is not uniform. The highest pressure is near the bolts and sometimes very light pressures exist in the center of the modules.

- The primary purpose of a thermal interface material is to replace the air gap with a higher conductivity material. Any arrangement resulting in air gaps will compromise thermal performance

- Thermal Impedance includes the effect of material thickness. Thus, thermal impedance must be accompanied by a thickness of interface. Thinner interfaces are better than thick if no air gaps are present

- Large power modules can have non-flat baseplates. This non-flat profile should be

138

evaluated in both static temperature testing and thermal cycle testing. Wavy and concave baseplates have the greatest potential for thermal surprises. Testing using actual modules, TIM, and actual cold plates will mitigate the risk.

References

[1] M.H. Poech a, R. Eisele b , "A Modelling Approach to Assess the Creep Behavior of Large-Area Solder Joints", 2000, Microelectronics Reliability, Volume 40, Issues 8–10, August–October 2000, Pages 1653-1658.

[1] Garron K. Morris, Matthew Polakowski†, Lixiang Wei, Martin D. Ball, Mark G. Phillips, Craig Mosey, Richard A. Lukaszewski, 2014, "Thermal Interface Material Evaluation for IGBT Modules under Realistic Power Cycling Conditions".

[2] Yuichiro KONISHI, Keisuke HORIUCHI and Atsuo NISHIHARA, 2017," Visualization experiment on pump-out phenomena of thermal grease under thermal cycling", Transactions of the JSME, Vol.83, No.845, 2017,

Optimizing Structural Rigidity and Thermal Performance of Vapor Chambers via Pillar Distribution and Filling Ratio

Nawaf Rasheed[1], Dhruvil Prajapati[1], Chengyan Li[1], L. Winston Zhang[1,2]

[1]University of Illinois Urbana-Champaign, Urbana, Illinois, USA
[2]Novark Technologies, Inc., Shenzhen, Guangdong, China
lwzhang5@illinois.edu

Abstract

This study explores the critical balance between thermal performance and mechanical stress in vapor chamber (VC) optimization by examining pillar distribution and filling ratios. Using a combined experimental and numerical approach, the current research highlights the trade-offs between thermal efficiency and mechanical strain. The results demonstrate that reducing pillar size within the heating area improves thermal performance, even though at the cost of increased strain by two orders of magnitude, which could lead to VC flatness/assembly problem. Conversely, optimizing filling ratios achieves similar thermal improvement without a significant rise in strain. These findings provide insights into vapor chamber design, enabling more balanced approach to optimizing both thermal and mechanical performance.

Keywords

Vapor chamber, Optimization, Multiphase

Nomenclature

Symbol	Description
T_e	Temperature at evaporation surface
T_c	Temperature at condensation surface
T_w	Temperature at wall
$T_{cold\,plate}$	Temperature at cold plate surface
T_3, T_4, T_5	Temperature at measurement points
$\dot{q}$	Heat flex
S_M	Mass source term
S_E	Energy source term
ρ	Density
$\boldsymbol{u}$	Velocity vector
μ	Dynamic viscosity
P	Pressure
E	Copper Young's modulus
υ	Copper Poisson's ratio
α	Coefficient of thermal expansion
k_c	Copper thermal conductivity
$k_{cold\,plate}$	Cold plate thermal conductivity
k_v	Water vapor thermal conductivity
φ	Porosity
K	Permeability
c_{pv}	Vapor specific heat
c_{pl}	Liquid specific heat
I_c	Inertial resistance factor
η	Filling ratio
m	Mass
V	Volume
σ	Standard deviation
R_{vr}	Vertical thermal resistance
R_{hr}	Horizontal thermal resistance

1. Introduction

With the increasing computing demands and the rapid advancement of microelectronics technology, effective thermal management has become critical [1–3]. High-performance electronic components generate significant localized heat, making conventional cooling methods, such as air-cooled heat sinks, less viable due to space limitations and lower heat transfer capabilities. This has positioned vapor chamber as a promising solution for high-power electronics, leveraging phase-change cooling mechanisms to uniformly spread heat across two dimensions. However, optimal VC design is essential to fully utilize their thermal management potential [4, 5].

The thermal performance of a vapor chamber has been extensively studied in the literature, with numerous parameters identified for optimization, including wick structure, filling ratio, vapor core height, and, notably pillar size and distribution [6–8]. Among these, pillar distribution has drawn particular attention due to its dual role in enhancing both thermal performance and structural rigidity. Studies indicate that increasing the number of pillars alone may not always reduce thermal resistance; instead, an optimized pillar distribution can guide vapor flow more effectively, thereby improving the VC's performance. Effective pillar distribution facilitates improved vapor flow pathways within the VC, enabling faster vapor transport from the evaporator to the condenser while assisting liquid return through capillary action. Moreover, a well-organized pillar pattern can prevent thermal resistance buildup at higher heat loads by ensuring even temperature distribution across the evaporator surface [9–11].

While much of the prior research has focused on optimizing VC geometry, such as pillar configuration and wick structures, comparatively little attention has been given to the impact of the working fluid filling ratio [12]. Recent studies highlight the filling ratio as a critical factor influencing temperature uniformity and thermal conductivity [13]. The filling ratio directly affects the temperature homogenization characteristics of VCs by balancing vapor flow and liquid supply. Lower filling ratios promote stable and homogeneous temperature distributions, minimizing thermal resistance by ensuring a consistent liquid supply to the wick structure. Conversely, higher filling ratios can lead to excess liquid accumulation, constraining effective vapor flow and causing uneven temperature distribution, particularly near the chamber's edges [13]. However, excessively low filling ratios may decrease the critical heat flux (CHF) of the vapor chamber [14].

Another key aspect of VC design, beyond thermal performance, is structural rigidity. Structural rigidity is crucial for maintaining the shape and flatness of the VC, as deformation during vacuum sealing or installation can compromise surface flatness. This deformation increases thermal contact resistance by reducing effective adhesion to heat sources, ultimately diminishing overall VC performance [15, 16]. Literature shows that strategically positioned pillars can significantly reduce strain, achieving up to a 91% reduction in some configurations [17]. Pillar placement and design are critical; strategically arranged pillars not only enhance the chamber's resistance to external forces but also maintain a stable internal vapor core space, which is essential for effective vapor flow and liquid return pathways. Additionally, wall thickness and material choice impact strain, with thicker walls reducing deformation under thermal load. Copper, due to its superior rigidity, is often preferred over aluminum for maintaining a stable chamber structure under operational conditions [17]. Additionally, copper is compatible with water, preventing the generation of non-condensable gases, which makes it an ideal material for heat pipes and vapor chambers.

This paper aims to address existing research gaps by conducting a comprehensive investigation integrating thermal performance and structural rigidity analyses to optimize vapor chamber design. Using a combination of experimental and numerical methods, we analyze various pillar distribution configurations and filling ratios to evaluate their effects on thermal efficiency and structural rigidity.

2. Experimental and Numerical Methods

2.1. Experimental Method

A series of experiments are conducted. Figure 1 illustrates the experimental setup, which features two cold plates. The VC was positioned between a heater block (cartridge heater) and the cold plates, which served to cool the VC during the experiment. Data from the cold plates, including mass flow rate, power, and temperature, were recorded via flow meter and thermocouples respectively.

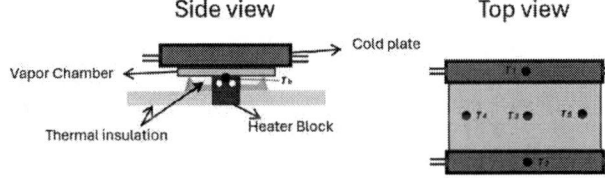

Figure 1: Schematic of the experimental setup showing VC, heater block, cold plate.

Six measurement points were distributed across the VC, which measured 80 mm × 45 mm (length × width), using T-type thermocouples (model name: TT-T-30-SLE) for temperature measurement, as shown in Figure 1. The experiment (Figure 2) was repeated three times, with error bars in the results plot representing one standard deviation (σ) from the mean.

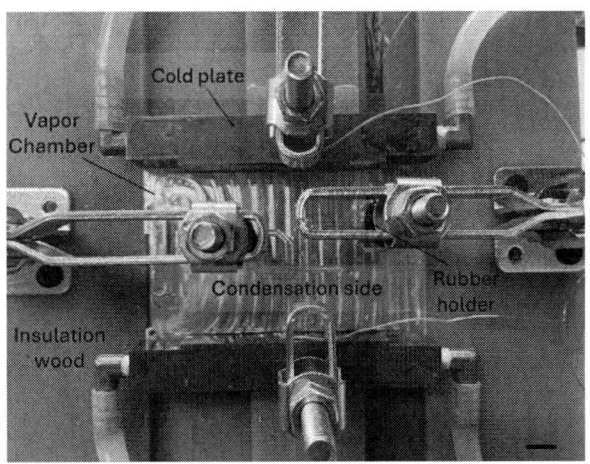

Figure 2: Experimental testing setup for the Vapor Chamber in top view (dimensions: 80 mm × 45 mm). Scale bar: 5 mm.

The experiment was conducted on a design configuration referred to as "Configuration Design A" as shown in Figure 3a with internal structure shown in Figure 3b. Testing began with a 20 W heat input, which was incrementally increased to 70 W over a heat source area of 4.84 cm².

To determine the filling ratio in the vapor chamber, we measured its initial weight after full assembly without any working fluid in VC. Then water was then injected into the vapor chamber through the filling pipe, a vacuum was applied to remove air, and the filling pipe was sealed via tail welding (Figure 3c and d). The final weight of the vapor chamber was subsequently measured. The water content was calculated as the difference between the final and initial weights:

$$m_{water} = m_{wet} - m_{dry} \qquad (1)$$

Finally, the filling ratio (η) was determined using the known wick volume from the VC design [6].

$$\eta = \frac{V_{water}}{V_{wick}} = \frac{\frac{m_{water}}{\rho_{water}}}{V_{wick}} \times 100\% \qquad (2)$$

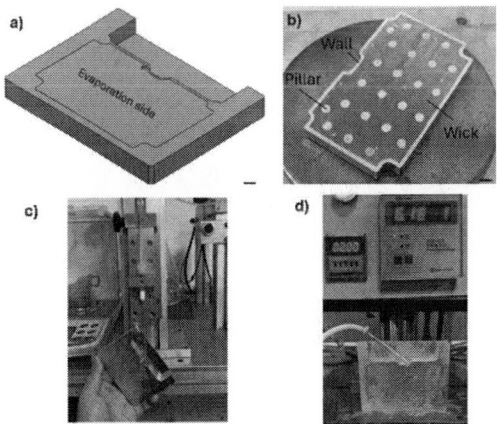

Figure 3: Design A configuration: a) CAD model of vapor chamber. b) Actual vapor chamber showing internal structure. Scale bar 5 mm. c) Water filling process. d) Vacuum process.

2.2. Numerical Methods

The numerical model of the vapor chamber is a 3D representation that replicates the geometry of the experimental vapor chamber. The model comprises five layers: the outer wall (solid copper), wick, vapor, wick, and the opposite wall, respectively. Pillars connect the two wick regions through the vapor core, as illustrated in Figure 4.

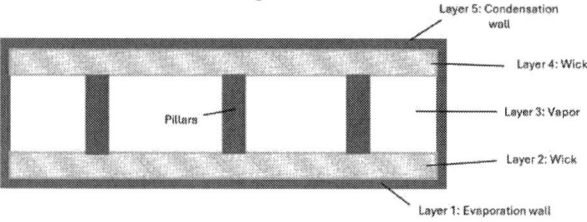

Figure 4: Schematic of vapor chamber numerical model layers

The following assumptions were made in the model. The flow is considered laminar and compressible, with phase changes occurring within the thin film evaporation layer at the liquid-vapor interface, effect of two cold plate can be simplified to be symmetric. Both the wicks and pillars are treated as homogeneous and isotropic porous media, and the effects of gravity are neglected.

The boundary conditions for the model are illustrated in Table 1, where k is thermal conductivity of copper, T is the temperature at specific surface, $\dot{q}$ is the heat flux.

Table 1: Thermal numerical model boundary conditions

Evaporation side	$-k_c \dfrac{\partial T_e}{\partial z} = \dot{q}$	(3)
Condensation side	$-k_c \dfrac{\partial T_c}{\partial z} = -k_{cold\,plate} \dfrac{\partial T_{cold\,plate}}{\partial z}$	(4)
Adiabatic wall	$-k_c \dfrac{\partial T_w}{\partial x} = -k_c, \dfrac{\partial T_w}{\partial y} = -k_c \dfrac{\partial T_w}{\partial z} = 0$	(5)

The model tracks mass and energy transfers at the liquid-vapor interfaces using source terms derived from the Lee model, as described in the literature [18]. The continuity equation and momentum equation are represented as follows:

$$\nabla \cdot (\rho \boldsymbol{u}) = S_M \qquad (6)$$

$$\nabla \cdot (\rho \boldsymbol{uu}) = -\varphi \nabla P + \mu \nabla^2 \boldsymbol{u} - \left(\frac{\mu}{K}\boldsymbol{u} + \frac{\rho}{2}I_c \mid u \mid \boldsymbol{u}\right) \quad (7)$$

Where ρ is density, $\boldsymbol{u}$ is velocity vector, S_M is mass source term, P is pressure, μ is dynamic viscosity, φ is porosity, K is permeability, I_c is inertial resistance factor.

The energy equations for each layer are given as:
Wall:

$$\nabla \cdot (kc\nabla T) = 0 \qquad (8)$$

Vapor Core:

$$\nabla \cdot \left[(\rho c_p)\right]_v \boldsymbol{u}T = \nabla \cdot (k_v \nabla T) \qquad (9)$$

Wick:

$$\nabla \cdot \left[(\rho c_p)\right]_l \boldsymbol{u}T = \nabla \cdot (k_c \nabla T) + S_E \qquad (10)$$

Where S_E is energy source term, ρc_p is volumetric heat capacity of vapor/liquid. The physical and material properties/parameters are listed in Table 2.

Table 2: Material properties/parameters

Property/Parameter	Symbol	Value	Unit
Copper Young's modulus	E	1.1e11	Pa
Copper Poisson's ratio	υ	0.33	-
Coefficient of thermal expansion	α	1.8e-5	m°C^{-1}
Copper thermal conductivity	k_c	385	W m^{-1} K^{-1}
Cold plate thermal conductivity	$k_{cold\,plate}$	385	W m^{-1} K^{-1}
Water vapor thermal conductivity	k_v	0.026	W m^{-1} K^{-1}
Porosity	φ	0.56	-
Permeability	K	6.1E-11	m^2
Vapor specific heat	c_{pv}	2014	J kg^{-1}K^{-1}
Liquid specific heat	c_{pl}	4182	J kg^{-1}K^{-1}
Inertial resistance factor	I_c	8800	m^{-2}
Dynamic viscosity water liquid	μ_l	0.001	kg m^{-1}s^{-1}
Dynamic viscosity water vapor	μ_v	0.0001	kg m^{-1}s^{-1}

The simulations were carried out using ANSYS Fluent [19], [20]. The finite element structure analysis model uses three-dimensional solid elements which structure analysis couples with thermal analysis. To simplify computation, symmetry boundary conditions are applied by modeling only a quarter of the structure with symmetric displacements in ANSYS. Two loads are considered in structure analysis: the hydrostatic pressure due to the working fluid and the thermal stresses derived from the thermal analysis.

The thermal performance of the vapor chamber was evaluated by horizontal thermal resistance (R_{hr}) and vertical thermal resistance (R_{vr}) [14]:

Vertical Thermal Resistance:

$$R_{vr} = \frac{T_b - T_{avg(3\sim5)}}{Q} \qquad (11)$$

Horizontal Thermal Resistance:

$$R_{hr} = \frac{T_3 - T_{avg(4\sim5)}}{Q} \qquad (12)$$

Here T_b is the temperature at the center of evaporator's outer surface near to heater block (cartridge heater), T_3 is the temperature at the center of condenser outside surface, and T_b, T_3, T_4, T_5 are the temperature at four measurement points as shown in Figure 1 for the experiment setup.

Figure 5 illustrates variations in VC geometry. Design A features 24 pillars arranged in a uniform grid with zero angular offset between rows. Design B contains 20 pillars with rows offset by 45°. Design C reduces the number of pillars to 15 by removing those within the heat source region, creating a central void. Design D further reduces the pillar count to 10, leaving a single pillar in the heat source region by eliminating two rows.

Designs A and B exhibit uniform pillar distribution, which is a common configuration in VC designs. In contrast, Designs C and D are asymmetrical, which affects temperature uniformity. These design configurations are chosen because it can provide a comparative analysis, revealing the interplay between thermal performance and mechanical stress distribution.

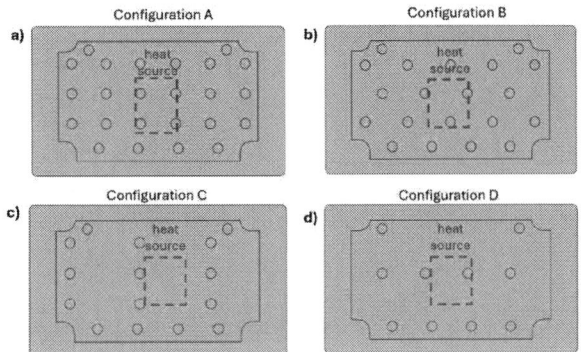

Figure 5: Vapor chamber designs with pillar arrangements relative to heat source region: a) Design A: 24 uniformly distributed pillars. b) Design B: 20 pillars with a 45° offset between rows c) Design C: 15 pillars, excluding pillars in the heat source region d) Design D: 10 pillars, with only one pillar remaining in the heat source region

3. Results and Discussion

3.1. Results

To validate the numerical model developed, Figure 6 compares the numerical and experimental temperatures on the condenser surface at various heat flux levels, measured at different distances from the center. Error bars represent one standard deviation from the mean, (number of reading (n) = 3). Additionally, Figure 7 compares the numerical and experimental temperatures on both the evaporator and condenser surfaces, highlighting the percentage error at thermocouple locations with a heat input of 20 W, the experiment values are the mean of that point.

The model requires tuning, with key parameters including the inertial resistance factor, porosity, and permeability, which should be set within reasonable bounds. The final values for these parameters are provided in Table 1. Although some differences exist, the results demonstrate that the numerical predictions generally align well with the experimental measurements.

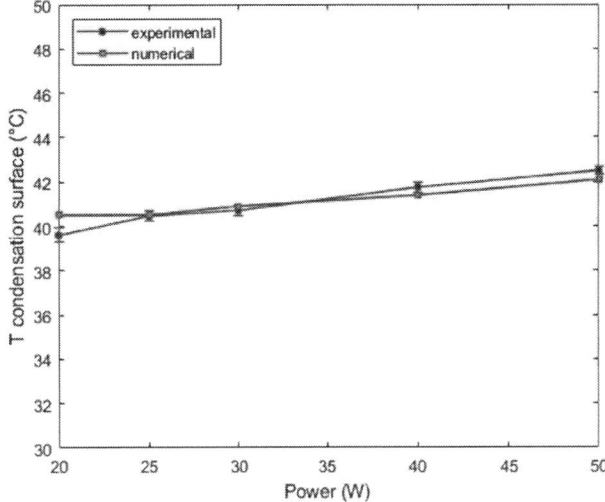

Figure 6: Comparisons of numerical and experimental temperatures at condensation surface

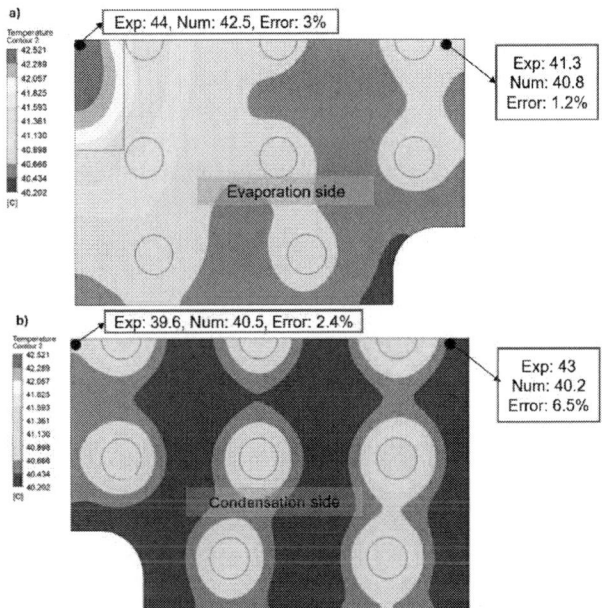

Figure 7: Comparison of exterior wall temperatures between numerical and experimental results: a) Evaporation surface temperature b) Condensation surface temperature

After tuning and validating the numerical model with experimental data, a comparison of all design configurations is shown in Figure 8. Among the configurations, Design Configuration C outperforms the others in thermal performance, having the lowest number of pillars in the heat source region. The comparison criterion used is the maximum temperature rather than temperature uniformity across the condensation surface, as the maximum temperature more directly correlates with stress levels.

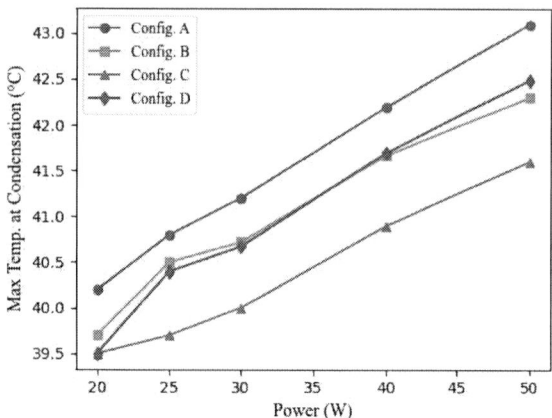

Figure 8: Maximum temperature vs power for all design configurations calculated numerically.

This analysis demonstrates that the thermal performance of the vapor chamber (VC) can be improved around ± 1°C by reducing the pillar density within the heater area, as demonstrated by Design Configuration C. This finding aligns with existing literature, which emphasizes the benefits of optimizing pillar distribution for enhanced heat dissipation. However, this study extends beyond thermal optimization by investigating the coupled effects of stress and strain on system performance, providing a more comprehensive understanding of design trade-offs.

To explore these trade-offs, stress and strain in the top-right quarter of the VC were analyzed for all design configurations. Figure 9 presents stress distributions, showing that Designs C and D experience the highest stress levels. Similarly, strain levels along the x-axis direction were analyzed, with Figure 10 illustrating the results.

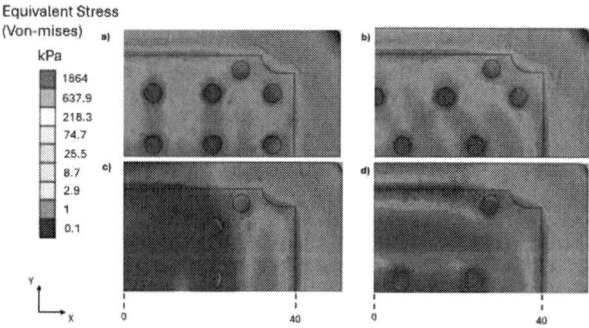

Figure 9: Equivalent stress distribution all design configurations: a) Design configuration A. b) Design configuration B c) Design configuration C d) Design configuration D

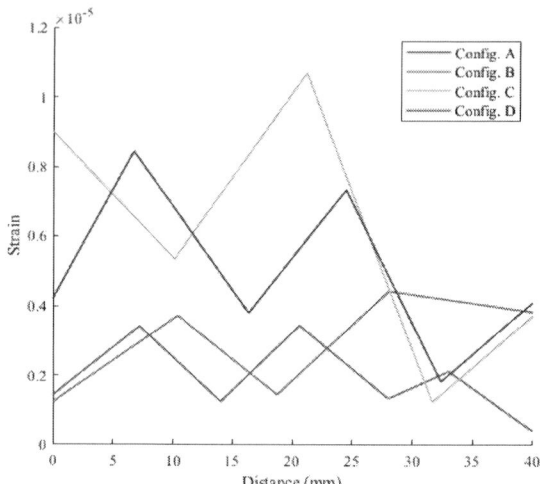

Figure 10: Strain stress distribution all design configurations

The stress and strain results are well below the stress limit of copper, ensuring structural safety. However, significant differences are observed between Configurations A & B (with more pillars) and Configurations C & D (with fewer pillars). In Configurations C & D, the maximum stress is distributed across the thin walls of the evaporation and condensation surfaces, while in Configurations A & B, the pillars bear most of the maximum load. This design approach helps prevent buckling and enhances the flatness of the evaporation and condensation surfaces, ensuring better contact with the heat source and sink.

Additionally, the absence of pillars in Configurations C & D results in higher strain levels, whereas Configurations A & B show significantly lower strain. Notably, strain levels differ by nearly two orders of magnitude between these configurations. Higher strain levels can compromise surface flatness, reducing contact efficiency with the heat source and sink, and potentially degrading thermal performance over time.

The trade-off between pillar density and thermal performance lies in balancing stress distribution and strain-induced deformation. Designs with fewer pillars (Configurations C & D) achieve better thermal performance by reducing thermal resistance, as the thin walls distribute stress more evenly. However, this improvement comes at the cost of significantly higher strain levels, which can compromise surface flatness over time.

To enhance the VC performance without increasing strain levels, an alternative approach involving optimization of the fluid filling ratio was analyzed experimentally. Water was used as the working fluid with filling ratios of 46% (1.6 cc) and 55% (1.9 cc). Figure 11 shows the temperature distribution of the upper surface (condensation surface) under four power conditions: 25 W, 30 W, 40 W, and 70 W. Since this VC operates in an anti-gravity configuration, heat tends to accumulate at the bottom due to gravitational effects, resulting in consistently higher bottom temperatures.

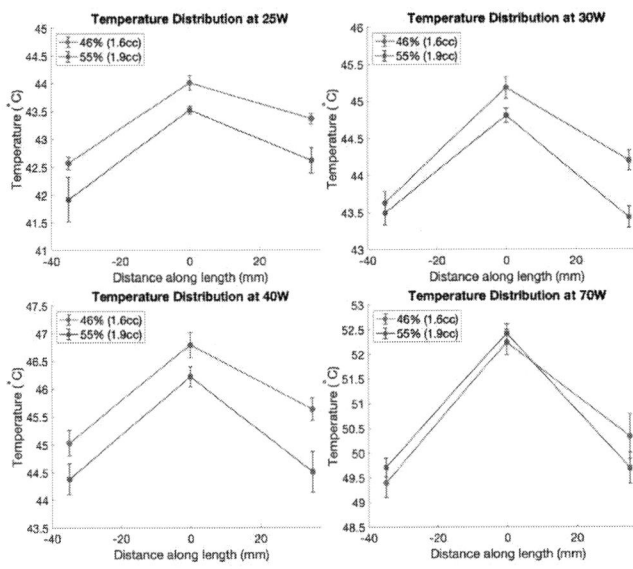

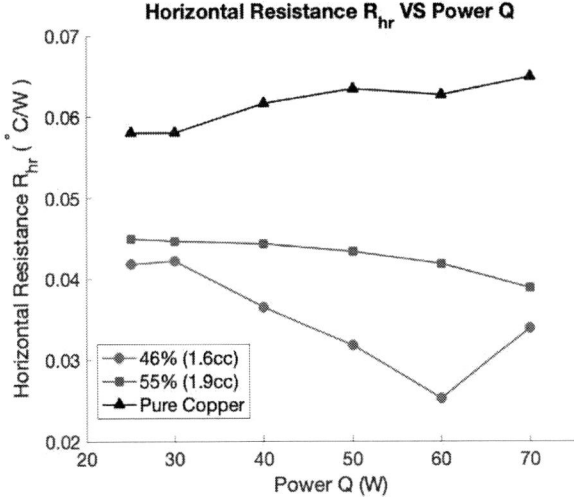

Figure 13: Comparison of horizontal thermal resistance under different power levels

Figure 11: Temperature distribution for vapor chamber under different filling ratios

At low power conditions, the design with a filling ratio of 46% (1.6 cc) effectively transfers heat to the upper surface, resulting in a higher and more uniform temperature distribution than the 55% (1.9 cc) design. However, at high power conditions (70 W), the 46% filling ratio exhibited significant dry-out phenomena, leading to reduced performance. This suggests that while a lower filling ratio performs well at low power, it struggles to maintain stability and efficiency under high power conditions due to inadequate fluid replenishment.

For vertical thermal resistance (R_{vr}), which represents typical VC thermal resistance, Figure 12 shows that the 1.6 cc design reduces R_{vr} by about 50% compared to the pure copper benchmark, while the 1.9 cc design achieves a reduction of about 35%. However, at 70 W, the 1.6 cc design shows a significant increase in thermal resistance due to dry-out phenomena, suggesting that a lower filling ratio leads to a lower critical heat flux limit and reduced VC stability.

For horizontal thermal resistance (R_{hr}), which reflects temperature uniformity, Figure 13 indicates that R_{hr} continuously decreases as power increases. Both the 1.6 cc and 1.9 cc designs reduce R_{hr} by 33% and 25%, respectively, compared to the benchmark, demonstrating significant improvements in horizontal heat transfer efficiency.

In summary, the filling ratio critically affects VC performance. A lower filling ratio reduces thermal resistance and improves heat transfer efficiency at low power levels but suffers from a lower critical heat flux (CHF) and stability issues at higher power due to dry-out phenomena. Conversely, a higher filling ratio improves CHF and stability at the expense of higher thermal resistance.

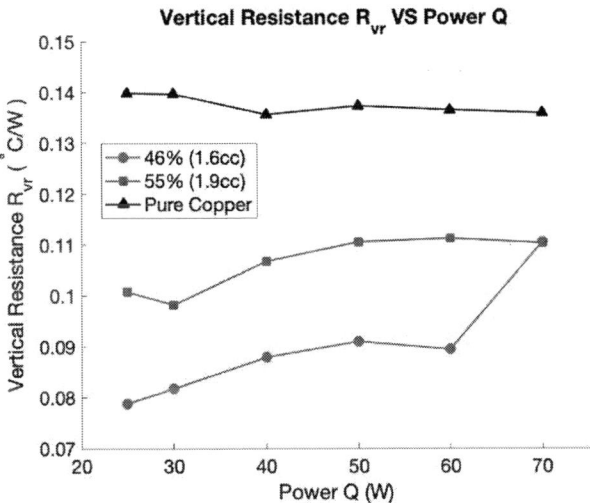

Figure 12: Comparison of vertical thermal resistance under different power levels

4. Conclusions

This study explored optimizing VC structural rigidity and thermal performance through pillar distribution and filling ratio. Reducing pillar density within the heater area enhanced thermal performance but caused a two-order-of-magnitude increase in strain, compromising VC flatness and thermal performance over time. Reducing the fluid filling ratio improved thermal performance without significantly affecting strain levels but exhibited limitations under high heat flux, leading to dry-out and reduced stability. The key takeaway from this study is that optimizing the fluid filling ratio plays a more critical role in achieving sustained thermal performance over time compared to geometric optimization of the VC structure.

Future work will include long-term experiments to analyze the relationship between stress and thermal performance degradation over time quantitatively, providing deeper insights into the interplay between mechanical strain and heat transfer efficiency.

Acknowledgments

The authors thank Mr. Jinji Yu and Ms. Alice He from Novark Technologies for their insightful discussions and assistance during the experiments.

References

[1] H. Tang *et al.*, "Review of applications and developments of ultra-thin micro heat pipes for electronic cooling," *Appl. Energy*, vol. 223, pp. 383–400, Aug. 2018, doi: 10.1016/j.apenergy.2018.04.072.

[2] P. Chen, S. Chang, K. Chiang, and J. Li, "High Power Electronic Component: Review," *Recent Pat. Eng.*, vol. 2, no. 3, pp. 174–188, Nov. 2008, doi: 10.2174/187221208786306270.

[3] L. C. Lv and J. Li, "Micro flat heat pipes for microelectronics cooling," *Recent Pat Mech Eng*, vol. 6, no. 3, pp. 169–184, 2013.

[4] Y. Koito, H. Imura, M. Mochizuki, Y. Saito, and S. Torii, "Numerical analysis and experimental verification on thermal fluid phenomena in a vapor chamber," *Appl. Therm. Eng.*, vol. 26, no. 14–15, pp. 1669–1676, 2006.

[5] K. Rezk, M. A. Abdelrahman, A. A. A. Attia, and M. Emam, "Thermal control of temperature-sensitive electronic components using a vapor chamber integrated with a straight fins heat sink: An experimental investigation," *Appl. Therm. Eng.*, vol. 217, p. 119147, Nov. 2022, doi: 10.1016/j.applthermaleng.2022.119147.

[6] D. Xie, Y. Sun, G. Wang, S. Chen, and G. Ding, "Significant factors affecting heat transfer performance of vapor chamber and strategies to promote it: A critical review," *Int. J. Heat Mass Transf.*, vol. 175, p. 121132, Aug. 2021, doi: 10.1016/j.ijheatmasstransfer.2021.121132.

[7] Y. Rahmatullah and T.-Y. Wen, "Optimization of thermal-hydraulic performance for pillar-reinforced vapor chamber using Lee model based numerical scheme and analysis of variance," *Int. Commun. Heat Mass Transf.*, vol. 152, p. 107319, Mar. 2024, doi: 10.1016/j.icheatmasstransfer.2024.107319.

[8] S. Zhu and Y. Li, "Sensitivity and optimization analysis of pillar arrangement on the thermal–hydraulic performance of vapor chamber," *Appl. Therm. Eng.*, vol. 240, p. 122246, Mar. 2024, doi: 10.1016/j.applthermaleng.2023.122246.

[9] M. Muneeshwaran, Y.-F. Lin, L. Lin, V. Lin, and C.-C. Wang, "A parametric study on the performance of vapor chamber in association with pillar distribution," *Appl. Therm. Eng.*, vol. 207, p. 118217, May 2022, doi: 10.1016/j.applthermaleng.2022.118217.

[10] L. Cong and J. Qifei, "Influences of copper columns on thermal hydraulic performance of the vapor chamber," *Heat Mass Transf.*, vol. 55, pp. 3065–3075, 2019.

[11] G. Huang, W. Liu, Y. Luo, T. Deng, Y. Li, and H. Chen, "Research and optimization design of limited internal cavity of ultra-thin vapor chamber," *Int. J. Heat Mass Transf.*, vol. 148, p. 119101, Feb. 2020, doi: 10.1016/j.ijheatmasstransfer.2019.119101.

[12] A. A. Attia and B. T. El-Assal, "Experimental investigation of vapor chamber with different working fluids at different charge ratios," *Heat Pipe Sci. Technol. Int. J.*, vol. 3, no. 1, 2012.

[13] J. Han, H. Bae, and W. Joung, "Effect of the Working Fluid Charge Ratio and Heat Flux on the Temperature Homogenization Characteristics of a Vapor Chamber-Type Heat Spreader," *Int. J. Thermophys.*, vol. 43, no. 11, p. 168, Nov. 2022, doi: 10.1007/s10765-022-03088-9.

[14] Y. Yang, J. Li, H. Wang, D. Liao, and H. Qiu, "Microstructured wettability pattern for enhancing thermal performance in an ultrathin vapor chamber," *Case Stud. Therm. Eng.*, vol. 25, p. 100906, Jun. 2021, doi: 10.1016/j.csite.2021.100906.

[15] Y. Chen, B. Li, X. Wang, Y. Yan, Y. Wang, and F. Qi, "Investigation of heat transfer and thermal stresses of novel thermal management system integrated with vapour chamber for IGBT power module," *Therm. Sci. Eng. Prog.*, vol. 10, pp. 73–81, May 2019, doi: 10.1016/j.tsep.2019.01.007.

[16] U. Vadakkan, G. M. Chrysler, and S. Sane, "Silicon/water vapor chamber as heat spreaders for microelectronic packages," in *Semiconductor Thermal Measurement and Management IEEE Twenty First Annual IEEE Symposium, 2005.*, San Jose, CA, USA: IEEE, 2005, pp. 182–186. doi: 10.1109/STHERM.2005.1412176.

[17] Y.-H. Hung, S.-W. Kang, and W.-C. Tsai, "STRAIN ANALYSIS OF VAPOR CHAMBER HEAT SPREADERS," *J. Mar. Sci. Technol.*, vol. 18, no. 2, Apr. 2010, doi: 10.51400/2709-6998.2327.

[18] Y. Rahmatullah and T.-Y. Wen, "Lee model based numerical scheme for steady vapor chamber simulations," *Int. J. Heat Mass Transf.*, vol. 201, p. 123636, Feb. 2023, doi: 10.1016/j.ijheatmasstransfer.2022.123636.

[19] I. Ansys, "ANSYS FLUENT theory guide," *Canonsburg Pa*, vol. 794, 2011.

[20] I. Ansys, "ANSYS FLUENT user's guide," *Canonsburg PA*, vol. 15317, 2011.

Transforming Battery Thermal Management through the Design and Optimization of High-Heat Transfer Lithium-Ion Batteries

Alfred J. Piggott, Jeffrey S. Allen, Ahmad A. Pesaran

Applied Thermoelectric Solutions LLC, 24875 Novi Rd., Novi, 48376, Michigan, USA

+ 1 (231) 753-8201, Alfred@ThermoelectricSolutions.com

Abstract

The battery industry has identified key battery performance metrics requiring significant improvements. Enhanced battery thermal management (BTM) can improve these key metrics. Herein, reducing battery thermal resistance as an alternative approach to BTM is investigated. A new reduced thermal resistance, high heat transfer (HHT) battery technology is introduced. An equivalent resistance battery thermal model is developed and validated against an experiment and model in the literature. Maximum heat removal rates are compared between the target conventional battery and the HHT battery using a parametric study with air, liquid, and refrigerant boundary conditions over a range of battery aspect ratios. The HHT battery shows a heat removal rate ranging from 1.3 to 20.7 times higher than a conventional battery. For the refrigerant cooled instance, the HHT battery has a maximum heat removal rate of 362 watts compared to the conventional battery's 17.5 watts. Unlike conventional batteries, HHT batteries have the potential to achieve a heat transfer rate that enables extreme fast charging (XFC). Even the lower-performing air-cooled HHT batteries demonstrated significant improvements, with the potential to reach a heat transfer rate that supports a 15-minute charging time. The next step for this research is to collaborate with an industry partner to build prototypes and run bench tests. HHT batteries are a promising technology with potential to enable increased energy density, extreme fast charging, increased battery life, reliability, safety, reduced power loss in cold weather, and decrease costs.

Keywords

Battery Thermal Management, Thermal Resistance, High heat transfer battery (HHT), Extreme Fast Charging (XFC)

1. INTRODUCTION

The United States is working to decarbonize the transportation sector by increasing the market share of zero emission electric vehicles (EVs) [1]. The DOE, ARPA-E and industry have identified key EV battery improvements that are needed to boost the market share of electric vehicles. These improvements include increasing battery energy density, capacity, life, reliability, safety, decreased battery costs, increased battery charge rate, and improved range retention of the battery in cold weather [2]. Improved battery thermal management (BTM) is the essential solution for all of these issues [3] [4]. To enhance battery thermal management, certain battery thermal management system (BTMS) needs must be addressed. These include increased heat removal rates [5], reduced battery temperature gradients [5], faster thermal response[4], lower energy consumption [4], Lower

complexity BTMS design [4], size and weight reduction[4], lower cost [4] and potentially a fundamentally new approach.

Mathematically the heat removal rate of a battery can be described by the equation $\dot{Q} = \Delta T_{batt}/R_{th\,batt}$ as shown in Figure 1. The equation indicates that there are two methods to enhance the rate of heat removal from the battery. Either increase the temperature delta, ΔT_{batt} or reduce the battery thermal resistance, $R_{th\,batt}$. Historically, the predominant approach to battery thermal management has been to use an increased ΔT_{batt} approach. This approach is generally accomplished by increasing the heat transfer coefficient, h at the battery surface, and / or increasing the contact area of the battery A that h is acting upon. Some examples of these methods include natural convection [6], forced air [7], pumped single phase liquid [8], solid to liquid phase change (PCM) [9], vapor compression refrigeration [10], dielectric fluid immersion [11], pumped two phase fluid [12] and jet impingement [13]. The increased ΔT_{batt} approach has its limitations. The United States Advanced Battery Consortium (USABC) advises keeping ΔT_{batt} to no more than 3°C [14]. Even at this level, a 3°C ΔT_{batt} can lead to a 300% rise in battery degradation compared to a battery with a uniform temperature [15]. Furthermore, with ΔT_{batt} constrained; the heat transfer rate is limited by $\dot{Q} = \Delta T_{batt}/R_{th\,batt}$. As a result, any strategies aimed at enhancing h or A to subsequently increase ΔT_{batt} become largely indistinguishable in terms of heat removal rate.

In contrast to the increased ΔT_{batt} approach, the research herein focuses on reducing $R_{th\,batt}$. The thermal resistance of a three-dimensional rectangular cuboid, such as certain battery formats, is defined by the equation $R_{th\,batt} = L/kA$. As illustrated in Figure 2, L represents the battery's thickness, while A is the area perpendicular to the direction of heat flow. Due to the layered construction of batteries, their thermal conductivity, k is highly anisotropic, being approximately 32 times more conductive when heat flows parallel to the layers (in-plane) compared to the conductivity in the cross-plane or through the layers.

Typically, batteries rely on the lower thermal conductivity of the cross-plane heat transfer path to dissipate heat. In this context, L is minimized and A is maximized to reduce $R_{th\,batt}$. But as evidenced by the needs of battery thermal management today, this approach has not had the desired impact. Research has compared high k in-plane thermal management with cross-plane methods [16], and while in-plane methods demonstrated several outstanding advantages for battery thermal management, those studies did not focus on the impact of reducing the battery's thermal resistance. To summarize, current batteries utilize improved L and A to reduce $R_{th\,batt}$ using low k cross-plane heat transfer, and some investigations have explored utilizing high k in-plane heat transfer. However, no studies have focused on simultaneous optimization of L, A for a battery using the high k in-plane heat path to reduce $R_{th\,batt}$ and in-turn increase $\dot{Q}$.

The aim of this work is to quantify the effect on $\dot{Q}$ of employing optimized L and A for a battery that takes advantage of high-k in-plane heat transfer. The in-plane heat transfer for this work is enabled by a mass-producible and innovative heat transfer design. This enhanced heat transfer design will be referred to herein as a high-heat transfer (HHT) battery [17].

The HHT battery facilitates in-plane heat transfer by employing aluminum "thermal connectors" that are ultrasonically welded to the compressed current collectors of the battery. These connectors are electrically isolated while being thermally conductive. They create a thermal bridge to the battery case and occupy space within the battery that would otherwise be empty, so energy density is not impacted.

To accomplish the research aim, an A-B comparison modeling study is developed to show the performance difference between a conventional battery and HHT battery over a wide range of aspect ratios and varied cooling heat transfer coefficients.

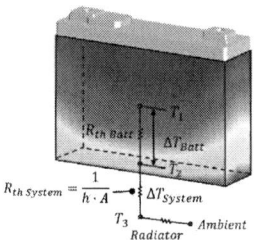

Figure 1: Battery and System Thermal Resistance Diagram

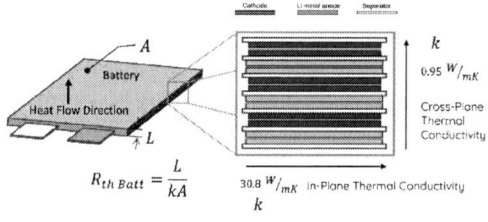

Figure 2: Battery cell showing layered structure with anisotropic thermal conductivity

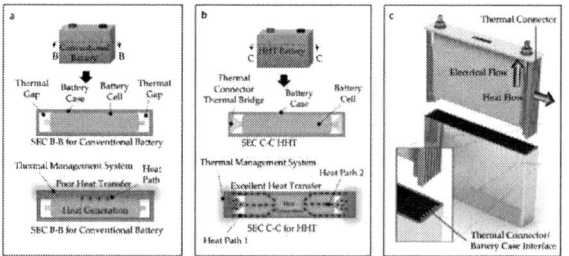

Figure 3: (a) Schematic cross section of a conventional battery thermal path. (b) Schematic cross section of a HHT battery thermal path. (c) Three-dimensional view of a HHT battery showing thermal connector thermal path, electrical flow direction and the

2. METHODS

A steady-state equivalent resistance electrical-thermal analogy model (Figure 4) is developed for a conventional prismatic battery. The target conventional battery studied herein is a 25Ah automotive battery cell by SANYO PANASONIC. The equivalent circuit model is solved symbolically for voltages which are equivalent and analogous to temperatures. The model was validated against the result of a laboratory test and transient CFD model [18] with good agreement. A comprehensive resistive circuit for the thermal connector technology and other associated thermal resistances is added to the validated conventional model to create the HHT battery model.

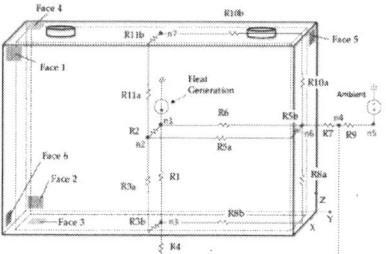

Figure 4: Equivalent resistance electrical-thermal analogy model

To compare the thermal performance of the conventional and HHT batteries, each resistance in the equivalent resistance circuit

is

modeled as dependent on the aspect ratio of battery. This allows the thermal performance to be modeled at many aspect ratios for both the conventional and HHT batteries at constant internal case volume. Boundary conditions applied are base cooling (face 3, Figure 3b) for the conventional battery and cooling two sides (face 5 and face 6, Figure 3b) for the HHT battery. Side cooling is chosen for the HHT battery as it minimizes the thermal resistance path and provides the best thermal performance. For the conventional model, base cooling is chosen because it is a widely used and recognized method of cooling prismatic cells and thus a comparison with HHT batteries that could be put into perspective and recognized by a wide audience.

3. RESULTS

Figure 5 summarizes the results of the maximum heat removal rate study. This maximum heat removal rate is defined as the rate of heat flow out of the battery (or heat generation rate) that produces no more than $60°C$ battery center temperature or a ΔT_{batt} in the cell in the x, y, or z-dimension that does not exceed $20°C$ under the stated cooling boundary conditions. The cooling boundary conditions were air, liquid, and refrigerant with heat transfer coefficients of 25, 390 and 1740 W(m2K)$^{-1}$ respectively. In Figure 5, the overlayed red horizontal lines are the heat generation rates of various battery charging rates. The conventional battery is unable to reach the target cooling rate for a 15-minute charge with any cooling method. However, the HHT battery with a typical aspect ratio has the possibility for a 15-minute charge

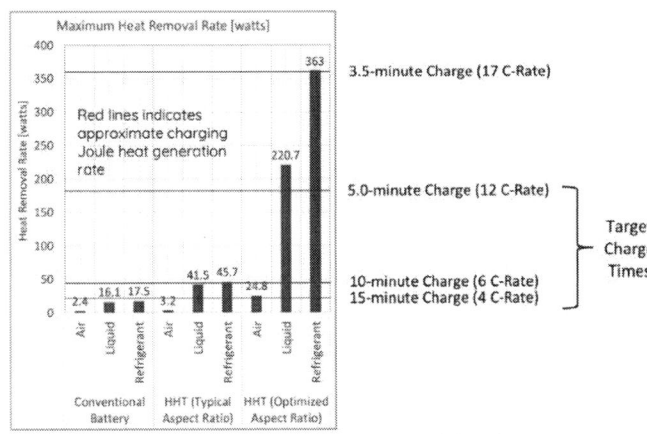

with liquid cooling and 10-minute charge with refrigerant

Figure 5: Maximum possible heat removal rates with overlay of battery joule heat generation during various fast charging rates.

cooling. The heat transfer capability of an HHT battery with an optimized aspect ratio and refrigerant cooling could be capable of a 3.5-minute charge from a thermal perspective and interestingly, a 15-minute charge with a simple and low-cost air-cooled system.

4. CONCLUSION

The battery industry has pinpointed essential performance metrics that require enhancement to boost the market share of electric vehicles. Improved battery thermal management is crucial for advancing these performance metrics. To improve battery thermal management, specific needs of the battery thermal management system (BTMS) must be addressed. The most widely used thermal management approach today involves increasing ΔT_{batt}, which is typically accomplished by enhancing the heat transfer coefficient at the battery surface. However, this method effectively restricts the heat transfer rate from the battery and accelerates its degradation.

The strategy for reducing $R_{th\ batt}$ is examined, and an innovative solution, the high heat transfer (HHT) battery is presented. To evaluate the heat transfer capabilities of the reduced $R_{th\ batt}$ approach and HHT batteries, a comparison between the HHT battery and a conventional battery is conducted using A to B thermal modeling. To support this comparison, a steady-state equivalent resistance thermal model is developed and validated against experimental results and a 3D transient CFD model found in the literature.

The HHT battery demonstrates an improvement in heat removal rate of 1.3 to 20.7 times that of a conventional battery, depending on the battery aspect ratio. From a thermal perspective, this performance increase has potential to enable extreme fast battery charging in 3.5 minutes.

This breakthrough finding has the potential to revolutionize battery thermal management and support the key metrics needed to boost the market share of electric vehicles, including higher energy density, longer battery life, improved reliability and safety, reduced power loss in cold weather, and lower costs. Future work will explore how HHT batteries can facilitate each of these metrics.

The next steps include working with industry partners to prototype and benchmark HHT batteries alongside conventional batteries.

HHT technology has the potential to transform battery thermal management, not just for electric vehicles, but also for other sectors such as aerospace, consumer electronics, and military applications, addressing long-standing battery performance challenges.

References

[1] T. W. House, "FACT SHEET: President Biden Announces Steps to Drive American Leadership Forward on Clean Cars and Trucks," The White House. Accessed: Feb. 07, 2024. [Online]. Available: https://www.whitehouse.gov/briefing-room/statements-releases/2021/08/05/fact-sheet-president-biden-announces-steps-to-drive-american-leadership-forward-on-clean-cars-and-trucks/

[2] "EVs4ALL | arpa-e.energy.gov." Accessed: Feb. 07, 2024. [Online]. Available: http://arpa-e.energy.gov/technologies/programs/evs4all

[3] S. Liu, G. Zhang, and C.-Y. Wang, "Challenges and Innovations of Lithium-Ion Battery Thermal Management Under Extreme Conditions: A Review," *ASME Journal of Heat and Mass Transfer*, vol. 145, no. 080801, Mar. 2023, doi: 10.1115/1.4056823.

[4] Y. Ortiz, P. Arévalo, D. Peña, and F. Jurado, "Recent Advances in Thermal Management Strategies for Lithium-Ion Batteries: A Comprehensive Review," *Batteries*, vol. 10, no. 3, Art. no. 3, Mar. 2024, doi: 10.3390/batteries10030083.

[5] A. Kumar Thakur *et al.*, "A state-of-the art review on advancing battery thermal management systems for fast-charging," *Applied Thermal Engineering*, vol. 226, p. 120303, May 2023, doi: 10.1016/j.applthermaleng.2023.120303.

[6] O. Kalkan, A. Celen, and K. Bakirci, "Experimental and numerical investigation of the LiFePO4 battery cooling by natural convection," *Journal of Energy Storage*, vol. 40, p. 102796, Aug. 2021, doi: 10.1016/j.est.2021.102796.

[7] T. Wang, K. J. Tseng, J. Zhao, and Z. Wei, "Thermal investigation of lithium-ion battery module with different cell arrangement structures and forced air-cooling strategies," *Applied Energy*, vol. 134, pp. 229–238, Dec. 2014, doi: 10.1016/j.apenergy.2014.08.013.

[8] A. A. Pesaran, S. Burch, and M. Keyser, "An Approach for Designing Thermal Management Systems for Electric and Hybrid Vehicle Battery Packs," Jan. 1999.

[9] S. A. Khateeb, M. M. Farid, J. R. Selman, and S. Al-Hallaj, "Design and simulation of a lithium-ion battery with a phase change material thermal management system for an electric scooter," *Journal of Power Sources*, vol. 128, no. 2, pp. 292–307, 2004, doi: 10.1016/j.jpowsour.2003.09.070.

[10] J. Kim, J. Oh, and H. Lee, "Review on battery thermal management system for electric vehicles," *Applied Thermal Engineering*, vol. 149, pp. 192–212, Feb. 2019, doi: 10.1016/j.applthermaleng.2018.12.020.

[11] M. Suresh Patil, J.-H. Seo, and M.-Y. Lee, "A novel dielectric fluid immersion cooling technology for Li-ion battery thermal management," *Energy Conversion and Management*, vol. 229, p. 113715, Feb. 2021, doi: 10.1016/j.enconman.2020.113715.

[12] Y. Zhu, Y. Fang, L. Su, and Y. Fei, "Experimental study on thermal performance of a pumped two-phase battery thermal management system," *International Journal of Energy Research*, vol. 44, no. 6, pp. 4664–4676, 2020, doi: 10.1002/er.5247.

[13] M. Keyser, "Thermal Implications for Extreme Fast Charge," presented at the Vehicle Technologies Office (VTO) Annual Merit Review and Peer Evaluation, Washington, D.C., Jun. 06, 2017.

[14] "USABC – USCAR." Accessed: Oct. 21, 2024. [Online]. Available: https://uscar.org/usabc/

[15] S. Li, C. Zhang, Y. Zhao, G. J. Offer, and M. Marinescu, "Effect of thermal gradients on inhomogeneous degradation in lithium-ion batteries," *Commun Eng*, vol. 2, no. 1, pp. 1–14, Oct. 2023, doi: 10.1038/s44172-023-00124-w.

[16] I. A. Hunt, Y. Zhao, Y. Patel, and G. J. Offer, "Surface Cooling Causes Accelerated Degradation Compared to Tab Cooling for Lithium-Ion Pouch Cells," *J. Electrochem. Soc.*, vol. 163, no. 9, p. A1846, Jul. 2016, doi: 10.1149/2.0361609jes.

[17] A. Piggott, "ParaThermic® Battery Thermal Management." [Online]. Available: https://thermoelectricsolutions.com/parathermic-battery-thermal-management/

[18] J. Kleiner, L. Komsiyska, G. Elger, and C. Endisch, "Thermal Modelling of a Prismatic Lithium-Ion Cell in a Battery Electric Vehicle Environment: Influences of the Experimental Validation Setup," *Energies*, vol. 13, no. 1, Art. no. 1, Jan. 2020, doi: 10.3390/en13010062.

Experimental Evaluation of Thin Oscillating Heat Pipes: A Comparison with Commercial Thermal Solutions

Corey Wilson

ThermAvant Technologies, LLC

2508 Paris Road, Columbia, MO 65202

573-321-1567, Corey.Wilson@ThermAvant.com

2-Page Abstract

Abstract

The oscillating Heat Pipe (OHP) is a highly efficient thermal management technology which is particularly suited for space constrained, high-performance applications such as electronics cooling and aerospace systems. Unlike conventional heat pipes, OHPs use the pressure gradient between the heat source and heat sink to push the working fluid throughout the channel and transport heat. While OHPs are ideal for space constrained applications, development of sub-2 mm thick OHPs would enable a broader range of applications.

At these thicknesses, the current commercial solution for passive heat transport is thin vapor chambers, flattened heat pipes, and pyrolytic graphite. Thin vapor chambers and flattened heat pipes are both established technologies, offering efficient heat spreading however the total heat load and transport length are limited and they do not provide structural integrity to the system. Pyrolytic graphite sheets have high in-plane thermal conductivity and are lightweight, however they are also limited in terms of heat transport capacity across long distances.

This experimental study will analyze the performance of sub-2 mm thick OHPs and how they compare with commercial off the shelf vapor chambers. The results will also be compared with published data on flattened heat pipes and pyrolytic graphite sheets to provide a comprehensive analysis of the strengths and limitations of each technology.

1. INTRODUCTION

The oscillating Heat Pipe (OHP) is a highly efficient thermal management technology which is particularly suited for space constrained, high-performance applications such as electronics cooling and aerospace systems. Unlike conventional heat pipes, OHPs use the pressure gradient between the heat source and heat sink to push the working fluid throughout the channel and transport heat. While OHPs are ideal for space constrained applications, development of sub-2 mm thick OHPs would enable a broader range of applications.

At these thicknesses, the current commercial solution for passive heat transport is thin vapor chambers [1,2], flattened heat pipes [2], and pyrolytic graphite [3]. Thin vapor chambers and flattened heat pipes are both established technologies, offering efficient heat spreading however the total heat load and transport length are limited and they do not provide structural integrity to the system. Pyrolytic graphite sheets have high in-plane thermal conductivity and are lightweight, however they are also limited in terms of heat transport capacity across long distances. This paper will build upon the theoretical analysis presented in the 2024 Semi-Therm, "Evaluation of Sub-Millimeter Thick Oscillating Heat Pipes for Space Constrained Electronics Applications".

2. EXPERIMENTAL/NUMERICAL METHODS

This experimental study will measure the performance of sub-2 mm thick OHPs and similar sized sub-2 mm thick commercial off the shelf (COTS) vapor chambers and pyrolytic graphite sheets. The three heat transfer devices will be selected to have a similar length, width, and thickness, however due to availability there will be some variation. One thermal test bed will be used to measure the performance of all three devices to minimize variability in testing conditions. The devices will be tested in horizontal and tilted orientations and at a range of heat loads and rejection temperatures.

3. RESULTS

The thermal performance of the three devices will be compared with each other and they will be compared against published data on flattened heat pipes and encased pyrolytic graphite heat spreaders to provide a comprehensive analysis of the strengths and limitations of each technology.

4. CONCLUSIONS

The expected operational bounds of each type of heat transfer device will be presented to provide guidance to thermal engineers on how OHPs fit with current thin heat spreaders.

5. REFERENCES

[1] ATS - Advanced Thermal Solutions, Inc. | Innovations in Thermal Management, (n.d.). https://www.qats.com/ (accessed July 26, 2022).
[2] Fujikura's Heat Pipes and Vapor Chambers, Fujikuras Heat Pipes Vap. Chamb. (n.d.). https://thermal.fujikura.jp/ (accessed July 25, 2022).
[3] Advanced Cooling Technologies, (n.d.). https://www.1-act.com/.

High Performance 3U-Form Factor Pulsating Heat Pipe Heat Spreader

Sai Kiran Hota, Kuan-Lin Lee, Ramy Abdelmaksoud, Srujan Rokkam
Advanced Cooling Technologies, 1046 New Holland Ave Building 2
Lancaster, PA. USA. 17601
saikiran.hota@1-act.com; 717-205-0679

Abstract

Next generation electronics cards subjected to high heat fluxes must be effectively managed for reliable operation. To enable efficient thermal management, an additively manufactured Pulsating Heat Pipe (PHP) was investigated as a two-phase heat transfer electronics heat spreader for modular 3U form factor electronics. High performance was achieved with ammonia as the working fluid. Peak thermal performance determined for the PHP heat spreader was determined to be greater than 3.5 W/°C, which was about 3.3X higher than conduction plate. The PHP heat spreader was able to transport more than 450 W (~70 W/cm^2) without dry-out at lower condenser temperatures of -10°C. Additionally, it was estimated that the PHP heat spreader could yield up to 8% mass savings compared to conduction plate, which is critical for space applications.

Keywords

Heat spreader, thermal management, pulsating heat pipe, electronics cooling, two-phase cooling

Nomenclature

C	Thermal conductance (W/°C)
C_{pl}	Specific heat capacity in liquid form, (J/kg.K)
h_{fg}	Enthalpy of vaporization (J/kg)
M_{PHP}	Merit Number
Q	Heat (W)
R	Gas constant
T	Temperature (°C)
Z	Compressibility
σ	Surface tension (N/m)
μ	viscosity (Pa.s)
ϕ	Volume fill ratio
ρ	Density (kg/m^3)

1. Introduction

Recent trends in semiconductors have realized high computing electronics cards with small footprints. These high heat fluxes, must be effectively handled to maintain the electronics below 75-85°C for safe and reliable operation of the electronics card [1]. "Commercial" solutions employ conduction-based heat spreaders based on aluminum, copper, or even diamond, based on system limitations. However, the performance of these conduction cards is limited by material thermal conductivity, and thus are less feasible for application with next generation electronics dissipating very high heat fluxes, >> 10 W/cm^2. For a 1-inch x 1-inch footprint, the corresponding heat power is more than 65 W. In lieu of these thermal management challenges, two-phase heat spreaders, like Embedded copper-water Heat Pipes (EHP) [2], have been

developed and fielded for electronics cooling. The EHPs have demonstrated between 2-8 times improvement in thermal performance for small form factor to larger 6U form factor electronics cards [2, 3, 4]. However, the freezing challenges with water as the heat pipe working fluid limits the application in situations where the heat spreader is exposed to temperatures ≤0°C like space environments. While freeze tolerant mixtures are explored, they are not reliable. Alternate working fluid like methanol cannot handle very large heat fluxes at low temperatures. A Pulsating Heat Pipe (PHP) heat spreader is a good alternative for such applications. PHP with appropriate working fluid enables efficient heat spreading of electronics cards in both near room temperature to low temperature environments. Here, the working fluid is typically a refrigerant like propylene, ammonia, or R134a [5, 6, 7, 8].

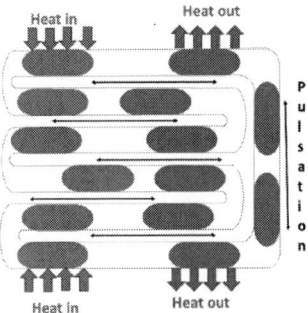

Figure 1. Schematic illustrating the operation of a PHP [9].

A PHP is a form of heat pipe where heat transfer occurs by means of liquid-vapor phase change and two-phase flow of the working fluid in capillary sized channels. Figure 1 shows the schematic of the operation of the PHP. The operation of the PHP is as follows [10, 11]: When heat is added (evaporator section), nucleation occurs through the liquid slug in the channels next to the heated wall resulting in an increase in the vapor pressure. Consequently, at the other end (condenser), the heat is delivered to the sink by the condensing vapor, which results in the reduction of vapor pressure. This dynamic interplay of the fluid pressures results in movement of the two-phase fluid in the capillary sized channels. PHPs have been investigated for use in varying extreme operating temperatures from cryogenics to ultra-high temperatures [12, 13] for applications in solar desalination [14], thermal management of electrical transformers [15], cooling of superconducting magnets [16], etc. Likewise, investigations have been ongoing on PHPs for electronics cooling like PHP-based heat spreaders. Similar to EHPs, research on PHPs based electronics cooling have 2 to 8 times improvement in small form factor electronics cards [17].

The PHP technology can be implemented for electronics cooling by integrating the fluid channels within the heat spreader plate. A PHP heat spreader, additively manufactured with Al-6061 RAM2 for 3U form factor electronics is presented here, with ammonia as the working fluid. The manuscript is organized as follows: Section 2. describes fluid selection criteria; section 3. provides description of the PHP heat spreader along with test method; section 4. discusses experimental results; section 5. discusses ammonia-PHP (vessel) transportation requirements; and section 6. provides the conclusion.

2. PHP fluid selection

The PHP working fluid must satisfy the following primary constraints to be deemed as a suitable working fluid:

- Satisfy fluid channel diameter consideration by satisfying Bond number limitation, which when rearranged to determine the maximum diameter is given as [18]:

$$d_{cirt}(mm) = 2000 * \sqrt{\left(\frac{\sigma}{g(\rho_l - \rho_v)}\right)} \quad \text{(eq.1)}$$

Where, σ is the surface tension, ρ_l is the liquid density and ρ_v is the vapor density. The chosen PHP diameter was 1.52 mm.

- The working fluid must have high $\frac{dP}{dT}$ (at saturation) to enable high driving force within the fluid with low temperature drop.

- High PHP merit number to yield low thermal resistance [19].

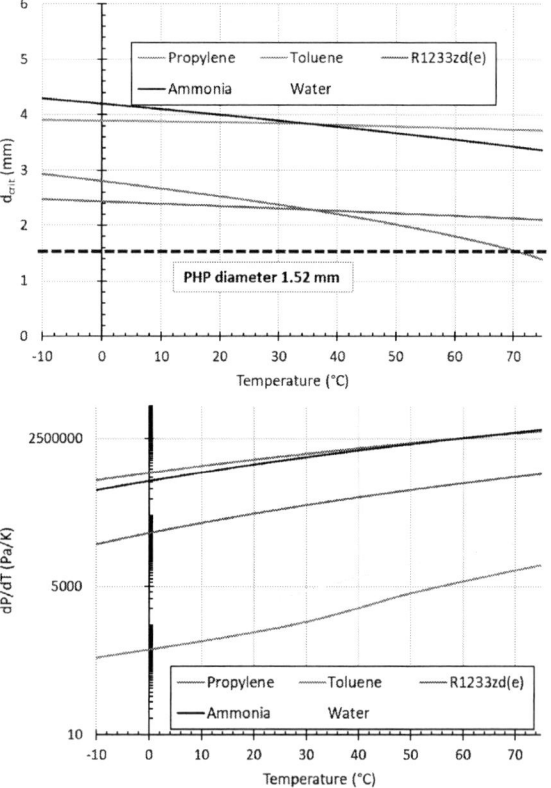

Figure 2. Critical diameter and dP/dT at saturation for PHP working fluids

Figure 2 shows the critical diameter and dP/dT of different classes of working fluids – refrigerants, toluene (organic compound) and water. Excluding propylene at very high temperatures, the working fluids easily meet the critical diameter requirement. However, the driving force potential by toluene and water are the lowest. Propylene, which was previously investigated as the working fluid for electronics cooling has the dP/dT on the order 1E6 Pa/K. Ammonia, likewise, has similar values as that of propylene. Based on these two aspects, both propylene and ammonia are suitable working fluids. However, fluid properties influencing the pulsation, as discussed below, indicate relative comparison of propylene and ammonia selection.

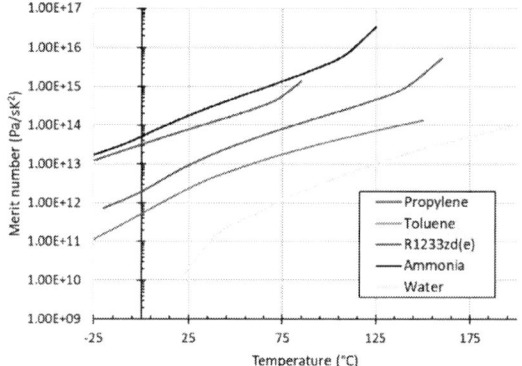

Figure 3. Merit number of PHP working fluids *[20]*

Figure 3 shows the merit number of PHP working fluids based on the performance (not heat transport limit). In addition to the driving force dP/dT, the merit number accounts for fluid density, specific heat capacity (C_{pl}), enthalpy of vaporization (h_{fg}), viscosity (μ_l), etc. all of which influence the pulsation (or circulation) of the working fluid within the channels as fluidic forces. The merit number is given as [19]:

$$M_{PHP} = \frac{\rho_l C_{pl} \left(\frac{dP}{dT}\right) ZRT}{\mu_l h_{fg}} \quad \text{(eq.2)}$$

Based on the above calculation, it was determined that the merit number of ammonia was higher than propylene, while, the merit number of all other fluids were several orders of magnitude lower. In addition to the above three selection criteria, typically, the heat transfer limits of the working fluids were estimated to further support fluid selection. Heat transfer limits are often calculated to estimate the maximum heat load the two-phase technology can deliver. This often varies with temperature and the fluid being used. When the fluid is near the freezing point, often viscous and sonic limit dictate the maximum heat transfer rate, however, in the current context they do not feature. For PHP, maximum heat transfer rate is associated with vapor-inertial limit and corrected with swept-length limit at higher operating temperature range. Vapor-inertial limit occurs when the vapor plug starts to penetrate into the slug causing the breakage of slug-plug pair. Swept-length limit is attributed to the relationship between the heated length and the liquid slug-length to cover the heater section to maintain the fluid nucleation rate for boiling/ vaporization. In the PHP heat transfer limits proposed by Drolen and Smoot [21], it was noted that identifying appropriate number of turns in a PHP heat spreader within a highly conductive plate like aluminum is difficult and could result in heat transfer limits discrepancy. However, from preliminary calculations, it was computed that the heat transfer limits for propylene varied from ~250W at 10°C to below 100W at 40°C, to below 10W at 60°C. Some of the operational data points and dry-out values were correlated to the heat transfer limits experimentally with reasonable accuracy [20]. Using these models, the estimated heat transfer limits of ammonia were determined to be higher than propylene, reaching a value significantly greater than 500W at a temperature above 40°C. Investigating these heat transfer limits the following inferences can be drawn: the typical maximum heat transfer limit of PHP given by the vapor-inertial limit is higher with ammonia compared to propylene; and the influence of the swept length limit is witnessed in temperature ~ 45°C with propylene, while, for ammonia, it does not appear until above 70°C. With regards to heat transfer potential, this indicates ammonia is a highly suitable PHP working fluid for electronics cooling in comparison to alternative options like propylene.

3. Description of heat spreader and testing methodology

A 3U form factor PHP heat spreader was fabricated for 3U form factor electronics cards, of dimensions 160 mm x 100 mm x 3.38 mm. The capillary PHP fluid channels were formed with a diameter of 1.52 mm. Figure 4 shows the geometric information on the additively manufactured PHP heat spreader. The material used for fabrication was Al-6061 RAM2. To facilitate fabrication and charging of the working fluid, powder removal ports were integrated within the plate. After fabrication, excess powder was removed and welded shut. The working fluid was charged through the fill tube port. The performance of the PHP heat spreader was compared against solid aluminum-conduction plate of similar dimensions. The conduction plate was made fabricated by CNC machining with Al-6061. The theoretical thermal conductivity of the heat spreader was 167 W.mK, which returns a thermal conductivity of 1.14 W/°C. Experimental thermal conductance of the conduction plate was 1.1±0.1 W/°C.

Quasi steady state test method was adopted for performance testing of the PHP heat spreader using a central heat in with two edge heat-out method. A 25.4 mm x 25.4 mm (1-inch x 1-inch) aluminum block with cartridge heater rod inserts was used to simulate the electronics heat source. Three thermocouples were attached to the heat spreader at the heater block interface using plunger thermocouples with springs providing constant pressure. Two cold plates attached to the two edges of the heat spreader through wedge-lok card retainers [22] were used as heat sinks for heat PHP heat rejection. Liquid Nitrogen (LN) was circulated through the cold plate for heat removal. A solenoid valve was used to continuously obtain feedback from the PHP condenser thermocouple to maintain constant temperature (referred to as operating temperature here). The overall test setup is shown in Figure 5. The test fixture was insulated using fiberglass insulation on the heat spreader and the heater block and a 1-inch thick garolite foam insulation was put around the test block.

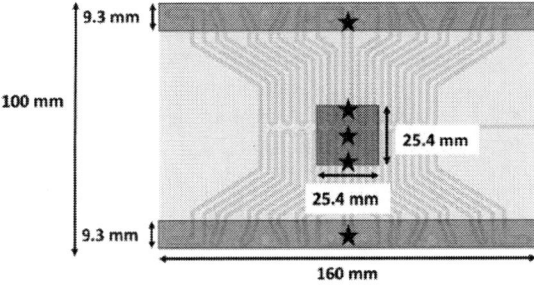

Figure 4. Channel layout geometry, and testing method for characterizing performance of PHP heat spreader. The bottom of the schematic shows heater block with plungers holding thermocouples at the heater-PHP heat spreader interface to record evaporator temperature.

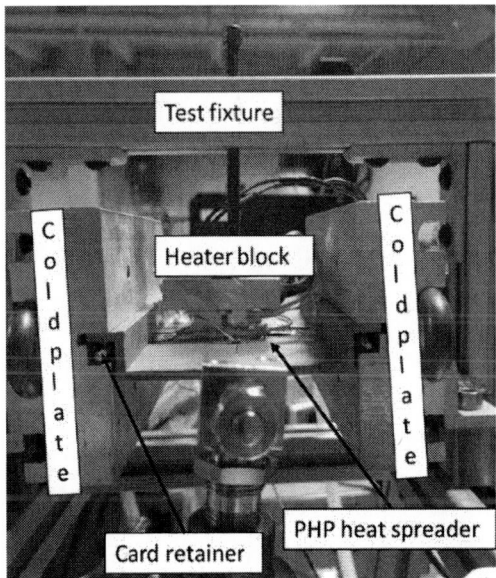

Figure 5. PHP heat spreader test fixture

The thermal performance was determined as the thermal conductance (C), calculated as:

$$C = \frac{Q}{\Delta T} \qquad \text{(eq.3)}$$

Where, Q is the heat input and ΔT is steady state temperature difference between the average evaporator temperature and the condenser temperature. The experiments were performed three times at each data point for repeatability. At highest heater power of ~450W, the maximum variation in the heat input was ±3.5W, and estimated power loss through the insulation using equivalent heat transfer coefficient of ~5 W/m²K was 6.4W. Total uncertainty in heat input was 2.2% at highest heat input. The uncertainty in thermocouples was ±0.5°C. The variation in uncertainty of thermal conductance based on the formulation presented by [23] was calculated to be 2.5%.

The thermal performance testing of the PHP heat spreader was undertaken with fluid fill ratios of 50% and 75% (reference at 22°C) with condensers maintained in the temperature range of -10°C to +30°C in horizontal and vertical configuration. The PHP was charged with ammonia at -23°C to reduce effects of high-pressure fluid in the charging tubes at room temperature.

4. Thermal performance of PHP plate heat spreader

4.1 Thermal performance of PHP at 50% fill ratio

The PHP heat spreader was tested at 50% fill ratio (@22°C) in quasi-steady state method. The temperature profiles Figure 6 and Figure 7 shown below corresponds to condenser temperatures of -10°C and +20°C, respectively.

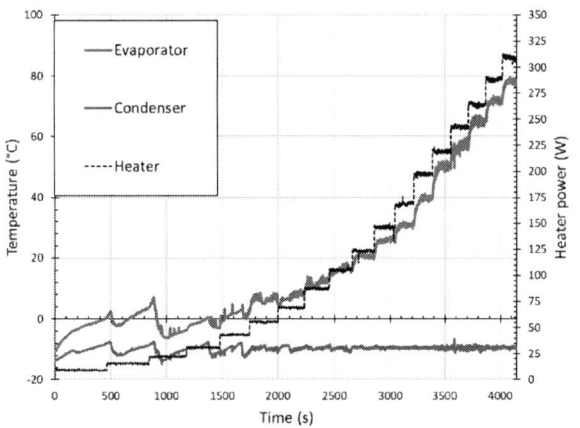

Figure 6. PHP heat spreader temperature profile with condenser at -10°C

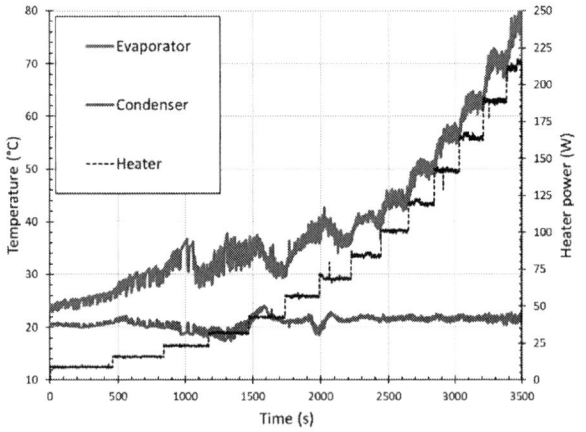

Figure 7. PHP heat spreader temperature profile with condenser at 20°C

Figure 6 and Figure 7 show the PHP wall temperature at condenser temperature of -10°C and +20°C, respectively in horizontal configuration. At lower heater powers, some variation in condenser temperature was observed due to delay in the PID learning curve of the solenoid valve and the high thermal mass discrepancy of cold plates. At low condenser temperature, a start-up power of ~30W was required for the two-phase operation of the PHP to initiate. As the heater power increased, especially above 50W, the pulsation on the wall evaporator temperature became more consistent with pronounced two-phase operation. The PHP could transport ~310W (48 W/cm²) without dry-out, with the evaporator reaching a temperature of about 80°C. As the condenser temperature increased to +20°C, easier start-up in the PHP was obtained. The PHP could deliver 210W (32.5 W/cm²) without dry-out. The corresponding evaporator temperature was 78°C. The testing was stopped above this point for equipment safety and to avoid very high pressures within the PHP.

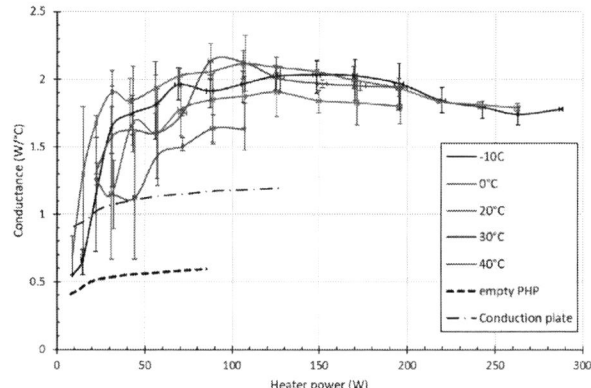

Figure 8. Thermal conductance of the PHP heat spreader at 50% volume fill ratio in horizontal configuration

Figure 8 shows the thermal conductance of the ammonia PHP heat spreader at 50% volume fill ratio. Additionally, testing was performed at condenser temperature of +40°C (for this data set only). The thermal conductance of the PHP increased gradually from below 1 W/°C (before or just after start-up) to above 2 W/°C (~2X conduction plate and ~3.9X empty PHP). At condenser temperatures of -10°C and 0°C, the thermal conductance was consistently above 2 W/°C until heater power of 200W. Above 200W heater power, a small decline in the thermal conductance was observed. This can be attributed to change in the fluid flow pattern (oscillation to circulation) or departure from slug to flow to other flow patterns. It can also be attributed to the fact that the "vapor quality" increases at higher powers, and the flow patterns vary with increasing vapor quality. High vapor quality leads to annular flow, which, in the case of PHP represents dry-out condition. In the case of PHPs, slug flow is vital for consistent operation of the PHP [24]. As the condenser temperature increased to +30°C or even +40°C, it was deduced that the peak thermal conductance decreased, and the performance improvement compared to conduction card was typically lower than 1.5X. This can be attributed to both changes in the vapor quality and dynamic change in the instantaneous fluid ratio because of changes in the liquid/vapor density of ammonia.

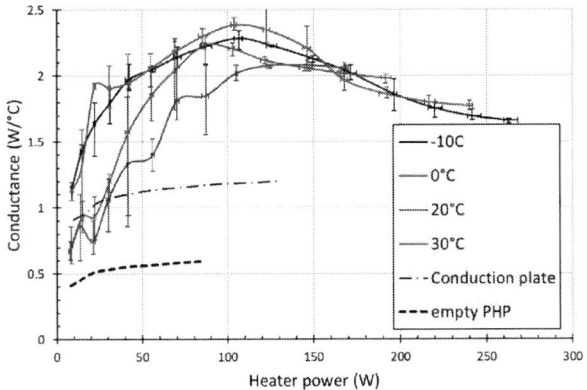

Figure 9. Thermal conductance of the PHP heat spreader at 50% volume fill ratio in vertical configuration

Likewise, performance testing of the PHP heat spreader was undertaken in vertical configuration. Figure 9 shows the thermal conductance of the PHP heat spreader in vertical configuration. At low heater powers, it was observed that the thermal conductance was marginally greater than that of the horizontal configuration. The peak thermal conductance greater than 2.2 W/°C was obtained at low condenser temperatures of -10°C and 0°C. However, as the heater power increased to above 175W, the thermal conductance reduced to a value of ~1.6-1.8 W/°C. In general, vertical orientation assisted with the PHP start-up by supporting fluid motion. However, at higher power, lower thermal conductance was observed in comparison to horizontal configuration due to the aforementioned reasons.

4.2 Thermal performance of PHP at 75% fill ratio

The PHP was tested at higher fluid fill ratio of 75% (@22°C) at similar condenser temperatures in the range -10°C to +30°C in both horizontal and vertical configuration.

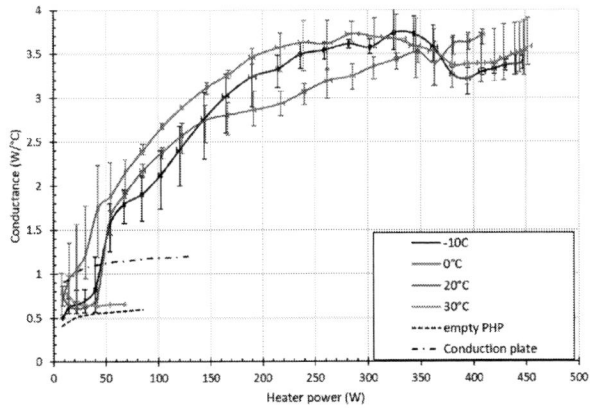

Figure 10. Thermal conductance of the PHP heat spreader at 75% volume fill ratio in horizontal configuration

Figure 10 shows the thermal conductance of the PHP heat spreader with ammonia charged at 75% fill ratio in horizontal

configuration. The thermal conductance increased from below 1W/°C to more than 3.5 W/°C at condenser temperatures of -10°C to +20°C. The thermal performance improvement of the heat spreader was about 3.3X conduction plate and 6.5X empty PHP. The PHP was able to deliver more than 400W (62 W/cm^2) and up to 450W (~70 W/cm^2) without dry-out. The maximum evaporator temperature was determined to be ~60°C. Further testing was limited by the heater power limitations of the cartridge heater. However, at a higher condenser temperature of +30°C, the PHP failed to operate. As mentioned in the previous section, this can be attributed to the instantaneous change in the fluid fill ratio with the corresponding PHP temperature. The highest performance was obtained for the PHP heat spreader with this condition. However, in one setting when the fluid fill ratio increased to 80%, the PHP did not operate. The influence of the fluid fill ratio is discussed in the subsequent sub-section.

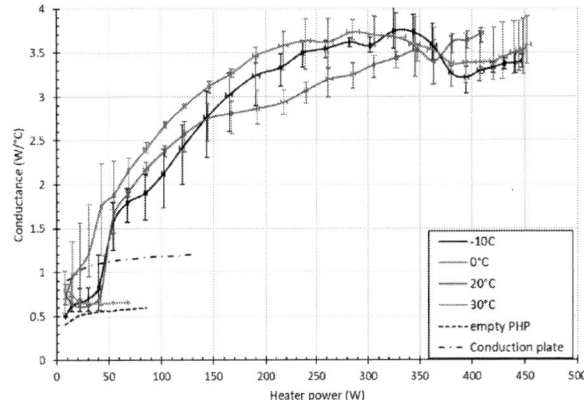

Figure 11. Thermal conductance of the PHP heat spreader at 75% volume fill ratio in vertical configuration

Figure 11 shows the thermal conductance of the PHP heat spreader in the vertical configuration. Interestingly, the peak thermal conductance was marginally lower in comparison to the horizontal configuration, demonstrating some of gravity in the PHP heat spreader thermal performance. This was furthermore elucidated by the non-operational PHP at 20°C condenser temperature, which was otherwise operational in the horizontal case. At higher fill ratio, the longer liquid slug experiences greater gravitational force in comparison to lower fill ratios, although the preferred-direction of fluid movement could not be ascertained here. The peak thermal conductance obtained here was ~3.3-3.4 W/°C, which was almost 0.3W/°C lower than the horizontal case. The vertical orientation hampered the thermal conductance to certain extent in this case. Although the PHP heat spreader was able to transport up to 450W heat, the maximum evaporator temperature obtained here was up to 80°C. The improvement in thermal performance over conduction plate and the empty PHP was 3X and 6X, respectively.

4.3 Evaluating influence of fluid fill ratio on PHP thermal performance

From above analysis, it was deduced that the fluid fill ratio of 50% experienced faster start-up, the peak thermal

157

conductance obtained was between 2.1-2.3 W/°C for chosen condenser temperature range from -10°C to +30°C in both horizontal and vertical configuration. Within the heat power range of 75-200W, the thermal conductance trend was almost similar and some variation in the performance, up to ±0.3W/°C or higher was observed at lower or higher heater powers for a given condenser temperature (from Figure 8 and Figure 9). As noted, at lower heater powers, at nominal fill ratio of 50%, gravity assists with fluid circulation and supports PHP start-up at lower condenser temperatures. Despite variation in thermal performance, the operation of the PHP heat spreader was less sensitive to gravity at 50% fill ratio.

However, at higher fill ratio of 75%, influence of gravity was noted at higher heater powers and at condenser temperature of +20°C and above. Higher fill ratio corresponds to longer liquid slugs, which experience higher gravitational force in comparison to other fluidic forces like viscous force or vapor expansion force, etc. At higher PHP temperatures resulted by higher heater powers, the weakening surface tension coupled with these forces further influences the fluid movement in vertical orientation in comparison to the horizontal configuration. While, high fluid velocity is desired, very high fluidic velocity is detrimental as it influences the pressure drop.

While, both 50% and 75% fill ratio showed significant improvement in PHP heat spreader performance, testing was performed at lower fluid fill ratio of 25% - smaller liquid slugs.

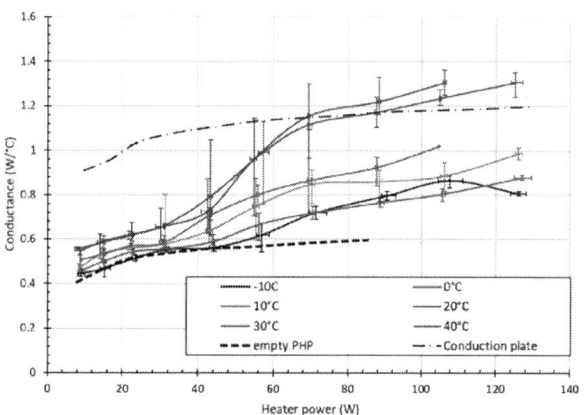

Figure 12. Thermal conductance of the PHP heat spreader at 25% volume fill ratio in horizontal configuration

Figure 12 shows the thermal conductance of the PHP heat spreader at low fill ratio of 25% (undercharged condition). As observed, although, two-phase operation was obtained, the uncertainty in the presence of the liquid slug and the instantaneous fluid velocity (and thereby the vapor quality) variation resulted in poor performance. Overall, the thermal conductance of the PHP heat spreader in the undercharged condition was poorer than or comparable to the conduction plate, based on the heater power and the operating condition.

While, undercharged fluid lends to poor PHP thermal performance, it was determined that higher fill ratio of 75% is suitable for PHP at low condenser temperatures. At higher

condenser temperatures, for any given power, if the instantaneous fill ratio is higher than the maximum theoretical value, then the PHP does not operate. For example, the instantaneous fill ratio was 77% for PHP temperature of 30°C, but the theoretical maximum at the same temperature was 76.4%. The theoretical maximum (ϕ_{max}) was calculated as [25]: $\phi_{max} = \frac{1}{1+\left(\frac{kRT\rho_l}{K}\right)^{0.5}}$, Where, k is the adiabatic coefficient and K is the bulk modulus.

4.4 Comparison of thermal performance of ammonia PHP against literature data on propylene PHP and EHP

The thermal conductance of the PHP heat spreader with ammonia was compared against propylene PHP and EHP for similar geometry [6, 11]. The propylene PHP performance was tested at -10°C to +30°C and the EHP was tested at +20°C. The EHP had 4 U-shaped Cu-H$_2$O heat pipes.

Table 1. Thermal performance comparison of ammonia PHP against propylene PHP and EHP discussed in literature

Heat Spreader	Improvement in C over conduction plate	Maximum Power (dry-out/ tested) [W]	Relative mass change
Conduction plate	-	150	-
Propylene PHP (50%	~1.3-1.4X	210	-8.5%
Propylene PHP (75%)	1.5-2X	260	-8%
EHP	2X	260	+2.2%
Ammonia PHP (50%)	1.7-2X	260	-8.3%
Ammonia PHP (75%)	3-3.3X	450	-7.75%

Table 1 shows the thermal performance comparison of ammonia PHP (discussed here) against propylene PHP in the temperature range -10°C to +20°C and EHP at +20°C. In general propylene PHP at higher fill ratio and EHP showed up to 2X improvement against conduction plate [6, 11]. The ammonia PHP at nominal 50% fill ratio presented similar performance improvement, but higher fill ratio of 75% showed pronounced thermal conductance improvement by 3.3X over conduction plate with capability of delivering more than 450W heat (70 W/cm^2 heat flux) without dry-out. Another advantageous feature of PHP is mass savings. The PHP, with the current channel design saves up to 8% in mass in comparison to conduction card (~184.2 grams). This is an important criterion for space electronics cards. In contrast, EHP showed a slight increase in mass by 2.25%.

5. Transporting ammonia-PHP (heat spreader)

Ammonia PHP was determined to be high performance thermal heat spreader for electronics cooling. However, ammonia is classified as a toxic gas by the Department of Transportation (DOT) under hazard class 2.3 (UN 1005). The transportation of the heat spreader from the manufacturing facility to the recipient must meet stringent guidelines. The heat spreader envelope treated as a pressure vessel must meet following pressure safety criteria [26, 27, 28].

- The fluid container must be designed to minimum packaging pressure no less than 1.5 times vapor pressure at 46°C, which is 2750 kPa.

- ASME recommends safety pressure check up to 4 times the maximum fluid pressure, which is expected to be ~10350 kPa.

- DOT-SP 11818 recommends testing up to 1400 Psi (9650 kPa).

The saturation temperature for ammonia at 2750 kPa is 62°C. The PHP heat spreader was repeatedly tested to temperature above 80°C in several instances. In addition, at the start, the PHP heat spreader was tested with nitrogen gas up to a pressure of 10350 kPa.

Every manufacturer maintains a record of components with ammonia fluid and the recipient can discuss requisite information as desired. Usually, ammonia vessels can only be shipped in cargo carriers following DOT permit authorization DOT-SP 11818. The safety measures in the permit indicate protocols for packaging and testing the heat spreader welds (ASTM E-1742, NAS1514, or equivalent).

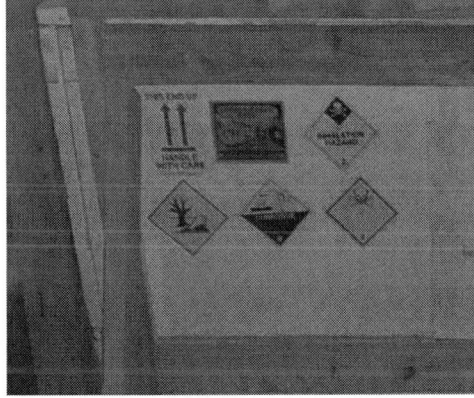

Figure 13. Warning signs and labels required to indicate the shipment contains ammonia.

The packaging box must properly contain the heat spreader within it and have precautionary labels and signs, like the photograph shown in Figure 13. The warning labels shown here indicate, inhalation hazard & toxicity category 3, cargo-aircraft only, environment hazard, and corrosion and skin irritation. Shipment destinations and modes mandate the use of some of these warning labels. Usually ground or ocean shipment is preferred, but in air transportation, the ammonia filled heat spreader (container) can only be transported in cargo. For domestic (within U.S.A) transportation, non-flammable gas green class 2.2 sticker is used. International shipment must contain corrosion and inhalation hazard stickers. Poison/toxicity stickers are particularly required for ocean transportation.

6. Conclusions

Thermal performance characterization of ammonia charged PHP heat spreader at fill ratios of 50% and 75% is presented at various condenser temperatures in the range -10°C to +30°C. The dimensions of the heat spreader were 160 mm x 100 mm x 3.38 mm with fluid channel diameter of 1.52 mm. Tests were performed using a central 25.4 mm x 25.4 mm heater block with two edge condensers through a cold plate interfaced with wedge-lok card retainer. Following inferences were drawn:

At nominal fill ratio of 50%, peak thermal conductance up to 2.1±0.1 W/°C was obtained in the heater power range 75-200W. This was about 2X compared to conduction plate. At higher power loads, a slight reduction in thermal conductance to 1.8 W/°C was obtained. The thermal conductance of the PHP heat spreader in both horizontal and vertical orientation showed similar trend with small differences. The PHP heat spreader operated effectively in all condenser set points from -10°C to +30°C.

At higher fluid fill ratio of 75%, the thermal conductance of the PHP heat spreader further pronounced to values > 3W/°C. Peak thermal conductance >3.5W/°C was obtained in horizontal orientation. The increase in thermal conductance compared to the conduction plate was 3.3X. The PHP was able to deliver more than 450 W (70 W/cm^2) without dry-out at operating (condenser) temperatures of -10°C and 0°C. In vertical orientation, some slight influence of gravity was observed, potentially due to larger liquid slug length, coupled with high instantaneous fill ratio for a given heat load. However, the vertical orientation also demonstrated high thermal conductance, up to 3X more than conduction plate. In the horizontal orientation, the PHP heat spreader did not operate at +30°C set point, while in vertical orientation, the PHP heat spreader did not operate at +20°C and +30°C condenser set points.

While, significant improvement in thermal conductance was obtained with PHP heat spreader, it was deduced that ammonia as working fluid was suitable for condenser temperatures below +30°C. While, the PHP showed improvement in performance at +40°C at 50% fill ratio, the improvement in thermal conductance compared to the conduction plate was less than 1.5X.

When compared to past published literature on similar geometry, ammonia as the working fluid showed more suitability compared to propylene. Furthermore, it was determined that PHP could yield mass savings in the electronics card, which is an important criterion for consideration for space applications.

Acknowledgments

This work was performed with the support of NASA's Small Business Innovation Research (SBIR) Phase II contract #80NSSC22CA205. The authors are grateful to Dr. Sergey Y.

Semenov for his support. The authors express their gratitude to the engineering technician, Mr. Eugene Sweigart and Mr. Phil Texter for their support with fabrication and experiments. The authors also thank ACT's Operations team for their support with welding tasks. Authors thank Mr. Brad Hartzler, shipping and receiving manager at ACT.

References

[1] S. Murshed and C. de Castro, "A critical review of traditional and emerging techniques and fluids for electronics cooling," *Renewable and Sustainable Energy Reviews,* vol. 78, pp. 821-833, 2017.

[2] K.-L. Lee, S. K. Hota, A. Lutz and S. Rokkam, "Advanced two-phase cooling system for modular power electronics," in *51st International Conference on Environmental Systems*, St. Paul, MN, 2022.

[3] S. K. Hota, K.-L. Lee, G. Hoeschele, R. W. Bonner and S. Rokkam, "Performance investigation on different form factor embedded heat pipe and pulsating heat pipe heat spreaders," in *Proceedings of 17th International Heat Transfer Conference*, Cape Town, South Africa, 2023.

[4] S. K. Hota, K.-L. Lee, G. Hoeschele, R. Bonner and S. Rokkam, "Experimental comparison on thermal performance of pulsating heat pipe and embedded heat pipe heat spreaders," in *2023 39th Semiconductor Thermal Measurement, Modeling & Management Symposium (SEMI-THERM)*, San Jose, CA, 2023.

[5] Im, Y. Hoon, J. Y. Lee, Ahn, T. In, Youn and Y. Jik, "Operational characteristics of oscillating heat pipe charged with R-134a for heat recovery at low temperature," *International Journal of Heat and Mass Transfer,* vol. 196, p. 123231, 2022.

[6] S. K. Hota, Lee, Kuan-Lin, G. Hoeschele and S. Rokkam, "Investigation on Pulsating Heat Pipe (PHP) Heat Spreader Plate for Electronics Cooling," in *2024 40th Semiconductor Thermal Measurement, Modeling & Management Symposium (SEMI-THERM)*, San Jose, CA, 2024.

[7] N. Van Velson, R. Kumar, C. Tarau and G. F. Moroni, "Thermal Management of Large Area Heat Loads Using Multi-Pass Cryogenic Loop Heat Pipe," in *AIAA SCITECH 2024 Forum*, 2024.

[8] Hu, Lingfeng, Y. Chen, C. Wu and H. Zhang, "Optimization design and performance analysis of ammonia heat pipe air heater under low ambient temperature conditions," *Applied Thermal Engineering,* vol. 225, p. 120202, 2023.

[9] S. K. Hota, K.-L. Lee, T. McFarland, G. Hoeschele, J. Weyant and S. Rokkam, "High performance heat spreader for thermal management of high heat flux optical and electronics systems," in *Next-Generation Optical Communication: Components, Sub-Systems, and Systems XIII*, San Francisco, CA, 2024.

[10] V. S. Nikolayev and M. Marengo, "Pulsating heat pipes: basics of functioning and modeling," in

ENCYCLOPEDIA OF TWO-PHASE HEAT TRANSFER AND FLOW IV: Modeling Methodologies, Boiling of CO₂, and Micro-Two-Phase Cooling Volume 1: Modeling of Two-Phase Flows and Heat Transfer, World Scientific, 2018, pp. 63-139.

[11] S. K. Hota, K.-L. Lee, G. Hoeschele, T. Mcfarland, S. Rokkam and R. Bonner, "Experimental comparison of two-phase heat spreaders for space modular electronics," in *2023 International Conference on Environmental Systems*, 2023.

[12] L. Kossel, J. Pfotenhauer and F. Miller, "Thermal performance and stability experiments of a 1.75-meter-long helium pulsating heat pipe," in *IOP Conference Series: Materials Science and Engineering*, 2024.

[13] D. Takouda and T. Inoue, "Heat transport characteristics of a sodium oscillating heat pipe: thermal performance," *International Journal of Heat and Mass Transfer,* vol. 196, p. 123281, 2022.

[14] H. Abad, M. Ghiasi, S. Mamouri and M. Shafii, "A novel integrated solar desalination system with a pulsating heat pipe," *Desalination,* vol. 311, pp. 206-210, 2013.

[15] R. Abdelmaksoud, J. Diebold, S. K. Hota, K.-L. Lee, S. Sinha and D. Maksimovic, "Development of ceramic pulsating heat pipes for medium-voltage power electronics," in *International Electronic Packaging Technical Conference and Exhibition*, San Diego, CA, 2023.

[16] K. Natsume, T. Mito, N. Yanagi, H. Tamura, T. Tamada, K. Shikimachi, N. Hirano and S. Nagaya, "Heat transfer performance of cryogenic oscillating heat pipes for effective cooling of superconducting magnets," *Cryogenics,* vol. 51, no. 6, pp. 309-3014, 2011.

[17] S. K. Hota, K.-L. Lee, B. Leitherer, G. Elias, G. Hoeschele and S. Rokkam, "Pulsating heat pipe and embedded heat pipe heat spreaders for modular electronics cooling," *Case Studies in Thermal Engineering,* vol. 49, p. 103256, 2023.

[18] M. Fazli, S. Mehrjardi, A. Mahmoudi, A. Khademi and M. Amini, "Advancements in pulsating heat pipes: Exploring channel geometry and characteristics for enhanced thermal performance," *International Journal of Thermofluids,* vol. 22, p. 100644, 2024.

[19] J. Kim, Kim and S. Jin, "Experimental investigation on working fluid selection in a micro pulsating heat pipe," *Energy Conversion and Management,* vol. 205, p. 112462, 2020.

[20] S. K. Hota, K.-L. Lee, G. Hoeschele, T. McFarland and S. Rokkam, "Development of Modular 3U Form Factor Pulsating Heat Pipe Based Plate Heat Spreader for Electronics Cooling," in *53rd International Conference on Environmental Systems*, Louisville, USA, 2024.

[21] B. Drolen and C. Smoot, "Performance limits of Oscillating Heat Pipes: Theory and Validation,"

Journal of Thermophysics and Heat Transfer, vol. 31, no. 4, 2017.

[22] https://www.1-act.com/thermal-solutions/embedded-computing/ice-lok/?srsltid=AfmBOopZ_isd2wp3-DmvjQoLOuBjarpGaDUOzLXaNfSZdCOyEYOrZiFv.

[23] D. Liu, J. Liu, K. Yang, F. Shang, C. Zheng and X. Cao, "Evaluation of the Heat Transfer Performance of a Device Utilizing an Asymmetric Pulsating Heat Pipe Structure Based on Global and Local Analysis," *Energies,* vol. 17, no. 22, 2024.

[24] Z. Li, "Study of a Hybrid of Pulsating Heat Pipe and Distributed Jet Array," *Journal of Thermophysics and Heat Transfer,* vol. 34, no. 2, 2020.

[25] D. Yin, H. Rajab and H. Ma, "Theoretical analysis of maximum filling ratio in an oscillating heat pipe," *International Journal of Heat and Mass Transfer,* vol. 74, pp. 353-357, 2014.

[26] https://www.ecfr.gov/current/title-49/subtitle-B/chapter-I/subchapter-C/part-173/subpart-G/section-173.315.

[27] https://info.thinkcei.com/think-tank/asme-standards#:~:text=Previous%20years%20of%20ASME%20Standards,can%20create%20a%20custom%20material..

[28] https://www.phmsa.dot.gov/hazmat/documents/offer/SP11818.pdf/offerserver/SP11818.

Effect of Surface Roughness on the Performance of Thermal Interface Materials

Roshan Sameer Annam, Loren Russell, Cara Rossetti, Dylan Shah, Navid Kazem
Arieca Inc
201 N Braddock Ave, STE #334
Pittsburgh, PA 15208, USA
dshah@arieca.com

Abstract

Many semiconductor applications suffer from inefficient thermal interfaces that impede heat transfer and require designers to reduce the power throughput to avoid overheating. Much effort is put into developing thermal interface materials (TIM) for such interfaces, by adding thermally conductive fillers to polymer materials. However, there has been relatively less research into the effect of surface properties on thermal performance. In this study, we investigate thermal and adhesive performance as a function of the roughness of the target surfaces. The test specimens for this study were three-layer specimens of copper-TIM-copper "sandwiches", made with copper coupons with varying surface roughness (Ra) values. We focus primarily on liquid metal embedded elastomers, with a gallium-based liquid metal filler and silicone matrix, while comparing to a popular thermal grease TIM as a baseline. Tests were conducted using an ASTM D5470-compliant Thermal Interface Material Analyzer (TIMA). In addition, we compare the results using a laser flash analyzer. In all configurations, we find that higher roughness leads to higher thermal resistance.

Keywords

Liquid metal, thermal interface material, surface roughness

1. Introduction

Modern power electronics and semiconductor devices dissipate considerable heat during use, resulting in overheating and degraded performance. However, designing effective thermal management systems is often thwarted by inefficiencies at the interfaces between each layer in the stack. Air gaps need to be filled with advanced thermal interface materials (TIM) to fill the air voids and bond the two surfaces together, thereby reducing the interface thermal resistance (ITR) between the surfaces.

ITR is a measure of the performance of the TIM and is known to be a function of the operating pressure and temperature [1]. While there have been studies for several decades focusing on the interfacial thermal resistance or contact resistance [2, 3], the effects of surface roughness on this phenomenon have been relatively understudied. Some studies show surface roughness increased contact resistance, such as for brass-on-brass and aluminum-on-aluminum interfaces [4]. Other studies found no effect of roughness, suggesting no benefit of utilizing costly precision machining methods in an attempt to improve the contact thermal resistance between such surfaces [3]. Given such ambiguity paired with the ubiquity of metallic heat spreaders in electronics, it is important to understand how thermal resistance varies with surface roughness, to help designers choose the appropriate surface finish to balance thermal resistance with ease of manufacture and adhesion strength.

The ideal interface for TIM and power electronics would have low thermal resistance for electrical efficiency, and high adhesion to survive the reliability requirements such as the popular AQG 324 standard [6]. In our previous study, we investigated the use of liquid metal embedded elastomers (LMEE [7]) as an alternative to the popular sintered silver materials used in power electronics [8] and found that maintaining adhesion was critical in maintaining low thermal resistance after thermal shock tests (TST). Building on this, here we investigate the effects of surface roughness on thermal resistance and adhesion strength of LMEE on copper substrates while benchmarking against a popular thermal grease to allow for comparison to other studies.

2. Experimental Methods

2.1. Sample Preparation

In this study, we used various grits of sandpaper (Allied High Tech) ranging from P60 to P1200, to polish cylindrical copper coupons (13 mm diameter and thickness 2 mm) to 4 surface roughness (Ra) values spanning from ~0.5 µm to ~4 µm respectively, covering a range commonly found in

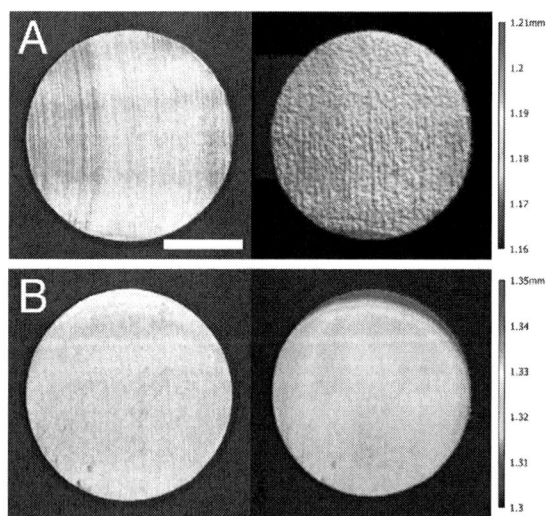

Figure 1. Microscopy (left) and profilometry (right) of copper test specimens polished using (A) P120 sandpaper, achieving $R_a \sim 1.13\ \mu m$; compared to (B) P1200 sandpaper, achieving $R_a \sim 0.40\ \mu m$. Scalebar in A is 5 mm and applies to all images. Colormaps in A-B have full scale range of 50 µm.

commercial heat sinks. First, each side of each coupon was pre-polished using P120 grit polishing paper to remove rough edges from manufacturing.

Next, one side of each coupon was polished to its target roughness, using polishing paper attached to a 152.4 mm determined the appropriate grit size and measured the surface roughness with a white-light optical profiler (VR-6100) using appropriate high-pass filters to eliminate the effects of warpage (approximately 40 microns) across all the grit sizes (**Figure 1**).

LMEE was then applied to 5 coupons of each roughness, in a cross-shaped dispense pattern, and then another coupon with matching roughness was placed on top (total 20 sandwiches). Next, we applied uniform pressure to the sandwich using glass slides and spring-steel binder clips (**Figure 2**). We then used Q-tips to remove LMEE that squeezed out of the interface and reapplied the binder clips. Finally, the sandwiches were cured at 150 °C overnight to cure the TIM.

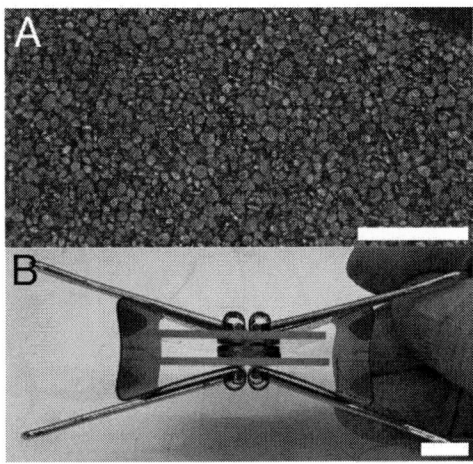

Figure 3. LMEE sample preparation. A) Microscopy of LMEE dispensed by hand onto a glass slide. Scale bar, 500 μm. B) Applying pressure to copper-TIM-copper sandwiches in preparation for curing. Scale bar, 10 mm.

To investigate whether there is a simple way to fill in the micro-grooves in the copper coupons, we additionally made samples where we used a Q-tip to imbue the surface with liquid metal (from 5N+, Inc.) applied after sanding but before the LMEE was applied (5 samples each, for roughness 0.4 and 0.85 μm). Additionally, to serve as a non-LMEE control, one sandwich of each roughness was prepared with non-curing thermal paste (Arctic MX6) as the interface, instead of LMEE. These samples were prepared identically as the LMEE, except no curing step was necessary.

2.2. Test Setups

The prepared copper sandwiches were measured in an ASTM D5470 [7] compliant machine (TIMA5, Nanotest, GmbH, **Figure 2**). To provide a standardized, reworkable interface between the test heads and the sandwiches, we applied thermal paste (MX6, Arctic GmbH) to both test heads. Then, the sandwich was applied to the bottom test head and the top testhead was moved down at 0.5 mm/s to 200 μm, to finish sample preparation. Next, the TIMA test was begun with control speed of 1 μm/s while it ramped up to 0.2 MPa applied pressure, with the heater at 100 °C and chiller at 15 °C. Once the system reached the desired pressure and thermal equilibrium, the system began recording the thermal resistance measurements. Each measurement therefore gave a repeatable

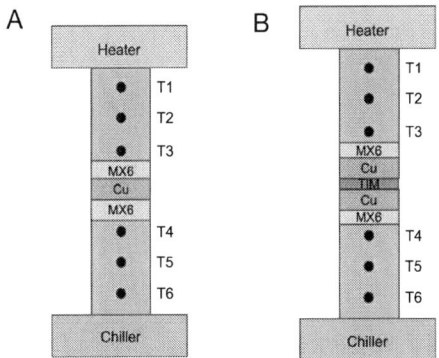

Figure 2. Schematic of thermal characterization in TIMA 5 test setup, along with thermal resistances. A) Baseline test setup used to measure the thermal resistance of the copper coupons. B) Full thermal stack for testing the TIM sandwiches

estimate of the full stack thermal resistance, which could be compared to a baseline to estimate the desired interface, which was between the two sandwiched copper coupons.

For the laser flash measurements, we used a LFA 467 Hyperflash by Netzsch, with the 13 mm circular diameter sample holder. To prepare the sandwiches for laser flash measurements, we first coated them with a graphite film using a graphite film spray. Next, we entered the graphite coating and sandwich thickness into the software and applied an appropriate measurement template. We used the thick sample template for most samples, which had a thickness greater than 2.5 mm. For sandwiches with a thickness lower than 2.5 mm, we chose a high diffusivity template of 20-100 mm/s^2.

To measure adhesion strength of the samples, we tested the samples under uniaxial tension. First, we cleaned the samples with simple green and fine-grit sandpaper (150 grit) and adhered the sandwiches to a rigid acrylic base plate. Then, we attached the top copper test piece to a loading block that was tied to a steel cable, to reduce effects of misalignment and get a more "normal" tensile force. Finally, we pulled the cable at 10 mm/min using a materials testing machine (Mark-10 F305-EM). We recorded the maximum observed force and calculated the adhesion strength by dividing it by the interface area.

2.3. Mathematical Model

The full-stack thermal model for the test setups (**Figure 2**) can be visualized as a 1D heat flow model, where the heat travels from the heater to the chiller. Thermal resistances exist at the interfaces of the different materials within the stack, along with the bulk material (Arctic MX6, copper, and TIM). Upon further inspection, we can identify appropriate experiments and normalizations to isolate the influence of thermal resistances caused by the copper disc and its interface with Arctic MX6 thermal grease. We assume that the system is under 1D heat flow from the hot side to the cold side through multiple layers, resulting in a linear (in-series) thermal resistance network. Here, we consider two cases: a single copper coupon and the copper sandwiches with the TIM, described in section B.

From **Figure 2a**, we can determine the thermal resistance of the stack, with one copper coupon (with a known thickness) whose sides were both roughened by P120 grit paper. This condition includes the contact resistance between the TIMA test heads to Arctic MX6, and Arctic MX6 to the copper coupon, with both interfaces being duplicated on the top and bottom surface. The thermal resistance of this stack is then taken as R_{stack1}. **Figure 3b** gives us the thermal resistance of the complete stack with the copper sandwich in it. This is denoted by R_{stack2}. Developing the resistance network for both the conditions, we get the following equations:

$$R_{stack1} = R_{C1,Th-MX6} + R_{MX6} + R_{C2,MX6-Cu} + R_{Cu} \\ + R_{C3,Cu-MX6} + R_{MX6} + R_{C4,Mx6-TH} \quad (1)$$

$$R_{stack2} = R_{C1,Th-MX6} + R_{MX6} + R_{C2,MX6-Cu1} + R_{Cu1} \\ + R_{C3,Cu1-TIM} + R_{TIM} + R_{C4,TIM-Cu2} \\ + R_{Cu2} + R_{C5,Cu2-MX6} + R_{MX6} \\ + R_{C6,MX6-TH} \quad (2)$$

where for example the term $R_{C1,Th-MX6}$ represents the contact resistance between the copper test head attached to the heater and Arctic MX6. The number after the letter C represents the order in which contact resistance is experienced by the heat flow from the hot to the cold test head. Other terms, without the letter C, represent the bulk resistance of the materials that are there in the stack. Since there are all independent resistances within a network, rearranging and combining certain terms will not affect the result. Therefore, the final expressions for the 2 equations are as follows:

$$R_{stack1} - R_{Cu} = (R_{C2,MX6-Cu} + R_{C3,Cu-MX6} + R_{MX6}) + \\ (R_{C1,Th-MX6} + R_{MX6} + R_{C4,Mx6-TH}) \quad (3)$$

$$\text{and,} R_{stack2} = (R_{C1,Th-MX6} + R_{MX6} + R_{C6,MX6-TH}) + \\ (R_{C5,Cu2-MX6} + R_{C2,MX6-Cu1} + R_{MX6}) + (R_{C3,Cu1-TIM} + \\ R_{TIM} + R_{C4,TIM-Cu2}) + R_{Cu2} + \\ R_{Cu1} \quad (4)$$

In the case of the two stacks, we know that,

$$(R_{C1,Th-MX6} + R_{MX6} + R_{C6,MX6-TH})$$

$$+(R_{C5,Cu2-MX6} + R_{C2,MX6-Cu1} + R_{MX6}) = (R_{C2,MX6-Cu} + \\ R_{C3,Cu-MX6} + R_{MX6}) + (R_{C1,Th-MX6} + R_{MX6} + \\ R_{C4,Mx6-TH}) = R_{stack1} - R_{Cu} \quad (5)$$

because they represent the same resistances from stack 1. Therefore, the equation can be written as,

$$R_{stack2} = R_{LMEE} + (R_{stack1} - R_{cu}) + R_{Cu2} + R_{Cu1} \quad (6)$$

$$R_{stack2} - (R_{stack1} - R_{cu}) - R_{Cu2} - R_{Cu1} = R_{LMEE} \quad (7)$$

where $(R_{C3,Cu1-TIM} + R_{TIM} + R_{C4,TIM-Cu2})$ is considered to be the resistance of the LMEE for the copper sandwich under study, denoted by R_{LMEE}, and R_{Cu} is the bulk resistance of the copper coupon in R_{stack1}, and R_{Cu2} amd R_{Cu1} are the bulk resistances of the copper coupons in R_{stack2}. The bulk resistance of the copper coupons can be calculated using the following formula,

$$R_{Cu} = \frac{L}{k_{Cu}} \quad (8)$$

Here, L is the thickness of the copper coupon and k_{Cu} is the bulk thermal conductivity of copper, which was taken as 388 W/mK [10].

To investigate the thermal performance of the TIM at the interface of the sandwich with a non-contact technique, we subsequently measured the thermal resistance of the copper sandwiches using laser flash (LFA 467 Hyperflash by Netzsch). The method uses a Xenon lamp as a flash source and calculates the effective thermal diffusivity (α) of the sandwich based on the time it takes for the light flash to travel through the sample to the detector. This effective thermal diffusivity is then used to calculate the thermal resistance of the copper sandwich. First, we have the thermal resistance of the copper sandwich:

$$R_{Cu\,Sandwich} = \frac{L_{Cu\,Sandwich}}{k_{Cu\,Sandwich}} \quad (9)$$

where $L_{Cu\,Sandwich}$ is the thickness of the total sandwich and $k_{Cu\,Sandwich}$ is given by

$$k_{Cu\,Sandwich} = \alpha \rho_{Cu} C_{P,Cu} \quad (10)$$

where ρ_{Cu} [11] is the density of copper and $C_{P,Cu}$ [12] is the specific heat of copper. Here we assume that the addition of the TIM in the copper sandwich does not affect the density and the specific heat values of the copper sandwich. This is justified since the volumetric ratio of the TIM to the copper discs (γ) is approximately 10^{-5}, as can readily be verified from the rule of mixtures ($\rho_{sand} = \gamma \rho_{TIM} + (1 - \gamma)\rho_{cu}$). This simplifies the computation of the thermal resistance of the copper sandwiches:

$$R_{LMEE} = R_{Cu\,Sandwich} - R_{Cu1} - R_{Cu2} \quad (11)$$

3.   Results

First, we tested curing on a single roughness (P120, so Ra~1.13 um) using three pressures during curing (no binder

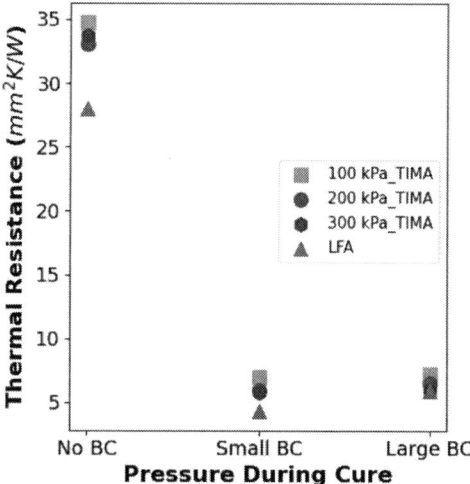

Figure 4. Thermal resistance of sandwiches cured under various conditions, then measured under various applied pressure. BC = Binder Clips.

clips, small binder clips, large binder clips; 2 samples each) and measured their thermal resistance in the TIMA at three pressures (100 kPa, 200 kPa, 300 kPa; **Figure 4**). We then used Equation 7 to estimate R_{LMEE} (**Figure 4**). The thermal resistance of the set with no binder clips was ~5x higher than the clipped sets. Surprisingly, both binder clips had comparable thermal resistance. Increasing pressures led to slightly lower LMEE thermal resistance. Using this as a screening study, we

164

selected the large binder clips and 200 kPa for the remainder of the tests, as mentioned in the Methods section.

Then, when testing the samples that were polished with various sandpaper grits, we observed negligible change in thermal resistance between the 0.4 μm and 0.83 μm samples, while there was a slight uptick at 1.13 μm and a statistically significant (P<0.05 compared to 0.4 μm) increase of ~100% for the samples with roughness 3.53 μm (**Figure 5**). This was observed for both the LMEE sandwiches and the Arctic MX-6 sandwiches. Correspondingly, the average BLT for the samples was 10 to 15 μm for the LMEE sandwiches across the range of surface roughness. Similarly, the BLT for Arctic MX6 sandwiches was around 10 to 15 μm. While we note that this matches what is expected from separate measurements made with just LMEE or MX6 in the TIMA at 200 kPa, there are likely some spots that are at a higher BLT due to imperfect surface flatness and coplanarity errors.

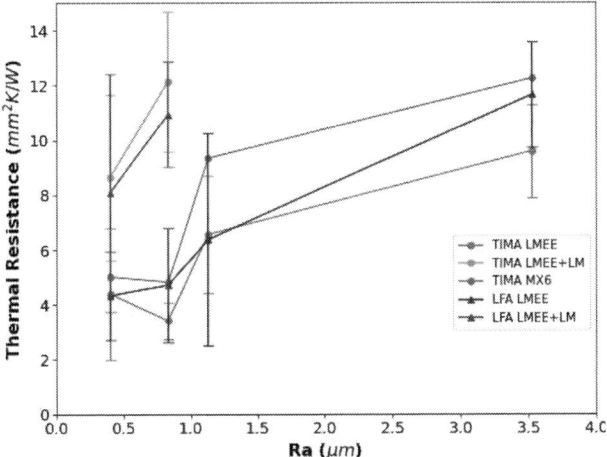

Figure 5. Thermal resistance measurements of TIM applied to copper with various surface roughness. N=5 for all tests (except only 1 for MX6), and error bars represent 95% confidence interval.

We originally hypothesized that the increase in thermal resistance is due to a combination of micro air voids remaining in the copper and having slightly higher effective BLT due to the surface texture at higher roughness. However, adding LM to fill in the micro voids surprisingly increased R_{th} (Fig. 5), suggesting that other properties may be at play, such as the bonding between the copper and the LMEE's polymer phase. Another effect that can be at play here is that the layer of LM prevents the applied TIM from conforming to the surface features by creating a physical barrier due to its poor wetting properties to copper, which leads to the inadequate filling of the microscale voids by LM and also with thermally conductive TIM, resulting in poor thermal performance for sandwiches with LM+LMEE at its interface.

The average R_{LMEE} measured using laser flash agrees well with the TIMA data, with maximum error of $2 \frac{mm^2 K}{W}$ for the 3.53 μm roughness samples (**Figure 5**). Importantly, the same trend was observed, where higher surface roughness above the 0.83 μm level led to higher interfacial thermal resistance.

Finally, all samples were cleaned and tested in uniaxial tension. The LMEE sandwiches had no statistically significant differences in adhesion strength, ranging from 140-240 kPa with standard deviations ~15-45 kPa. One interesting result that

deserves further study is the adhesion of the LMEE+LM samples was too high to use in the adhesion test setup. Our current hypothesis is that the LM alloyed with the copper and this alloy had improved adhesion to the polymer used in the LMEE.

4. Conclusion

In this study, we investigated how thermal resistance was affected by the surface roughness of copper test coupons, which are representative of heat spreaders and heat pipes used in modern electronics. The main findings were that increased roughness (above ~1 μm) increased thermal resistance, while the adhesion strength showed no statistically significant change. BLT measurements and surface profilometry are included to help with reproducibility. In addition, we provided further context of other variables that affect thermal resistance, namely pressure applied during curing and pressure applied during measurement.

Areas for further study include studying the effect of roughness on a wider range of materials, as well as obtaining tighter control of surface roughness. Regarding the former, it is unclear whether other classes of TIMs are affected by surface roughness variation. Relevant material classes include sintered metals, phase change materials, and pure liquid metals. In addition, the inclusion of other types of polymers (silanes, polyurethanes, etc.) and fillers into the LMEE architecture is expected to help counteract the effects of roughness and allow rougher heat spreaders to be used in applications.

To obtain tighter surface roughness, we noticed during sample preparation that the most challenging aspect was to keep the polished surface flat without inducing a significant tilt. This, in turn, affected our ability to keep consistent BLT, likely contributing significantly to the observed noise in thermal resistance. Possible routes to reducing sample variation include polishing entire sheets first (before punching or water-jetting out the copper disks) and utilizing milling and lapping instead of polishing. Finally, we note that our ongoing work extends the current study by testing effects of roughness and surface treatment on adhesion, using larger test specimens that simplify mechanical testing.

By demonstrating the effect of surface roughness on thermal resistance, we hope to aid thermal researchers in developing more efficient interfaces, while providing reproducible test methods for future studies. Our data suggests a general best practice is to polish the substrate and lid to below 1 μm Ra, with additional improvement down to 0.4 μm Ra. In some cases, we observed the thermal resistance reduced by up to 50% when reducing Ra from 3.5 down to 0.4 μm. Paying attention to surface roughness in the thermal design of advanced packages should therefore be one of the primary concerns, along with traditional considerations like thermal conductivity and bondline thickness. Optimizing these parameters in parallel should enable more efficient devices in a variety of industries, including electric vehicles, advanced packaging (heterogeneous integration, chiplets, etc.), and AI processors, where obtaining every performance improvement is critical for enabling further progress.

References

[1] Zhao, J.W., Zhao, R., Huo, Y.K. and Cheng, W.L., 2019. Effects of surface roughness, temperature and pressure on interface thermal resistance of thermal interface materials. International Journal of Heat and Mass Transfer, 140, pp.705-716.

[2] Madhusudana, C. & Fletcher, L. (1986). Contact heat transfer - The last decade. AIAA Journal. 24. 510-523. 10.2514/3.9298.

[3] S.S. Burde, M. M. Yovanovich., 1978. Thermal Resistance at Smooth-Sphere/Rough-Flat Contacts: Theoretical Analysis, Progress in Astronautics and Aeronautics: Thermophysics and Thermal Control, vol. 65, pp. 83-102

[4] M. Z. Abdullah, Y. C. Yau, Z. Z. Alauddin, and K. N. Seetharamu, "Effects of pressure on thermal contact resistance for rough mating surfaces," ASEAN Journal on Science and Technology for Development, vol. 18, no. 2, 2001.

[5] Zhang, P., Cui, T. and Li, Q., 2017. Effect of surface roughness on thermal contact resistance of aluminium alloy. Applied Thermal Engineering, 121, pp.992-998.

[6] ECPE Guideline AQG 324 - Qualification of Power Modules for Use in Power Electronics Converter Units in Motor Vehicles, 2019.

[7] M. D. Bartlett et al., "High thermal conductivity in soft elastomers with elongated liquid metal inclusions," PNAS, vol. 114, no. 9, pp. 2143–2148, Feb. 2017, doi: 10.1073/pnas.1616377114.

[8] Y. Mochizuki, T. Otsuka, N. Kazem, D. Shah, and K. Nakahara, "Large-Area Bonding with LMEE: Suppression of the Degradation of the Junction-to-Water Thermal Resistance in Power Modules," presented at PCIM 2024, Nuremberg, Germany, Jul. 2024. DOI: 10.30420/566262107

[9] Committee, Test Method for Thermal Transmission Properties of Thermally Conductive Electrical Insulation Materials. doi: 10.1520/D5470-12.

[10] Moore, J.P., McElroy, D.L. and Graves, R.S., 1967. Thermal conductivity and electrical resistivity of high-purity copper from 78 to 400 K. Canadian Journal of Physics, 45(12), pp.3849-3865.

[11] DAG G. ELLINGSEN, NINA HORN, JAN AASETH, CHAPTER 26 - Copper, Editor(s): Gunnar F. Nordberg, Bruce A. Fowler, Monica Nordberg, Lars T. Friberg, Handbook on the Toxicology of Metals (Third Edition), Academic Press, 2007, Pages 529-546, ISBN 9780123694133,

[12] Brooks, C.R., Norem, W.E., Hendrix, D.E., Wright, J.W. and Northcutt, W.G., 1968. The specific heat of copper from 40 to 920 C*. Journal of Physics and Chemistry of Solids, 29(4), pp.565-574.

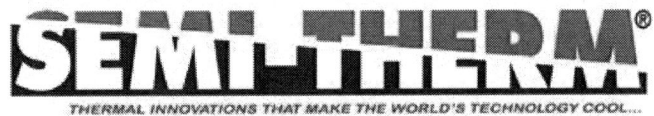

Vertically Aligned Flexible Insulation Thermal Conductive Sheet

Tomofumi WATANABE, Fumihiro MUKAI, and Hiroki NAITO
Bando Chemical Industries, Ltd.
6-6, Minatojima Minamimachi 4-chome, Chuo-ku, Kobe, 650-0047, Japan
E-mail address: shintaro.ozaki@bandogrp.com

Extended Abstract

Summary
We developed a thermal conductive sheet that significantly improves thermal conductivity in the thickness direction by controlling the orientation of shape-anisotropic fillers within a rubber matrix. This research focuses on using hexagonal boron nitride (h-BN) as the filler in an insulating type. The sheet with vertically oriented h-BN fillers demonstrates a thermal conductivity of 15 W/m·K and is capable of maintaining stable performance under varying pressures. This technology enables high thermal conductivity without the need for high-density filler packing, while also providing flexibility, allowing for efficient heat dissipation from the contact areas.

1.Introduction
In the rapidly evolving field of electronics, effective thermal management is crucial for maintaining performance and reliability. Thermal conductive sheets play an important role in facilitating heat transfer between components. Recent advancements in materials science have led to the development of innovative fillers (e.g., hexagonal boron nitride, h-BN) with excellent thermal conductivity and insulating properties. This paper aims to explore the use of shape-anisotropic fillers in a rubber matrix and improve the thermal performance of sheets produced using orientation control techniques that align the fillers in the thickness direction.

2.Experimental
Material Selection
Filler: Hexagonal boron nitride (h-BN) was selected for its high thermal conductivity (150–250 W/m·K) along the long axis and high electrical insulation properties. h-BN is in a flake-like form to facilitate orientation.
Matrix Resin: Silicone rubber was chosen as the matrix material due to its excellent heat resistance and electrical insulation properties, with a thermal conductivity of approximately 0.2 W/m·K.

Orientation control technology
The orientation of the fillers was controlled using a method that aligns the particles perpendicular to the sheet's thickness.

Evaluation of the Sheet's Properties
SEM Imaging: Scanning electron microscopy (SEM) was used to examine the microstructure of the sheet's cross-section and verify the vertical orientation of the h-BN fillers.
Thermal Conductivity Measurement: Based on ASTM D5470 standards, the thermal conductivity of the sheet was measured using a steady-state method. Thermal resistance values were plotted against the sheet's thickness to evaluate performance.
Testing at Different Compression Ratios: Measurements were conducted at various compression ratios (up to 50%) to assess their impact on thermal resistance. The TIM Tester Model 1300 was used to measure thermal resistance.
Hardness Measurement: The softness of the sheet was measured using an Asker C hardness tester.

Insulation Breakdown Voltage: The voltage at the point of insulation breakdown was measured for a 1mm thick sheet with a Φ20mm cylindrical electrode, under atmospheric conditions, using a DC voltage with a rising rate of 1kV/sec.

Volume Resistivity: The volume resistivity was measured by placing a 1mm thick sheet between surface and back electrodes and applying a DC voltage of 1000V using a high-resistance meter.

Flammability: Measured using the UL94 vertical burning test.

Data Analysis: The data obtained from thermal resistance measurements were analyzed to understand the correlation between applied compression ratio, thermal resistance, and thermal conductivity. Linear approximation was used to calculate the contact thermal resistance and thermal conductivity of the sheet.

3.Results

The results demonstrated that the application of h-BN as a shape-anisotropic filler in a silicone rubber matrix, along with effective orientation control, leads to significant improvements in thermal conductivity and stability. This indicates that the thermal conductive sheet is suitable for demanding thermal management applications in electronic devices, meeting the requirements for efficient heat dissipation without compromising durability.

Items	Sheet A	Sheet B Under development	Unit/Remarks
Filler	h-BN	h-BN	-
Thickness	0.3, 0.5, 1.0	1.0-2.0	mm
Thermal conductivity	9	15	W/m・K/ASTMD5470
Softness	55	30	Asker C
Dielectric breakdown*	No breakdown at 20	10	kV/mm
Volume resistivity	1×10^{14}	3×10^{13}	Ω・cm/JISK6911
Non flammability	V-0 equivalent	V-0 equivalent	UL94

* Test Piece:T=1mm, 1kV/s, Air Condition

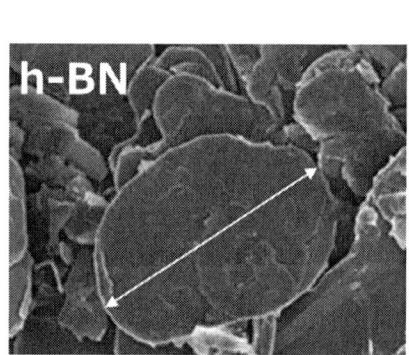

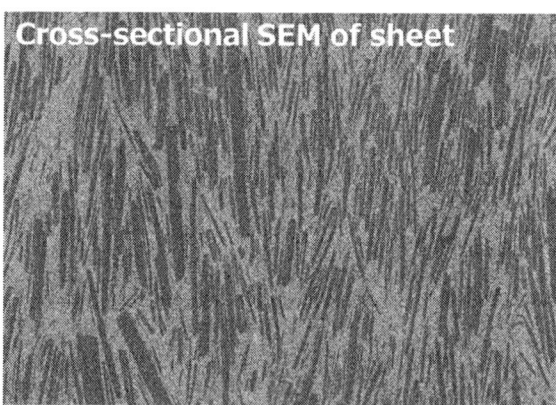

4. Conclusion

This paper discussed the importance of material selection, orientation control, and thorough testing in developing effective thermal conductive sheets. By achieving vertical orientation of h-BN fillers within a silicone rubber matrix, we were able to significantly enhance thermal conductivity while maintaining flexibility. This suggests that the developed sheets can ensure heat dissipation, stability, and durability in thermal management applications for modern electronic devices.

LIME: Liquid Metal Interconnections for Power Semiconductors

Nick Baker
University of Alabama, Tuscaloosa, Al, USA
nbaker2@ua.edu, +1 205 348 2623

Abstract

Almost all state-of-the-art power semiconductors use solid metal interconnects to connect the chip to other circuit components. Thermo-mechanical stress degrades solid metal interconnects and is the main cause of failure in power semiconductors. This paper uses liquid metals (LIME), which are inherently resistant to thermo-mechanical stress, to replace these interconnects. Preliminary results show that LIME increases the lifetime of power semiconductors by a factor of 80. This facilitates the miniaturization of power electronics, and both the cost and weight of power electronic systems can be reduced by up to 90%. Several iterations of liquid metal packages are shown, using both Silicon and Silicon Carbide chips. In the final iteration reported in this abstract, no solid interconnections remain within the module. Furthermore, the junction-to-ambient thermal resistance is improved by up to 10% using LIME in comparison to SAC305 solder and Aluminium wirebonds.

Synopsis

Almost all state-of-the-art semiconductor interconnects use solid metals welded together through thermo, thermo-sonic, ultrasonic, or thermo-compression bonding. These are most commonly wire-bonding, soldering, or sintering based technologies. Thermal cycling from power losses during semiconductor operation degrades these interconnects and eventually causes device and system failure. The lifetime of the power semiconductor is therefore a key design parameter for power electronic systems, and, according to representatives from the wind, solar, aircraft, automotive, and grid industries, power semiconductors are the component for which reliability is most critical [1].

The Young's Modulus, Coefficient of Thermal Expansion, and Mohs Hardness of the semiconductor chip impacts the thermo-mechanical robustness of power semiconductor modules. For Silicon Carbide, these parameters differ to that of Silicon. This leads to SiC MOSFET power cycling lifetimes that may be 20% - 30% of Si IGBTs when packaged with identical interconnect technologies [2]. As a result, interest in new packaging technologies for SiC devices are an area of interest [3]

In this paper, we replace both the die-attach (which is usually solder or sinter based) and active area interconnect (usually wirebonds) with liquid metals (LIME). Liquid metals are inherently resistant to thermo-mechanical stress. As a result, the power cycling lifetime of a SiC MOSFET devices can be increased. At present, results demonstrate a power cycling lifetime increase by a factor of 80x for SiC MOSFET chips packaged with LIME when compared to SAC-305 solder and Aluminium wirebonds.

This research has spanned 5 years with 5 prototype iterations constructed and tested. Figure 1 displays the average lifetime increase in the various iterations, and Table 1 shows the Min, Max, and layer descriptions for each prototype. Early prototypes showed several premature failures and low yield. However, these issues have been improved and premature failures are no longer observed. Figure 2a shows a photograph of the 2023 iteration, which was manufactured using a Silicon diode.

In the latest 2024 SiC MOSFET version, all interconnects have been replaced with liquid metal layers. No solid interconnects remain, and no encapsulation, barrier, recess, or floating structure, is used to contain the liquid metal. The final failure mechanism for the 2024 SiC MOSFET chip was observed to be corrosion based. Figure 2b displays an example of the type of corrosion observed on the topside of the chip during power cycling. The die-attach liquid metal layer did not observe significant corrosion. Figure 3 shows the evolution of the V_{ON} during the power cycling test for both the LIME and SAC305/Wirebonded SiC MOSFET.

Figure 4 displays the thermal resistance using LIME from 8 samples of a Si Diode. The mean junction-to-ambient thermal resistance is approximately 10% better than SAC305 and Al wirebonded samples. In addition, the variation across samples appears to be reduced. Thermal resistance is measured through a 120-second heating pulse, followed by a 120-second cooling period. The V_{FWD} is used to measure the diode junction temperature.

This paper will also estimate the impact of increased power cycling lifetime on the overall power electronic system. For example, a 20 kW Boost Converter (e.g., in an EV charger) using SiC MOSFETs, specified with a lifetime of 125,000 charge cycles, may have to be designed with a junction temperature swing of 53°C for Solder and Wirebond packaging. With an 80x lifetime increase using liquid metal, the allowable junction temperature swing can be increased to 180°C. The power losses during operation can therefore be increased. This could be achieved by increasing the switching frequency of the SiC MOSFET from 55 kHz to 425 kHz. As a result, the size of the filter components can be reduced, and a system weight and cost reduction of 80% and a 60% can be achieved, respectively, if assuming price parity with solder and wirebonded modules. These projections have been calculated through LTSpice simulation and are shown in Table 2.

Presentation

Further details of LIME interconnects will be given in the presentation. Explanations of the improvements made from early prototypes will be given, as well as the required future research to develop this technology for real-world use. Further results from more prototype variations will also be presented.

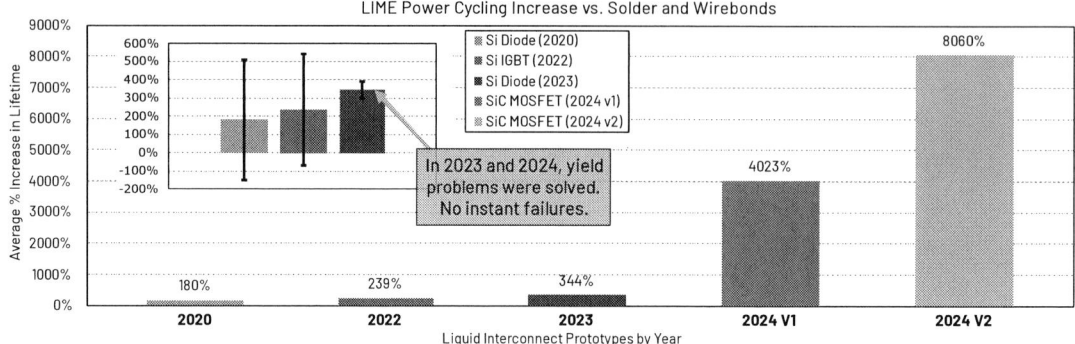

Figure 1. Increase in Power Cycling Lifetime of LIME based Power Semiconductors compared to Solder and Aluminium Wirebonded Power Semiconductors

Prototype Version	Chip Type	Interconnect Description			Lifetime Increase (%)			Samples
		Die Attach	Topside	DBC-Baseplate	Mean	Min	Max	
2020	200V Si Diode	SAC305	Ga-based LM	SAC305	180%	0%	658%	5
2022	600V Si IGBT	SAC305	Ga-based LM	SAC305	239%	0%	610%	5
2023	600V Si Diode	Ga-based LM	In-based LM	PCB with Dielectric	344%	296%	383%	8
2024 V1	1.2kV SiC MOSFET	Ga-based LM	In-based LM	SAC305	4023%	3745%	4300%	2
2024 V2	1.2kV SiC MOSFET	Ga-based LM	In-based LM	Ga-based LM	8060%	N/A	N/A	1

Table 1. Description of LIME Prototypes and Power Cycling Lifetime Increases Compared to Solder and Aluminium Wirebonded chips. Text in Green signifies liquid metal layers.

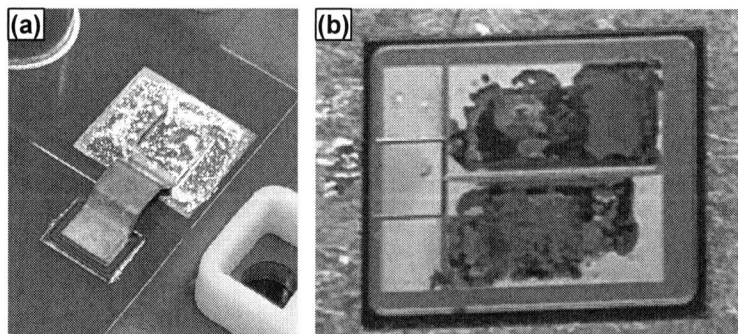

Figure 2. (a) A Silicon Diode manufactured using LIME on an Aluminium-clad PCB with a flexible dielectric. (b) Corrosion observed on a SiC MOSFET chip from the use of Liquid Metal interconnects. SiC MOSFET has been removed and the module disassembled to take the photo.

	SiC MOSFET LIME Packaging	SiC MOSFET Al Wire and SAC305	Si IGBT Al Wire and SAC305	US DoE 2035 Targets
Cost ($/kW)	10.9	28.0	55.2	16.67
Specific Power (kW/kg)	18.8	4.2	1.6	12.8
Power Density (kW/L)	19.4	8.0	3.7	14.8
Efficiency (%)	98.5	99.3	97.4	98.5
F_{SW} (kHz)	425	55	19	N/A
Allowable ΔT_J (°C)	180	53	78	N/A

Table 2. Performance metrics for a 20-kW Boost Converter with LIME vs. Aluminium wirebond and soldered power semiconductors *(figures are based on LTSpice simulation)*. US Department of Energy 2035 Targets are taken from "Electric Drive Technical Team Roadmap, March 2024". Green indicates meeting DoE 2035 targets, red indicates not meeting DoE 2035 targets.

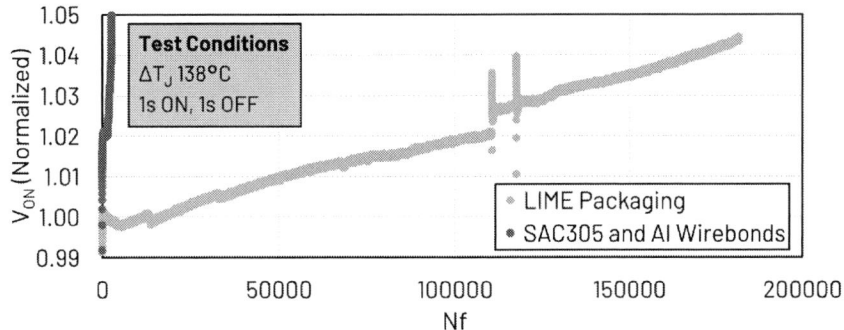

Figure 3. V_{ON} during a power cycling test of a LIME packaged SiC MOSFET and a SAC305 and Al wirebond package SiC MOSFET.

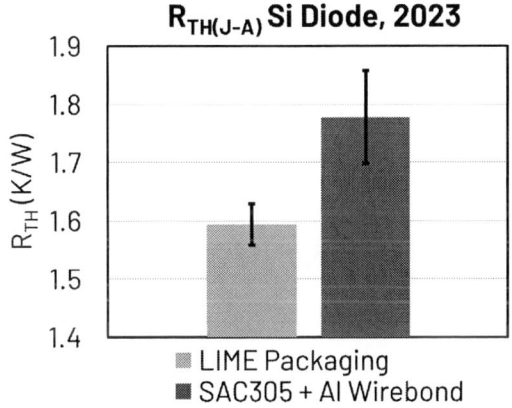

Figure 4. Junction-to-ambient thermal resistance of a Silicon Diode when packaged with LIME vs. SAC305 solder and Aluminium wirebonds. A total of 16 samples were tested (8 of each group).

References

[1] J. Falck et al, "Reliability of Power Electronic Systems: An Industry Perspective," in IEEE Industrial Electronics Magazine, vol. 12, no. 2, pp. 24-35, June 2018.

[2] F. Hoffmann, N. Kaminski and S. Schmitt, "Comparison of the Power Cycling Performance of Silicon and Silicon Carbide Power Devices in a Baseplate Less Module Package at Different Temperature Swings," 2021 33rd International Symposium on Power Semiconductor Devices and ICs (ISPSD), Nagoya, Japan, 2021, pp. 175-178.

[3] L. Wang, W. Wang, R. J. E. Hueting, G. Rietveld and J. A. Ferreira, "Review of Topside Interconnections for Wide Bandgap Power Semiconductor Packaging," in *IEEE Transactions on Power Electronics*, vol. 38, no. 1, pp. 472-490, Jan. 2023.

Non-thermal Plasma Activation to Increase Surface Area of Carbon-Rich Materials

Hector Gomez and Gerardo Diaz
Department of Mechanical Engineering, University of California, Merced
gdiaz@ucmerced.edu

Extended Abstract

SUMMARY

Biochar, a carbon-rich residue derived from agricultural waste, holds immense potential across various industries including food, water, agriculture, and energy, due to its highly porous structure and contaminant removal capabilities. Its application potential is attractive due to its carbon-negative production process, widespread availability, and inexpensive production from widely available biomass. In this work a plasma reactor containing biochar operates at significantly lower temperatures required for conventional physical activation methods, ranging between 200°C to 400°C. The plasma reactor uses steam as a carrier gas at a controlled flow rate (e.g., 3 mL/min), which interacts with the biochar in the plasma reactor. Plasma is generated within the reactor using high potentials (up to 12 kV), a frequency of 90 kHz, and an input power of up to 100 W. Experimental results demonstrate significant improvements in biochar properties post non-thermal plasma activation. For instance, at an activation temperature of 200°C, the surface area of biochar increased from 67.5 m²/g (without plasma) to 83.6 m²/g when utilizing 50 W of plasma power. With an increase in plasma power to 100 W, the surface area further rose to 103.4 m²/g. Processing temperature and plasma exposure have significant effects on the surface area and porosity distribution of the biochar when processed with the non-thermal plasma. This increase in surface area indicates enhanced porosity and thus improved potential effectiveness for industrial applications, while also reducing the energy-intensive nature of traditional activation methods.

1. INTRODUCTION

Porous materials are at the forefront of material research in light weight and high surface area characteristics used for various applications. A category of extensively studied porous materials used for various industrial applications including electrodes are carbon-rich materials [1-3]. They have a wide range of applications that have become more prevalent with human population growth and the need for energy, as well as technological and environmental adaptability. Some of the carbon-rich materials used for industrial applications include biochar and activated carbon.

Due to the relatively high porous structures of carbon-rich materials, and their ability to remove contaminants, they have a high potential for applications in the food, water, agricultural, energy, and semiconductor industries, to name a few. Biochar is obtained by taking biomass (agricultural waste) and exposing it to a thermo-chemical conversion process in an oxygen limited environment called pyrolysis. Consecutively, activated carbon is made by treating biochar in a process called activation. The activation process from biochar to activated carbon enhances the physical characteristics and absorption capacity of biochar by increasing the specific internal surface area and porous structure, making it suitable and desirable for a wider range of applications [4]. However, the conventional process of converting biochar to activated carbon by physical activation, consumes extensive energy, 44 - 170 MJ/kg. Recently, non-thermal plasma has been studied for material surface treatment, with the ability to alter the physical and chemical structure. Previous work treated organic carbon-rich material with non-plasma to tailor its physical and chemical properties [5-15]. Previous studies on the physical properties of biochar have shown inconclusive results, with some treatments even leading to deterioration. To address this, we propose enhancing the biochar activation process by incorporating a heating component. Specifically, we aim to heat the carrier gas at medium temperatures during plasma treatment to promote a more effective biochar activation method.

In this work a plasma reactor containing biochar operates at significantly lower temperatures required for conventional physical activation methods, i.e., operating temperatures between 200°C to 400°C instead of 800 to 1000°C for conventional physical activation. The plasma reactor uses steam as a carrier gas at a controlled flow rate (e.g., 3 mL/min), which interacts with the biochar in the plasma reactor. Plasma is generated within the reactor using high potentials (up to 12 kV), a frequency of 90 kHz, and an input power of up to 100 W.

2. EXPERIMENTAL/NUMERICAL METHODS

Biochar from ponderosa pine was used for the set of experiments due to its high availability in the Sierra Nevada of California. About 300 mg of biochar were used for each test. A schematic of the experimental assembly of the DBD plasma tubular reactor is shown in Figure 1. The outer electrode is composed of stainless steel sleeved into the dielectric barrier tube, made up of alumina ceramic. For the inner electrode, a stainless-steel coil spring was tightly screed into the ceramic tube with an inner diameter of 0.25 in (0.635 cm), making flush contact throughout the length of the outer electrode. The biochar was evenly placed between the coils at the bottom of the tubular reactor, contacting the inner electrode and dielectric barrier. For this experiment, power was measured from the electrical outlet source using a power sensor. The steam used was superheated via an in-house stainless steel coil heat exchanger using a natural gas burner and the temperature was collected using a Type K thermocouple. The temperature was controlled by the proximity of the natural gas burner to the heat exchanger coil. In the experimental procedure, the steam flowed for 10 min while being heated, interacting with the biochar, until reaching steady state and then treated with plasma simultaneously for 10 min, for a total run time of about 20 min.

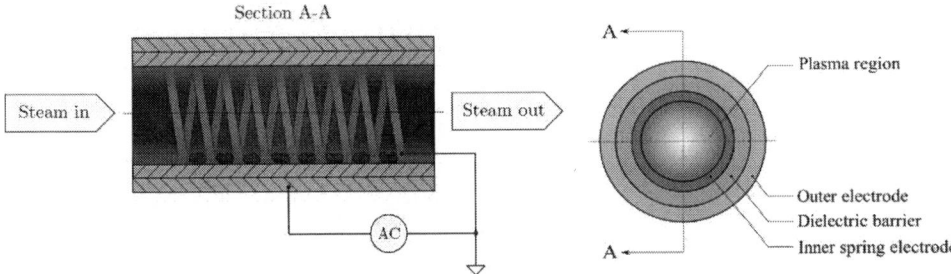

Figure 1: Plasma Reactor

3. RESULTS

For steam activation, two combination parameters were varied for the effects in the DBD plasma discharge, power and temperature. The first results in Table 1 are from the ponderosa pine biochar in a steam environment as we varied the total power used by the power supply. To see the effects that plasma has on the biochar, the reference was heated to 200°C without plasma treatment. The biochar was then treated with non-thermal plasma at 50 and 100 W of power. As we increased the power, both the Brunauer, Emmett and Teller (BET) surface area and micropore volume increased as the power was increased. The BET surface area has an increase of 24 % and 53 % when 50 and 100 W of plasma was applied, respectively. The increase in percentage was significantly higher when compared to other research conducted in atmospheric and/or vacuum conditions [5-15].

Table 1: Power Effects on Porous Structure

Experimental Name	Approximate Temperature (°C)	Total Plasma Power (W)	BET Surface Area (m^2/g)	Micropore Volume (cm^3/g)
Ponderosa Pine Sample 1	200	0	67.5	0.023
Ponderosa Pine Sample 2	200	50	83.6	0.026
Ponderosa Pine Sample 3	200	100	103.4	0.036

To see the effects of temperature on the biochar structure, the temperature was varied while keeping the plasma power constant at 100 W. Results are shown in Table 2. As with the power effects, as we increased the temperature, the BET surface area and micropore volume increased. The BET surface area increased by 17 % and 208 % as the temperature increased to 250 and 390°C at 100 W, respectively. The micropore volume increased by 23 % and 251 % when increased to 250 and 390°C, respectively. The power supply frequency treating these samples was 90 kHz.

Table 2: Temperature Effects on Porous Structure

Experimental Name	Approximate Temperature (°C)	Total Plasma Power (W)	BET Surface Area (m^2/g)	Micropore Volume (cm^3/g)
Ponderosa Pine Sample 4	200	100	103.4	0.036
Ponderosa Pine Sample 5	250	100	120.7	0.045
Ponderosa Pine Sample 6	400	100	318.6	0.127

4. CONCLUSIONS

This work presented a proposal for the improvement in the desired physical structure of biochar using plasma treatment in a developed medium temperature process. The results presented provided BET surface area and micropore volume consistently reaching higher values with plasma-assisted activation compared to original biochar sample. The increase in surface area and microporosity yielded the highest percent increase compared to the literature presented.

5. REFERENCES

[1] T. Huggins, H. Wang, J. Kearns, P. Jenkins and Z. J. Ren. Biochar as a sustainable electrode material for electricity production in microbial fuel cells. *Bioresour. Technol.*, 2014, 157, 114–119.

[2] K. Senthilkumar and M. Naveenkumar. Enhanced performance study of microbial fuel cell using waste biomass-derived carbon electrode. *Biomass Convers. Biorefin.*, 2023, 13, 5921–5929.

[3] J. Zhang, J. Li, D. Ye, X. Zhu, Q. Liao and B. Zhang. Tubular bamboo charcoal for anode in microbial fuel cells. *J. Power Sources*, 2014, 272, 277–282.

[4] H. Marsh, F. R. Reinoso, Activated Carbon. Elsevier Science, 2006.

[5] D. Lee, S. H. Hong, K.-H. Paek, W. T. Ju. Adsorbability enhancement of activated carbon by dielectric barrier discharge plasma treatment. Surface and Coating Technology 200 (7) (2013) 2277-2282.

[6] S. Kodama, H. Habaki, H. Sekiguchi, J. Kawasaki. Surface modification of adsorbents by dielectric barrier discharge. Thin Solid Films 407 (1) (2002) 151-155.

[7] G. Q. Wu, X. Zhang, H. Hui, J. Yan, Q. S. Zhang, J. L. Wan, Y. Dai. Adsorptive removal of aniline from aqueous solution by oxygen plasma irradiated bamboo based activated carbon. Chemical Engineering Journal 185-186 (2012) 201-210.

[8] B. C. Bai, H. U. Lee, C. W. Lee, Y. S. Lee, J. S. Im. N_2 plasma treatment on activated carbon fibers for toxic gas removal: Mechanism study by electrochemical investigation. Chemical Engineering Journal 306 (2016) 260-268.

[9] P. S. D. V. Maldonado, V. Hernandez-Montoya, M. A. Montes-Moran. Plasma-surface modification vs air oxidation on carbon obtained from peach stone: Textural and chemical changes and the efficiency as adsorbents. Applied Surface Science 384 (2016) 143-151.

[10] A. B. Garcia, A. Martinez-Alonso, C. A. Leon y Leon, J. M. D. Tascon. Medication of the surface properties of an activated carbon by oxygen plasma treatment. Fuel 77 (6) (1998) 613-624.

[11] L. Wu, W. Wan, Z. Shang, X. Gao, N. Kobayashi, G. Luo, Z. Li. Surface modification of phosphoric acid activated carbon by using nonthermal plasma for enhancement of cu(ii) adsorption from aqueous solutions. Separation and Purification Technology 197 (2018) 156-169.

[12] G. Z. Qu, J. Li, D. L. Liang, D. L. Huang, D. Qu. Surface modification of a granular activated carbon by dielectric barrier discharge plasma and its effects on pentachlorophenol adsorption. Journal of Electrostatics 71 (2013) 689-694.

[13] P. Pietrowski, I. Ludwiczak, J. Tyczkowski. Activated carbons modified by Ar and CO_2 plasmas - acetone and cyclohexane adsorption. Material Science 18 (2) (2012) 158-162.

[14] Q. Niu, J. Luo, Y. Xia, S. Sun, Q. Chen. Surface modification of biochar by dielectric barrier discharges plasma for hg removal. Fuel Processing Technology 156 (2017) 310-316.

[15] R. K. Gupta, M. Dubey, P. Kharel, Z. Gu, Q. H. Fan. Biochar activated by oxygen plasma for supercapacitors. Journal of Power Sources 274 (2015) 1300-1305.